Armorial Porcelain

Rachel L. Denyer · Morgan C. T. Denyer ·
Howell G. M. Edwards

Armorial Porcelain

The Genesis

Springer

Rachel L. Denyer
Thornton, West Yorkshire, UK

Morgan C. T. Denyer
The University of Bradford
Bradford, West Yorkshire, UK

Howell G. M. Edwards
The University of Bradford
Bradford, West Yorkshire, UK

ISBN 978-3-031-63744-5 ISBN 978-3-031-63745-2 (eBook)
https://doi.org/10.1007/978-3-031-63745-2

Cover illustration: Armorial plate from the Kemeys Tynte service, commissioned from Derby c.1820 to mark the marriage of Charles Kemeys Tynte and Elizabeth Swynnerton. Photo by Dr M.C.T Denyer: Private Collection

This Springer imprint is published by the registered company Springer Nature Switzerland AG
The registered company address is: Gewerbestrasse 11, 6330 Cham, Switzerland

Acknowledgements

The authors would like to record their grateful thanks and appreciation to several friends and colleagues who have supported them in their research: Fergus Gambon, Bonhams Antiques and Valuers, Bond Street London; Jo Fletcher-Lee, Tennants of Yorkshire Antiques and Fine Art Auctioneers Ltd, Leyburn, North Yorkshire; Sash Giles, Curator of the Devonshire Collections, Chatsworth House, Derbyshire. Also, Gwyn Jones, Oriel Plas Glyn-y-Weddw, Llanbedrog, Gwynedd, Martyn Downer, Kingschina Antiques, Mayfair, London, Indar & Aruna Pasricha, Kensington, London, Angela Howard and Errol & Henriette Manners, Kensington, London.

Finally, the authors most sincerely acknowledge their families and all the support and encouragement they have provided over the years. In particular, Howell would like to acknowledge the support of his daughter Kate and of his late wife Gill; Morgan and Rachel would like to acknowledge Sarah, Ian, Rachel, Verity, Tris and Gail as well as the wider family.

Contents

About the Authors

Rachel L. Denyer, M.A., MBCS has a History degree from The University College of North Wales in Bangor, Gwynedd. Since graduating, she has had a career in IT. She is a co-founder of Penrose Antiques Ltd, established in 2010, which specialises in English & Welsh ceramics, silverware and furniture. Between 2014 and 2018 she was one of four Directors of Rose Antiques Fairs, which organised a specialist annual antiques fair in York. In 2018, she was awarded a Masters degree in Art History, and her research led to the publication *A Century of Art: Clitheroe Artists Elijah and Frederick Cawthorne (1843–1940).* She was a contributing author to *The Billingsley Collection including Derby, Pinxton, Worcester, Swansea, Nantgarw and Coalport,* published in 2019. She has a strong interest in genealogical research and social history, and published *A Cocken Family History: The Descendants of William Cocken (born c1695) to 1920,* in 2020. A specialist interest in the history and development of food preparation and presentation, dining and social entertaining led to *1900s Store Cupboard Recipes for Cakes & Biscuits* published in 2021. She was a co-author of *The Pendock Barry Porcelain Service: A Forensic Evaluation,* published in 2023.

Dr. Morgan C. T. Denyer, B.Sc., Ph.D. was born in Yeovil, Somerset and was brought up in Carmarthenshire, South West Wales. He was educated at The University College of North Wales in Bangor, Gwynedd, where he gained a B.Sc. in Zoology and Ph.D. in Neuroscience. He then worked as a research scientist at The University of Glasgow and as a research fellow at RIKEN (Institute of Physical and Chemical Research), Wako, Saitama, Japan before becoming a lecturer at The University of Bradford. Morgan has authored over 90 peer-reviewed publications and patents. In 2010, he co-founded Penrose Antiques Ltd and later co-founded Rose Antiques Fairs. Morgan has published a number of works as author or editor, including: *Charles Horner: Short Guides* (2016); *William Billingsley—the Enigmatic Porcelain Artist, Decorator and Manufacturer* (2016); *Nantgarw Porcelain—The Pursuit of Perfection* (2017); *Swansea Porcelain—the Duck-Egg Translucent Vision of Lewis Dillwyn* (2017); *Derby Porcelain: The Golden Years,1780–1830* (2018); *A Century of Art: Clitheroe Artists Elijah and Frederick Cawthorne (1843–1940)* (2019); and *The*

Pendock Barry Porcelain Service: A Forensic Evaluation (2023). Morgan combines working in the antiques trade with a role as an honorary lecturer at The University of Bradford.

Howell G. M. Edwards, M.A., B.Sc., D. Phil., C.Chem., FRSC is a professor emeritus of Molecular Spectroscopy at the University of Bradford. He read Chemistry at Jesus College in the University of Oxford and after completing his B.A. and B.Sc. degrees he studied for his doctorate in Raman spectroscopy at Oxford and then became a Research Fellow at Jesus College, University of Cambridge. He joined the University of Bradford as a lecturer in Structural and Inorganic Chemistry, becoming the head of the Department of Chemical and Forensic Sciences, and was awarded a personal chair in Molecular Spectroscopy in 1996. He has received several international awards (Sir Harold Thompson Award; Charles Mann Award; Emanuel Boricky Medal; Norman Sheppard Award) in a spectroscopic career which has resulted in the publication of almost 1400 research papers in Raman spectroscopy and the characterisation of materials, along with six books on the application of this analytical technique to art, archaeology and forensic science. He has had a lifelong interest in the porcelains of William Billingsley, especially those from the Derby, Nantgarw and Swansea factories and has authored six books on porcelain: *Swansea and Nantgarw Porcelains: A Scientific Reappraisal, Nantgarw and Swansea Porcelains: An Analytical Perspective, Porcelain to Silica Bricks: The Extreme Ceramics of William Weston Young, 1776–1847, 18th and 19th Century Porcelain Analysis: A Forensic Provenancing Assessment, Porcelain Analysis and Its Role in the Forensic Attribution of Ceramic Specimens, "Raman Spectroscopy in Cultural Heritage Preservation"*, (with Philippe Colomban and Peter Vandenabeele, Springer, 2023) and *Welsh Armorial Porcelains: Nantgarw and Swansea Crested China*, published by Springer Publishing, Dordrecht, The Netherlands and two specialist texts, namely: *The Pendock Barry Porcelain Service: A Forensic Evaluation* (with R Denyer and M Denyer, 2023) and *The Farnley Hall Service: A Unique Survivor in Nantgarw Porcelain* (2023). He has also produced several monographs: *William Billingsley—The Enigmatic Porcelain Artist, Decorator and Manufacturer; Nantgarw Porcelain (with M.D. Denyer)—The Pursuit of Perfection; Swansea Porcelain—the Duck-Egg Translucent Vision of Lewis Dillwyn* and *Derby Porcelain: The Golden Years, 1780–1830.* Howell Edwards is an honorary scientific adviser to the de Brecy Trust on the scientific evaluation of their artworks and paintings.

Chapter 1
Historical Origins of Armorial Devices

Abstract This chapter sets the scene for the identification of the first recorded English armorial porcelain service on imported Chinese porcelain decorated in China and its forerunners of Chinese armorial porcelain artefacts and ornaments. The difficulties apparent for the correct representation and reproduction of Western heraldic coats-of-arms and devices by decorators in China are reviewed and the successful use of ex-libris bookplates in addressing this need is noted. Several specific examples of armorial services are considered and the input of genealogical research into the correct attribution of the heraldic devices to an armigerous family or person is reviewed. The underlying meaning of armorial devices used on ceramics is considered along with the crests and heraldic tinctures used.

Keywords Armorial services · *Pitt* service · *Carnatic* service · Heraldic genealogy · Crests · Heraldic devices · Ex-libris bookplates

> "To begin at the beginning".
> Sherlock Holmes, a Study in Scarlet.
> Sir Arthur Conan Doyle, 1887.

> "The painting of arms is now become a serious business and I must either lose or gain a great deal of business by it --- however, I must at all events come into it !"
> Letter from Josiah Wedgwood to Thomas Bentley. 12th September 1776.

1.1 General Introduction

The first example of a documented English armorial porcelain service for which specimens are still to be found is that of Sir Thomas Pitt, who was a governor of the HEIC at Fort St George in Madras, and who ordered an armorial service in Chinese porcelain in 1705. The distinct advantage possessed by an armorial service over single pieces is the number of artefacts that could have survived as an armorial service

R. L. Denyer et al., *Armorial Porcelain*, https://doi.org/10.1007/978-3-031-63745-2_1

could well have contained several hundred individual items and, in addition, duplicate services for dinner-dessert and tea-coffee are quite commonly encountered. Details of the *Pitt* armorial service will be discussed later, but it is intriguing that despite the multitude of armorial services which followed in the eighteenth century for English, Welsh, Irish and Scottish, and indeed American, families on Chinese porcelains there are no early examples of similar works from the English manufactories which were founded and grew in competition with the Chinese imports from the 1740s and onwards for many years. The reasons for this need to be examined critically.

These major themes will be explored in the current text, but the starting point of the research must necessarily be the consideration of the growth of armorial bearings historically from which the different heraldic parts that comprise a full coat of arms can be estimated and how these have developed over the ages into the forms used in the eighteenth century which provide the focus for the work here. Previous studies by the authors carried out into armorial bearings on porcelains have revealed that there is a lack of specificity which actually defines a ceramic armorial artefact and a debatably unacceptable degree of looseness which has pervaded the canon, giving rise to "exemplars" of armorial porcelains which should not be included definitively in any database (Edwards, *Welsh Armorial Porcelain: Nantgarw and Swansea Crested China*, 2022). In addition to the observation of manifestly incorrectly executed artwork existing on the artefacts concerned, there are noted cases of armorials which should not be included on the basis of a strict definition relating to an armorial piece, and these here have been labelled *pseudo-armorials*, which may originally have been designed to mislead a casual observer or to even to perpetuate a fraud. In some cases, an armorial porcelain service is found to not be really armorial at all upon applying our strictest definitions, yet it has been labelled as such by connoisseurs because the "heraldic device" portrayed on it has perhaps borne a superficial resemblance to the genuine bearings or achievements. Two examples of such ceramic *pseudo-armorials* which are still even now considered in the wider literature to be genuine armorial exemplars are the *Frog* service of Empress Catherine the Great of Russia, which was commissioned for her Chesma Palace in the 1770s from Josiah Wedgwood in his esteemed cream ceramic Queensware, and the *Carnatic* service in Chamberlain's Worcester porcelain commissioned in the 1820s by His Highness Willajah Nabob Azam Jah through their agents in Madras, Griffiths, Cooke & Co., in 1820. Pieces from both of these services are labelled as "armorial" nowadays but in fact they do not strictly exhibit any armorial bearings at all and really have a superficially decorative enamelled application that is reminiscent of an heraldic escutcheon, which is really not a genuine heraldic device, so enabling them both to be classified on our definitive scheme as *pseudo-armorials*. It is to be stressed that in no way does this degrade them in quality, and they must still be admired for the beautiful works of art that they are, but in the true definitive concept they should no longer be classified as armorials and in fact it seems that they never actually could have been commissioned as such! There now follows a more detailed account of several important services that have been mentioned here and the idiosyncrasies which are imparted to them by their armorial or pseudo-armorial bearings.

In this book we have set out to examine when armorials first were adopted as devices on English porcelains and to try and discover the reason(s) for their adoption and evolution. It is clear that the first armorial porcelains for British families were commissioned towards the end of the seventeenth century and the start of the eighteenth century: the earliest documentation refers to pieces of Chinese porcelain ordered by Sir Henry Johnson, of Blackwall, London, whose father, also Sir Henry Johnson, built East Indiamen ships for the HEIC first in Deptford and later in the EIC shipyard in Blackwall. These pieces of porcelain would have been enamelled in China towards the end of the seventeenth century (Howard, *A Tale of Three Cities: Canton, Shanghai and Hong Kong: Three Centuries of Sino-British Trade in the Decorative Arts,* 1997, p. 95). Specific details of these artefacts will be discussed later.

Le Corbellier and Freylinghuysen also make the comment that the East India Company regarded porcelain commissions as incidental to its main trade with China, preferring to deal in tea, spices, silver and silks, and that the porcelain imported to Europe amounted to only 6% of its trading business, despite the fact that David Howard has identified over 5000 armorial services alone decorated on Chinese porcelain for the British market, and these were themselves just a fraction of the more anonymous porcelain carried from China to London in their East Indiamen ships (Howard, *Chinese Porcelain*, Volumes I and II, 1974 and 2003).

An assessment needs to be made into the possible catalyst for the introduction of the concept of the placement of armorial bearings on ceramics. In many respects this opens up our basic theme of early armorial porcelains to the provision of armorial bearings on other artefacts such as silverware, library bookplates, furniture, carriages, mirrors and other household objects. In short, therefore, we can pose the question for analytical research: "What catalysed the use of armorial bearings on ceramics generally and porcelain specifically and did this arise independently of their adoption elsewhere around the homes of arms-bearing families or was it related to the more public demonstration of one's right to bear arms ?".

Specific points of consideration in our research will, therefore, need to impinge upon or address closely the following topics:

- When did the use of armorial bearings, which initially were designed for the differentiation of opposing forces and allies on the battlefield, become a matter for the emblazonment of one's personal effects around the home?
- What type of materials were used for the adoption of armorial bearings?
- What was the purpose of adopting the practice of engraving one's household artefacts such as silverware with personal crests and armorial bearings?
- When did armorial bearings start being used on porcelains or ceramics generally around the home: was this just for the landed gentry and the aristocratic "dedicated followers of fashion" and if so, who or what started it as an integral part of the "*contagion of china fancy*" as Dr. Samuel Johnson so aptly described it in the late eighteenth century?
- Armorial bearings are very specifically drawn up by the College of Arms in England and their equivalent heraldic overseers elsewhere: in that case, why are

there incorrect versions of armorial bearings in existence on porcelains which are, therefore, considered unacceptable both legally and socially - is there perhaps a hidden message being relayed here or does this just reflect the lack of entitlement of the user to bear arms?

What is also not clear is the role the College of Heraldry adopted when it became manifest that someone was using or had appropriated armorial bearings to which they were not entitled; although the registration of granted arms and their design approval needed to be lodged properly with the College of Heraldry, there are very few cases recorded where litigation was pursued against individuals for the misappropriation of armorial bearings. In the latter case, the corollary needs to be argued about the purpose of using an incorrect coat of arms or heraldic device to which one is not entitled in the first place, but there are many instances of where the illegal appropriation of a coat of arms or a partial coat of arms, such as an escutcheon or a crest, has in fact occurred without seemingly causing a punitive retribution from higher authority!

From the foundation of the Honourable East India Company in December, 1600, under Royal Charter from Queen Elizabeth I for trade with the Far East, the importance of the merchant classes in London grew steadily throughout the seventeenth century until the accession of King William III in 1688 saw a huge expansion in international trade with merchant traders providing the financial support for commerce and enterprise and giving rise to Daniel Defoe's comment (Defoe, *The Complete English Tradesman*, 1726) that "*Trade in England makes gentlemen and has peopled the nation with gentlemen, whose children are statesmen, parliamentarians, privy-counsellors, judges, bishops and noblemen*" (Howard, *From Canton to the City: Export Porcelain for London*, 2013). As Angela Howard has noted in her comprehensive text, this growth in merchant trade and its increasing influence in the City of London resulted in a large proportion of the Chinese porcelain imports at this time towards the end of the seventeenth century and the beginning of the eighteenth century being commissioned by those who were either directly involved with the business interests of the East India Company or who were closely associated with its trade, commerce and enterprise, and who occupied the highest financial and political positions nationally. It would be quite natural, therefore, that this class of armigerous gentry would certainly wish to emblazon their coats-of-arms or heraldic crests on the material benefits accruing from their newly acquired wealth, the most tangible of which would have been manifest on their desirable silver plate, carriages and porcelain. Despite this, the survival of armorial porcelain pieces from these very earliest commissions has been perhaps rather sparse: a major advantage of armorial porcelain is that the proper decipherment of the coat-of-arms thereon can provide an unambiguous and well-defined date range for its inception, particularly those that involve the impaled arms of a husband and wife when the genealogy is also known, unlike the more anonymous pieces of contemporary porcelain that can really be dated to only within a rather wider range of dates, which may transcribe twenty or more years within the identified Chinese Dynasty whence the porcelain was manufactured.

1.2 Sir Henry Johnson: The First Armorial Commission on Chinese Porcelain

The records indicate that Sir Henry Johnson Jr. was certainly one of the earliest English commissioners of Chinese porcelain imported via the East India Company towards the end of the seventeenth century. His father, Henry Johnson Sr., was apprenticed as a shipwright to the shipbuilder Phineas Pett in 1623 and in 1652 he moved from Deptford to Blackwall, purchasing the East India Company Shipyard and dry dock there in 1656 for the purpose of building and refurbishing East Indiamen vessels. These ships also doubled as Royal Navy third-rate ships- of-the-line (as heavily armed three-masted, two-deckers, carrying up to 42 eighteen-pounder cannon and a highly trained complement of gun crews): these were indeed formidable merchant vessels that could outrace and outgun any potential pirates in the South China Sea that may be encountered on their hazardous voyages there from Europe. Henry Johnson Sr. was knighted by King Charles II in 1679—he was a great friend of Samuel Pepys, and he features with entries in Pepys' Diary (Pepys, *Diary of Samuel Pepys*, transcribed 1893). Henry Johnson Sr. purchased Friston Hall, a sixteenth century house in Saxmundham, Suffolk, from Thomas Bacon in 1674 but he continued to reside mainly at the Gate House in the dockyard in Blackwall where he could oversee his business. His son, also Henry Johnson (born 1655), took over the shipyard business after his death in 1683: it prospered even more under him, and he too was knighted by King Charles II in 1685. At one stage, the younger Sir Henry Johnson owned shares in 39 merchant ships, some 32 of these trading with the Far East, and he was a man of considerable wealth. Sir Henry Johnson preferred the country life and moved out to the Saxmundham family estate from Blackwall in 1708, becoming MP for Aldeburgh in Suffolk: the dockyard went into decline some years later and could not then compete with the more modern yards that were being built further upriver (Hobhouse, *Blackwall Yard: Development to c. 1819*, 1994). It is this Sir Henry Johnson (who married first Anne Smithson in May 1686) who commissioned a set of six blue and white jardinieres, which were armorially decorated with his escutcheon displaying his coat-of-arms impaled with those of his second wife, Martha Lovelace, whom he married in March 1692. Martha Lovelace, daughter of John Lovelace, 3rd Baron Lovelace of Hurley, became Baroness Wentworth and Baroness Despenser *suo jure* (a phrase meaning, *in her own right* and independent of her husband) in May, 1697: it is clear, therefore, that these armorial artefacts cannot precede the date of 1692 and can therefore be dated provisionally to 1693–1698, to account for the fact that the time lag between the commissioning of Chinese porcelain and the delivery of porcelain ordered from China would normally take between two to three years because of the logistics of placement of the order at the port of Canton, the manufacture of the porcelains at Jingdezhen some 1000 km distant and the dependence of the sailing vessels on the prevailing monsoon winds for the outward and return sea journeys (Edwards, *Welsh Armorial Porcelain: Nantgarw and Swansea Crested China*, 2022). This timing would perhaps suggest that the porcelain commission was intended to celebrate the marriage of Henry and Martha and would have occurred on the granting of the new arms to the heiress Martha on her father's death in 1693. Sir Henry Johnson's second wife, Martha, who outlived him and died in 1745, brought with their marriage some estates in East Anglia and the South of England and he then became known as Sir Henry Johnson of The Gate House, Blackwall, Friston and Aldborough, Bradenham, Bucks., Water Eaton in Oxfordshire and Tuddington, Bedfordshire; estates that were together valued at over £8000 per annum (estimated current value equivalent to about £1 M per annum). Sir Henry Johnson in placing an order for several Chinese porcelain jardinieres was not alone in this respect: in 1697, the traveller Celia Fiennes noted *"that all sorts of pots of flowers and curious greens, fine orange citron and lemon trees"* occupied a place in porcelain jardinieres set in the dining room window of Woburn Abbey in Buckinghamshire, the country seat of the Duke of Bedford. Celia Fiennes (1662–1741) travelled around England on horseback between 1684 and 1703 *"to regain my health by variety and change of aire and exercise"* and visited most of the grand houses and estates in the process: her diary and account of her travels was intended for family purposes only but was later published (Fiennes, *Through England on a Side Saddle in the Time of William and Mary – Being the Diary of Celia* Fiennes, 1888) and provides an illuminating reflection on country life and society in the late 17th and early eighteenth centuries.

It is similarly recorded that Sir Henry Johnson commissioned a set of "flower pots" (also described variously as plant tubs or jardinieres) to contain small orange trees or shrubs as part of the interior decoration at his home in Suffolk: these pots were of a hexagonal and circular design, and now comprise six surviving pieces, four of which are hexagonal and two circular, according to David Howard, but in reality probably three hexagonal, two circular and one of square section (Angela Howard, personal communication). Two of these pots were discovered remaining at Friston Hall in Suffolk. Both are armorial cobalt blue and white decorated Chinese porcelains and the circular jardiniere is illustrated here in Fig. 1.1, depicting very finely executed armorial bearings in monochrome cobalt blue on a white ground: an enlargement of the escutcheon is shown in Fig. 1.2, which will serve to illustrate the detailed armorial discussion that is undertaken below. A hexagonal armorial jardiniere is now in the Bristol Museum and Art Galleries collection and the circular armorial jardiniere illustrated here is in the Metropolitan Museum of Art, New York (Le Corbellier & Freylinghuysen, *Chinese Export Porcelain*, 2003). Angela Howard has included details of the square sectioned armorial jardiniere in her article to be published later in 2023 (Howard, personal communication, 2023: and Howard in McGuire, *Four Centuries of Blue and White: The Freylinghuysen Collection of Chinese and Japanese Porcelains*, 2023).

The armorial escutcheon seems to be quite clearly defined on the jardiniere but the lambrequins or mantlings are rendered more as broad-leaved foliage, the helm is almost unrecognisable as such and the crest, comprising two stylised outspread wings, is mounted on what appears to be a straight torse. The arms on the escutcheon of the jardiniere (Fig. 1.2) comprising the impaling of Johnson with Lovelace have been discussed in detail by David Howard and John Ayers (*Chinese Porcelain and Decorative Art for Export over Three Hundred Years,* 1977) and can be summarised here as follows. Johnson: *Or, a water bouget sable on a chief of the second, three*

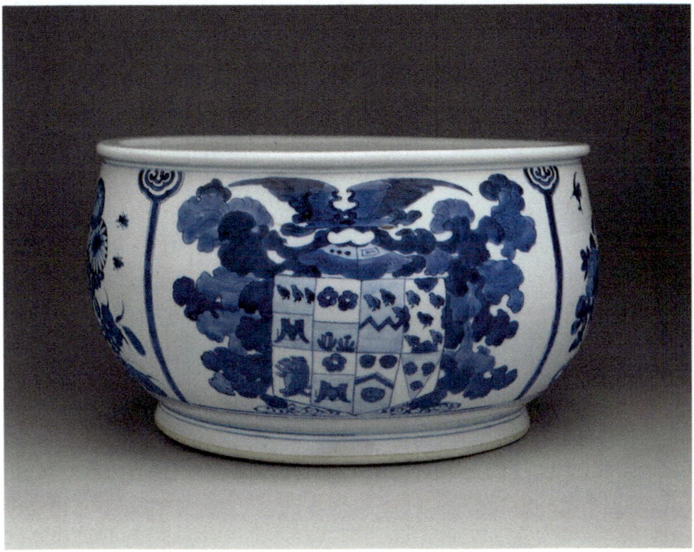

Fig. 1.1 Circular armorial jardiniere, Kangxi porcelain, 33 cm diameter, 19.7 cm high, ca. 1693–7, depicting the armorial bearings of Sir Henry Johnson impaled by those of his second wife Martha Lovelace, who were married in 1692. In 1697 Lady Martha Johnson became Baroness Wentworth and Baroness Despenser *suo jure* and her arms would have reflected this. Now in the Metropolitan Museum of Art, New York, the gift of Mr and Mrs Rafi Y. Mottahedeh, 1976, accession number 1976.112. *Public domain*

bezants (which here actually have been substituted by three martlets) with the crest of Smithson in the third quarter, a squirrel cracking a nut; crest, *out of a ducal coronet per pale argent and azure two wings expanded counter charged.* Quarterly, incorporating Lord: *argent on a fess between three cinquefoils azure, two pheons of the field.* Impaling Lovelace, *gules on a chief indented sable three martlets,* quarterly: Lovelace *azure on a saltire engrailed argent five martlets, or,* and Pye: *or on a pile azure, three escallops of the first.* Generally, the arms have been fairly well rendered by the Chinese artist but the rather curious representation of a martlet on the armorial porcelain must be commented on, this being more reminiscent of a small chick instead of a sleek swallow or swift, yet swallows and swifts are known in China, so it must be assumed that their heraldic significance was lost on the Chinese artist! The water bouget, which is an ancient heraldic symbol that appeared first in the thirteenth century and is meant to represent two leathern bags for water carried on a yoke by a soldier, is rather difficult to recognise as such and is here better defined as a distorted M. The pheons, which are heraldic arrowheads, are also unrecognisable and the cinquefoils are also absent. The most curious representation perhaps is the depiction of an heraldic squirrel cracking a nut which is rendered here as a squirrel being embraced by another animal resembling a bear! The inclusion in the arms of Martha's genealogy as daughter of her father, Baron Lovelace, and mother Margery,

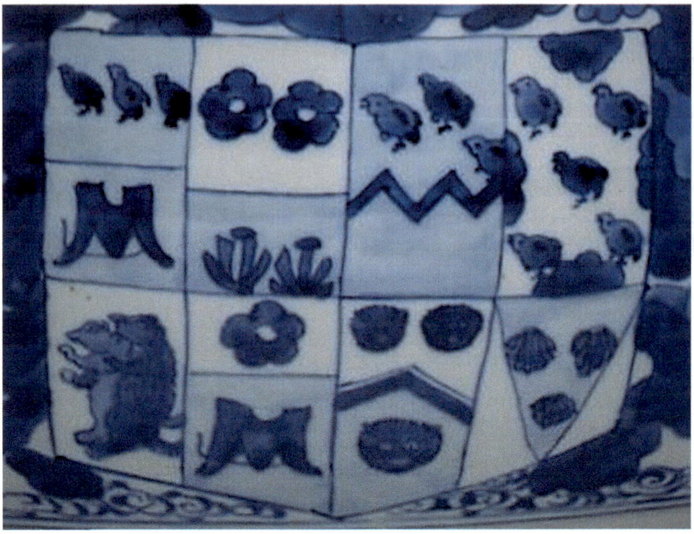

Fig. 1.2 Enlargement of the monochrome cobalt blue armorial bearing on the circular jardiniere shown in Fig. 1.1., depicting the armorial bearings of Sir Henry Johnson impaled by those of his second wife Martha Lovelace, who were married in 1692. In 1697 Lady Martha Johnson became Baroness Wentworth and Baroness Despenser *suo jure* and her arms would have reflected this. Now in the Metropolitan Museum of Art, New York, the gift of Mr and Mrs Rafi Y. Mottahedeh, 1976, accession number 1976.112. *Public domain*

daughter of Sir Edward Pye Bt., and granddaughter of Baroness Ann Wentworth, is to be noted.

1.3 The Sir Thomas Pitt Armorial Service

The earliest example of a surviving armorial Chinese porcelain dinner service for an English family for which examples of the artefacts are still extant is that of Sir Thomas Pitt, Governor of the HEIC Fort St George, Madras (1653–1726), who went to serve in India with the HEIC in the late seventeenth century and who placed an order for his service whilst employed there. This armorial service dates from 1705. It is of Kangxi, Qing Dynasty, Chinese porcelain decorated in the Imari style with predominantly red and blue pigments, in simulation of the imported Japanese Arita wares which were so popular in competition with their Chinese analogues in the mid-to-late seventeenth century, and the service bears the Pitt arms centrally. The escutcheon is noteworthy in being unusually placed heraldically above the enlarged crest of a stork and its torse; an example of a dinner plate from this *Pitt* service is shown here in Fig. 1.3. Kangxi (Xuanye) was the second of the Manchu Dynasty Emperors and reigned from 1661–1722, who succeeded to the Imperial throne following the interregnal and rebellious period in China after the collapse of the Ming Dynasty in 1644, a period that saw Japanese porcelain manufacturers seize the opportunity to market their Arita blue and white and polychrome wares in Europe and were assisted in the transportation of their porcelains to Europe by the VOC Dutch East India Companyan example of this so-called "kraak" blue and white porcelain is shown in Fig. 1.4. The heraldic supporters, helm, mantlings and family motto are also absent from the Pitt armorial service, so this then comes within our definition of being an armorial that possesses only a partial coat-of-arms, viz. bearing only the escutcheon and crest. The self-same "escutcheon" is also rather unusually depicted as a circular device surrounded by wreaths of blue foliage and comprises a chequy of azure and or with three gilt bezants, the correct heraldic assemblage: it is fair to comment here that this technically is not a standard escutcheon as we have defined above and perhaps its significance could escape a cursory examination, being of an oval or distorted elliptical shape, which is non-standard for English heraldry. Secondly, the crest is unusually and incorrectly located beneath the escutcheon and also is relatively of a remarkably large size compared with its accompanying escutcheon.

Thomas Pitt of Ewerne Stapleton, Dorset, made his fortune in India through the sale of a 127- carat diamond that he had acquired there to the French Regent of New Orleans, which was then incorporated as the Pitt Diamond into the Crown Jewels of the French monarchy. In common with many noble arms-bearing individuals and families, reference to Burke's Armory (Burke, *Burke's General Armory—The General Armory of England, Scotland, Ireland and Wales: Comprising a Registry of Armorial Bearings from the Earliest to the Present Time*, 1884) and cross-referencing with the descriptions and pictorial crests displayed in Fairbairn (Fairbairn, *Crests of the Families of Great Britain and Ireland and Their Kindred in Other Lands*, Volumes I and II, 1905) reveals the presence of a number of possibilities for the assignment of the arms and crest displayed on this *Pitt* service armorial plate to a particular family arms bearer. In all, 15 different arms-bearing individuals of different familial descent are registered as being entitled to bear the arms of Pitt in this literature, but there are some significant and yet some subtle differences observed between their displayed individual armorial bearings. Firstly, the input of genealogy can indicate that the award of arms to a particular individual or family may not match with the established chronological date of the porcelain service and it should be recognised that the title may have been an ancient one originally granted some centuries hitherto that is now deemed to be technically extinct and outside the time frame established for the commissioning of the porcelain service being investigated by the ceramic historian or analyst, yet the arms are still recorded. The first entry in *Burke's Armory* that is relevant here is that of Thomas Pitt of Boconnock, co. Cornwall, who was created Baron Camelford in 1784. His arms are seemingly described as they appear on the armorial plate shown in Fig. 1.3, but not precisely so: heraldically these comprise sable a fess chequy argent and azure between three bezants or. It should be noted that the chequy in the armorial plate shown here under discussion is actually tinctured for Pitt of Ewerne Stapleton as or and azure (yellow and blue) with gilt bezants. The crest is a stork argent as shown in Fairbairn (Fairbairn, *Crests of the Families of Great Britain and Ireland and Their Kindred in Other Lands*, Volumes I and II, 1905) and illustrated here in Fig. 1.5. The supporters of the escutcheon for

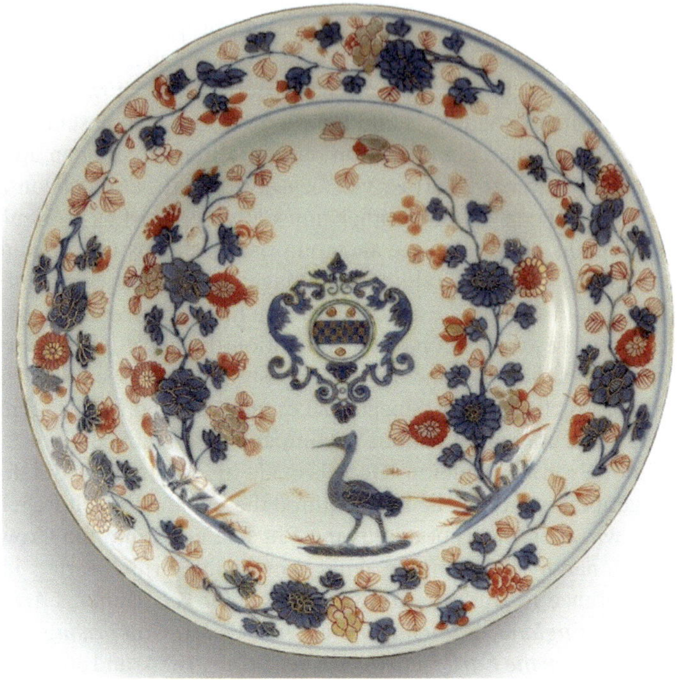

Fig. 1.3 Kangxi porcelain, Qing Dynasty, *ca.* 1705, Chinese armorial porcelain plate from the *Pitt* service, made for Sir Thomas Pitt of Ewerne—Stapleton, Dorset, the earliest surviving example of a Chinese armorial service ordered for the British aristocracy. Note the curious placement of the over-large crest below the escutcheon and the upwards curvature to the torse matching the curve of the cavetto of the plate, which should properly fit over the helm and require a downwards curvature. The oval escutcheon displays three bezants—a Byzantine gold coin discovered by the Crusaders *en route* to Jerusalem and in Latin called "*Bizantius aureus*", a gold bezant. The bezants depicted here are more akin to elliptical golden roses and should properly be shown as being plain and circular heraldically. *Private Collection. Reproduced with the permission of Kingschina Ltd, Mayfair, London*

Pitt of Ewerne Stapleton are given as two Cornish choughs regardant elevant proper. His motto is "*Per ardua liberi*", which translates as "*Freedom through adversity*". The younger brother of this Thomas Pitt was William Pitt, Earl of Chatham, the British Prime Minister, who has the same coat of arms, but the crest is now a stork which has its dexter claw resting on an anchor erect cabled or (Fig. 1.3, Fairbairn: Plate 224 /10). The supporters of the William Pitt escutcheon are a lion rampant gardant and a buck attired collared and chained with the motto "*Benigno numine*", which translates as "*By the Favour of the Heavens*". A third Pitt, namely Thomas Pitt, the Earl of Londonderry, was ennobled in 1726, but this became extinct upon his death in 1764, and the arms are quartered with those of Pitt having 1st and 3rd and the 2nd and 4th quarters being sable with two wings conjoined argent. The crest is now a stork argent, beaked and membered or, holding up its dexter claw (Fig. 1.3,

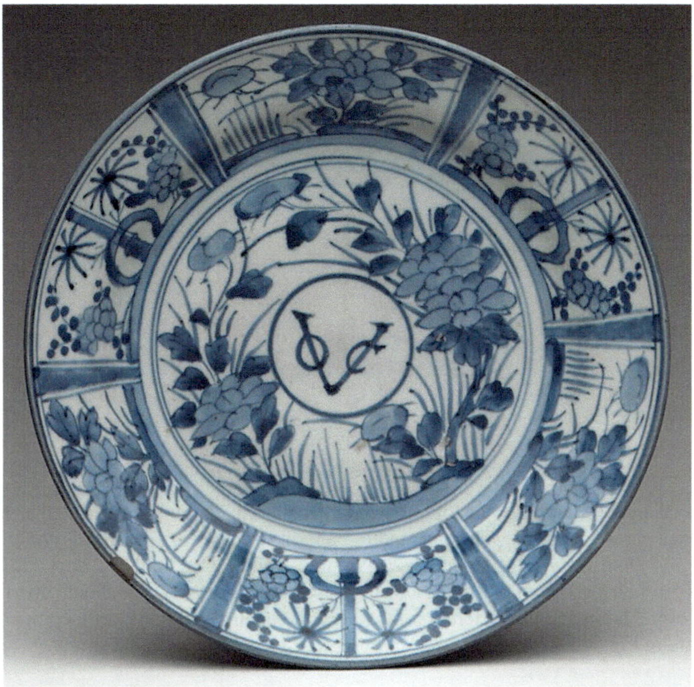

Fig. 1.4 Dutch East India Company *kraak* style large porcelain plate, diameter 12.5″ (31.3 cm), Japanese Arita ware, bearing the logo VOC in underglaze blue with a six-panelled geometric border, Edo period, *ca.* 1660. Metropolitan Museum of Art, New York, the bequest of Dr and Mrs Roger G. Gerry, Roslyn, New York, 2000, Accession number 2002.447.40. *Public domain*

Fairbairn: Plate 294 /11). The supporters of this Pitt escutcheon are two falcons sable with a chaplet of red roses barbed, with the motto: "*Amitie*" which translates as "*Friendship*". There are eleven other Pitts cited in Fairbairn with other notable differences in their escutcheons, supporters, crests and mottoes.

The choice of assignment to a family arms bearer now comes down to that between Pitt of Boconnock and Pitt of Ewerne Stapleton: the former had arms granted in 1784 whereas the latter had arms granted in 1604. Hence, it can be deduced logically that Pitt of Ewerne Stapleton is the one whose arms are depicted on the plate when the armorial service was commissioned chronologically in 1705. An interesting point emerges when the crest of Pitt of Ewerne Stapleton as depicted in *Fairbairn* (Fairbairn, *Crests of the Families of Great Britain and Ireland and Their Kindred in Other Lands,* Volumes I and II, 1905) (Fig. 1.5) is compared in detail with that painted on the armorial plate as shown in Fig. 1.3: the beak of the stork in the Pitt crest as appearing in Fairbairn's armory is clearly open and the tongue is exposed whereas the beak of the stork depicted on the crest enamelled on the plate is firmly closed. Also, the torse is displayed incorrectly on the armorial porcelain plate and curves upwards (i.e. it is of a convex profile), following the curvature of the edge of the

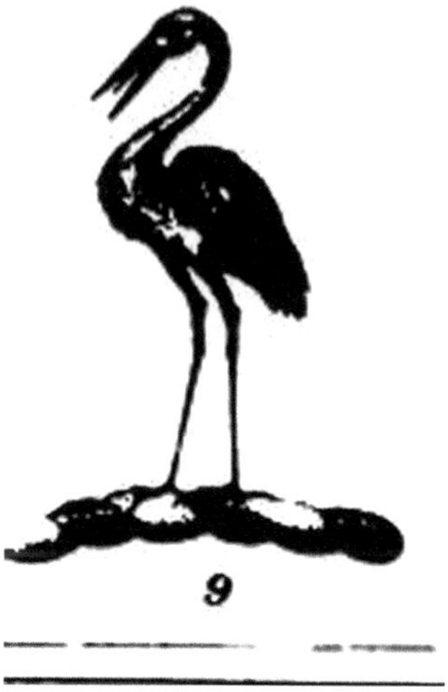

Fig. 1.5 Stork crest for Sir Thomas Pitt of Ewerne-Stapleton, co. Dorset, the titled arms-bearer of the *Pitt* service shown in Fig. 1.3, as depicted in Fairbairn (*Crests of the Families of Great Britain and Ireland and Their Kindred in Other Lands,* Volumes I and II, T.C. and E.C. Jack, Edinburgh and London, 1905, Plate 33/9). Note the torse curving downwards properly as it should do to encompass the contour of the helm and the different portrayal of the open beak of the stork compared with the armorial plate in Fig. 1.3

plate base, known as the *cavetto*, whereas a properly executed torse heraldically, if it is shown curved at all as in the crests that appear in Fig. 1.6 from the armory, should curve downwards to match the contour of the curvature of the top of the helm upon which it is designed to sit, as indeed it does as depicted in the armorial sketch from Fairbairn's armory in Fig. 1.5.

Several ceramic and armorial historians have commented hitherto that the Chinese depiction of British armorials enamelled on their export porcelain is sometimes lacking in exactitude and precision and that this would naturally depend critically upon the accuracy of the original copy drawing that was supplied by the arms-bearing individual or family to the HEIC agent (or supercargo) for enamelling in the workshops in Jingdezhen or Canton in China and also upon the specific knowledge and attention to detail given by the HEIC supercargo, who was responsible in Canton for inspecting and accepting the work carried out by the porcelain enamellers in China. This is probably the case that applied here and a small difference in the enamelled crest and torse has escaped the attention of the supercargo who was monitoring the enamelling being done, possibly on site in Canton, or more likely at this time in Jingdezhen and then presented to the HEIC supercargo as the finished article for shipment to London from Canton. In this particular case and at this time, the armorial decoration of the Pitt service probably would have been accomplished by enamellers in Jingdezhen and copied there from a template or drawing of the crest and associated arms, or perhaps even a bookplate from the library of Sir Thomas Pitt (although the first documentary use of a bookplate for this purpose probably dates form perhaps

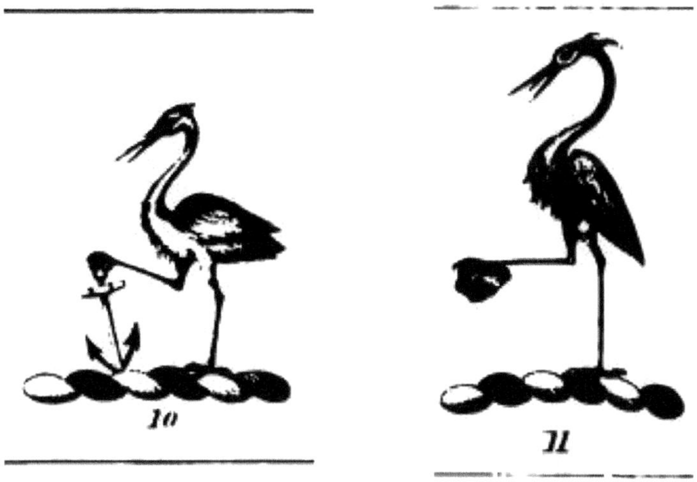

Fig. 1.6 Stork crests for some other Pitt individuals showing small variations within the crest as represented in Figs. 1.3 and 1.5: left, the crest of William Pitt, Earl of Chatham (Fairbairn: Plate 224 /10) and right, crest of Thomas Pitt, Earl of Londonderry (Fairbairn: Plate 294 /11). Depicted in Fairbairn (*Crests of the Families of Great Britain and Ireland and Their Kindred in Other Lands,* Volumes I and II, T.C. and E.C. Jack, Edinburgh and London, 1905)

a decade or more later, see below). It is also clear that the presence of the family motto in this case could perhaps have facilitated in the identification of the armorial service more easily by ceramic historians and would have acted as a confirmation of its correct attribution.

Although Pitt of Ewerne Stapleton has been identified unequivocally as the owner of this important historical service, our research indicates that he has also been matched elsewhere in some historical accounts with an incorrect mottohis arms there being accorded the motto of "*Amitie*" ("*Friendship*"), which is the actual motto of Pitt, the Earl of Londonderry, instead of his own true motto "*Per ardua liberi*" *("Freedom through adversity")*! The armorial researcher should be aware of the occasional occurrence of this sort of discrepancy and the inexactitudes that can consequently arise. In this case too, one should be aware that the selection of an individual or family made on the basis solely of the motto may not correlate with their coat-of-arms as displayed on the escutcheon and the presence of a motto does not always assist in the correct and unequivocal attribution of the arms being made and can even be a misleading side-track to the armorial investigation. The heraldic background is that the motto, unlike the coat-of-arms or the crest with which it has become associated, does not have to be formally registered with the College of Arms in England and can therefore be literally assumed by common usage in a coat of arms.

1.4 The Institution of Armorial Bearings and Coats-of-Arms

There is no doubt that the Normans brought the concept of heraldry and the bearing of heraldic devices to England in the eleventh and twelfth centuries and that these were used manifestly for the identification and rallying points of baronial forces on the battlefield and as colours for opposing teams at jousting competitions for their knights and their supporters. The main components of a full heraldic coat of arms are:

- The *escutcheon:* a shield which bore special figures and geometric designs which signified the personal descendancy and ownership of the armigerous person involved. The escutcheon at first was a direct image of the knight's shield that was borne in battle and took the form of an elongated tear-drop kite shape which often covered the body length of the person concerned and can be seen now in the stone effigies on the tombs of Norman knights buried in English cathedrals. This developed into the so-called "heater" shield which was of a smaller square-cut shape and more suited to cavalry and fighting on horseback. These practical designs gave way in the fifteenth and sixteenth centuries to further embellishments of a more rococo shape which included curved spandrels on the top of the shield and its sides and an embrasure, called a *bouche*, on the top of the shield at the dexter side designed to accommodate the lance. The shield differed materially for foreign European heraldic arms, such as those of France, and the states of modern Italy and Germany, and a greater variety of designs were adopted including a circular or elliptical "shield", see for example Fig. 1.10, which is never encountered formally in British heraldry. As the heraldic escutcheon was derived from its use by men on the battlefield, ladies entitled to bear arms in their own right do not display a shield, but their heraldic emblems and devices are contained in a diamond-shaped lozenge
- The *helm:* a representation of the metal helmet with slatted visor which protected the head and shoulders of the knight in battle. This may be shown full faced or in a side profile and with the visor open or closed. Generally, only royal knights were permitted to display their helm with a frontal aspect and with their visor open
- Above the escutcheon with its blazon may sit a crown or a coronet, with perhaps a chapeau (a soft cap), each of which marks the rank or noble standing of the bearer of the arms; the coronet indicates the status and rank in the English peerage, namely Marquess, Viscount, Earl and Duke, of the arms bearer through the presence, number and arrangement of small balls on spikes (known as "pearls") and strawberry leaves around the coronet (or a circlet for nobles). The numerical arrangement of leaves and pearls varies significantly in other national heraldries (Edwards, *Welsh Armorial Porcelain: Nantgarw and Swansea Crested China*, 2022). In the upper compartment of the coat-of-arms is also found the helmet (helm) or galero. For the regal status of an arms bearer the coronet is replaced with a jewelled crown. The helm comprises several types of open-faced or closed visor, again indicating the rank of nobility of the shield bearer—usually a knight

or a baronet will have a closed visor on the helm, whereas higher ranks in the peerage will have a barred or an open visor. The helm may also adopt a frontal or a side profile image, whether open or closed, but a regal status is always accompanied by a front-facing open visor. Sometimes the helm is accompanied by a prominent *bavier* (known as a "beaver"), or protective chin piece

- The helm is replaced for armigerous clerics by a galero: this is a broad-rimmed hat which is affixed with tassels, tinctured black in the Anglican Church and red in the Roman Catholic Church
- The *crest*: this sits atop the helm and was used as an additional identifier in battle. Initially, crests were adopted in the form of animals, birds, reptiles or tokens that designated bravery and valour such as a sword, harpoon, arrowhead (pheon) and lance. Later crests were less realistically depicted symbolically for wearing on a helm in battle and, for example, comprised depictions of the stern of a Spanish battleship (Admiral Lord Horatio Nelson) and a castellated fortified battlement tower (Pendock Barry)
- The crest sat on top of a *torse*: the torse originally was a silk entwined wreath, perhaps in two colours which facilitated the physical affixing of the crest to the helm. Hence it is properly represented in heraldry by concave twisted skeins which denoted its fixation to the crown of the helm
- The *supporters*: these had a great symbolism and significance in the selections of animals or objects used. They were situated to the right and left of the escutcheon (labelled *dexter* and *sinister* in heraldic terms, meaning right and left, respectively, to the person holding the escutcheon or shield). The animal chosen as a supporter popularly was a lion, leopard, wolf or bear, all designed to display valour and bravery and mythical beasts such as a unicorn, wyvern, gryphon or dragon were also popular as heraldic supporters. As real supporters at jousting tournaments protected the knights at their tent entrances so the supporters on their heraldic coats of arms always faced forward towards the observer and often displayed an aggressive stance, called *rampant* or *gardant* in heraldic terms. The use of supporters varied in overseas heraldic licence and in some cases, they were used singly or not used at all
- The *motto*: this was a statement, usually in Latin, English, French or Welsh which had a special significance for the arms bearer. The motto was situated in the lower field beneath the escutcheon and is sometimes accompanied by a badge of honour
- The *slogan:* alternatively derived from the battle-cry, a slogan was situated above the crest and does not appear on every coat of arms

Rarely does the full coat of arms with all the components listed above appear on a piece of armorial porcelain and in most cases a simple escutcheon, without supporters and helm, but possibly containing a motto will be used. The escutcheon may be accompanied by a crest, or the crest may feature alone or with a motto. Where a motto is cited, some care needs to be taken by the researcher since the motto in English heraldry is not part of the formal granting of arms by the College of Heralds: this means that several arms bearing families and descendants may have acquired different mottoes so that the same crest can have a different motto and, equally confusing, different escutcheons and crests may have the same motto for different branches of the armigerous family (Elvin, *A Handbook of Mottoes Borne by the Nobility, Gentry, Cities, Public Companies etc.*, 1860). Some examples of this will be cited later but the standard reference book for British armorial crests is that of Fairbairn (Fairbairn, *Crests of the Families of Great Britain and Ireland and Their Kindred in Other Lands,* Volumes I and II, 1905) who gives several thousand illustrated examples of crests and their descriptors and also lists more than six thousand mottoes, some of which have more than 15 families associated with their usage. The message for a definitive result in armorial assignment and attribution is quite clear: the researcher must be very careful in attributing an armorial assignment based on mottoes alone and an assiduous double-checking of the fine detail of the escutcheon and the crest if also provided in the armorial decoration is essential for the correct conclusion to be made.

1.5 The Use of Genealogy

In many cases it is essential to investigate the genealogy of the armigerous family for the correct assignment of the coat of arms or heraldic device on a piece of armorial porcelain. This is especially important where several possibilities can exist for the identification of a crest, an escutcheon or motto that has been cited. This is apparent for the proper assignment of the armigerous *Pitt* family, who commissioned the earliest known extant English armorial service of Chinese porcelain in 1705.

A second example is provided by the *Pendock-Barry* service (previously known as the *Barry-Barry* service) of Derby porcelain: for many years the origin and dating of this particular service has been debated by connoisseurs and ceramic historians but recent genealogical studies have now confirmed that this unique service was created at the Derby China Works in 1805 for Pendock Barry of Roclaveston Hall, Tollerton, bearing his escutcheon which formulated his descendancy from the Barry line and its premature usage on this service prior to its formal granting from the College of Heralds (Denyer, Denyer and Edwards, *The Pendock-Barry Service: A Forensic Re-Evaluation*, 2023) (Fig. 1.7). Reference to this book will reveal the critical and significant role of genealogical research in the correct attribution of the arms displayed in the escutcheon, which upon a cursory inspection may seem to be straightforward in assignment but in fact are not and which have now resulted in a re-naming and chronological reassignment of this important Derby porcelain service.

In other instances, too, the family genealogical input into the holistic evaluation of the origins of armorial services has had a significant effect upon the outcome. Two examples include the *Viscount Weymouth* service in Nantgarw porcelain (Fig. 1.8) and the *Clarke of Hereford* armorial service in Swansea porcelain (Figs. 1.9 and 1.10) (Edwards, *Welsh Armorial Porcelain: Nantgarw and Swansea Crested China,* 2022). Details of the genealogical input which exposed the previous assignment of these two services to be incorrect have been recounted elsewhere, as indicated above, but for many years the armorial connections were patently wrongly assessed to both of these

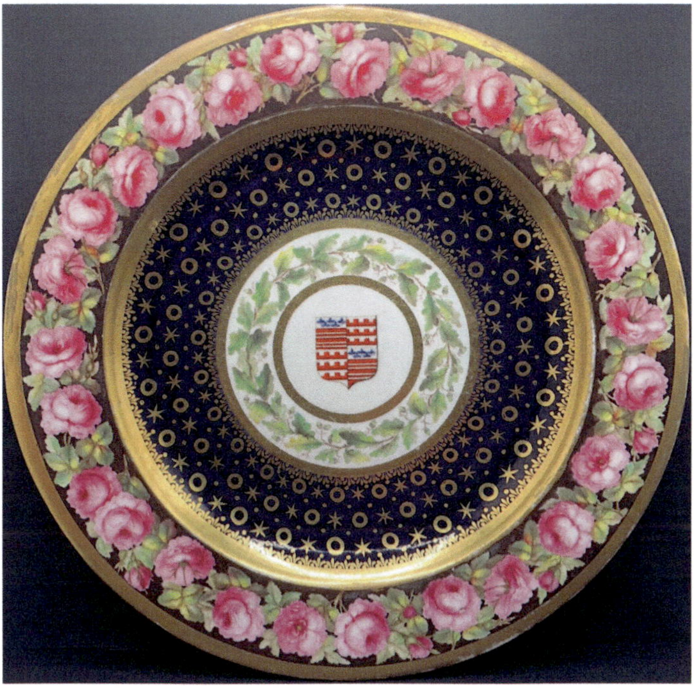

Fig. 1.7 Dessert plate from the *Barry-Barry* (*Pendock-Barry*) armorial service supplied from the Derby China Works to Pendock Barry Neale of Tollerton Hall, Nottinghamshire, in 1806/7. It is a beautiful example of William Billingsley's rose painting and was probably executed externally for Derby when he was resident as a freelance enameller in Brampton-in-Torksey. Many are marked in gold with the crossed swords, crown and two batons of the Derby China Works manufactory, this example does not carry the gold Derby mark. Note that this armorial exhibits the escutcheon only and much detailed genealogical research was undertaken from this to determine conclusively its correct attribution and chronology. *Private Collection*

and only recently have these been reattributed and confirmed that they should now be correctly entitled the *Thomas Thynne* service and the *Clarke of Newcastle -Parkinson of Kinnersley* service, respectively. It is also relevant to note here that both of these services possess unusual heraldic features which had hitherto escaped attention and masked their correct assignment and this reinforces the concept that the smallest details need to be appreciated to achieve the desired result. A case in point here is that the Swansea porcelain hitherto named *Clarke of Hereford* service had been assigned primarily on the basis of the escutcheon and supporting motto of *"Frangas non Flectes"*, which undoubtedly has the family association, but surprisingly the presence of the quartered scallop shell, which is an incorrect crest for this particular assignment, was ignored and which would have otherwise registered the correct attribution to Clarke of Newcastle impaling the arms of Parkinson of Kinnersley, Hereford, on the occasion of their marriage. The scallop shell crest can be clearly seen in Fig. 1.10.

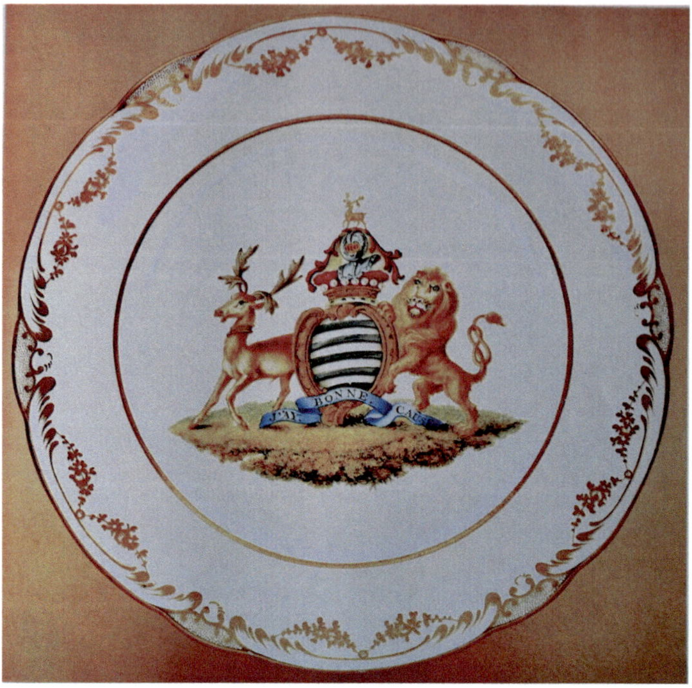

Fig. 1.8 Dinner plate from the Nantgarw China Works, *ca.* 1817–1820, impressed NANT-GARW C.W., bearing the full coat-of-arms of Viscount Weymouth, Lord Henry Frederick Thynne (1797–1837), who became the 3rd Marquess of Bath in 1837 just months before he died. The escutcheon is quartered: 1st and 4th, a barry of ten or and sable; 2nd and 3rd, or a lion rampant tail nowed gu. Crest: on a wreath of gu. and or, a reindeer statant or collared sable. Supporters: dexter, a reindeer or gorged with a plain collar sable; sinister, a lion, tail nowed and erect, gules. Motto: *"J'ay bonne cause"* which translates as *"I have good reason"*. Family seat, Longleat House, Wiltshire. Illustrated in Plate 56 John et al., *Nantgarw Porcelain Album*, 1975. *Private Collection*

1.6 A Genealogical Conundrum: Fawkes of Farnley Hall

To highlight the care that needs to be taken in researching heraldic coats-of-arms and symbols and the critical role that genealogical research can play in the elucidation of their commission and meaning a case study is presented here of the Chinese Qianlong armorial service bearing the arms of Fawkes. An interesting armorial service in Chinese Qianlong period porcelain is recorded for Fawkes of Farnley Hall, Otley, Yorkshire (Fig. 1.11), whose attribution will be discussed later. Walter Ramsden Hawkesworth Fawkes was a friend and benefactor of JMW Turner, who spent three months every year at Farnley Hall painting local scenes from 1796 until Walter Fawkes died in 1825: at one time Walter Fawkes had the largest collection of Turner oil paintings and watercolours in the country, some 206 in total it has been estimated. He was also an avid collector of porcelains and *objets d'art*.

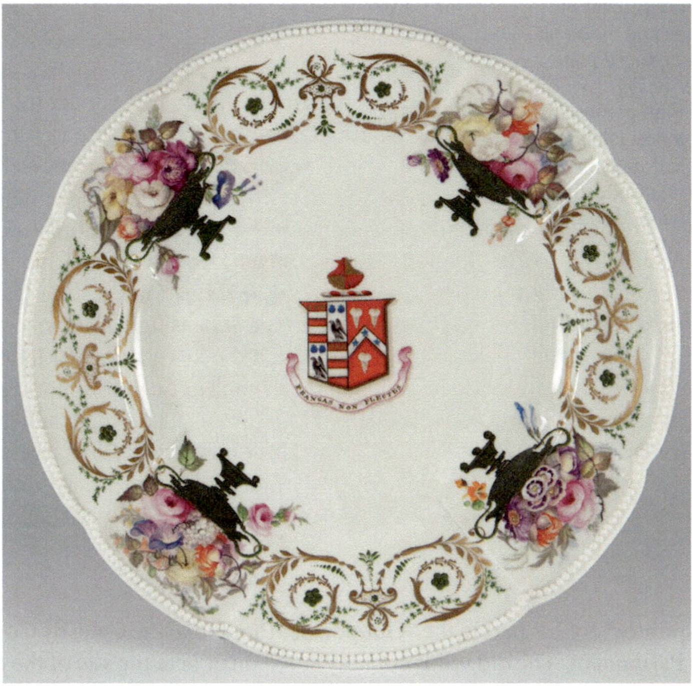

Fig. 1.9 This dinner plate of "trident" steatitic magnesian porcelain, with the impressed trident and SWANSEA mark, has applied small porcelain beads at the rim and is attractively decorated with a border of flowers, chrome green urns and chrome green enamel and gold scrolls painted locally at the factory by Henry Morris. The escutcheon and its quartering was definitively assigned to that of Clarke of co. Hereford impaling Parkinson of Kinnersley Castle, Hereford, as defined in *Burke's Armory*, However, the motto *Frangas non Flectes*", meaning "*Thou may'st break but shall not bend me", is* not that recorded for Clark of Hereford, which prompted Ernest Morton Nance (*The Pottery and Porcelain of Swansea and Nantgarw*, 1942) to assign the arms incorrectly to Abel Gower of Corsygedor, Merioneth. Now better assigned to Clarke of Newcastle impaling Parkinson of Kinnersley Castle on the basis of the escallop shell crest. Reproduced with permission from Gwyn Jones, Director of the John Andrews Charitable Trust Collection, Oriel Plas Glyn-y-Weddw, Llanbedrog, Abersoch, Gwynedd

Farnley is mentioned in the Domesday Book of 1086 as Fernelai (or Fernelie). The first mention of a Fawkes family in connection with Farnley occurs somewhat later in 1289 when Falcasius de Lindeley is reported as having successfully taken legal action against the Vicar of Leathley and 19 others for cutting down trees on his property. In the Subsidy Roll of 1378 there is a mention of "*John ffawkes and his wife and of their villa de ffarnlay*". A John Fawkes of Farnley, Steward of the Forest of Knaresborough, is mentioned in Dugdale (Dugdale, *Visitation of Yorkshire*, 1894, Part III, page 204: Sir William Dugdale was Norroy King-of-Arms in the seventeenth century) leading in to a Henry Fawkes, Esquire of the Sword and Lord Mayor of York between 1522 and 1549, whose son William (married to Ellen Haryngton) was the father of Guy Fawkes, who was born in 1570 in York and was executed in 1602 as

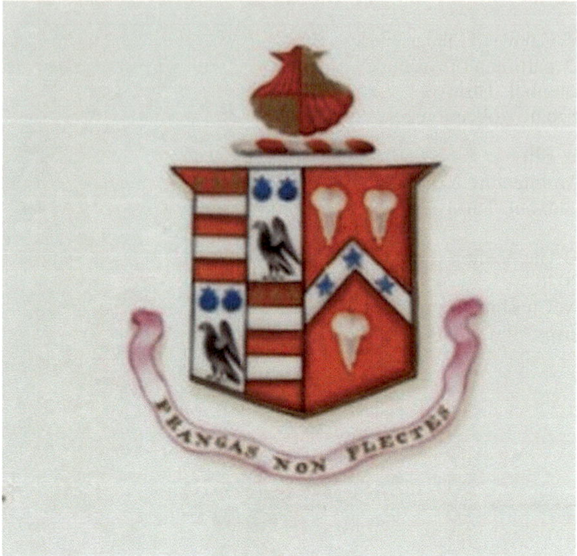

Fig. 1.10 Enlargement of the escutcheon and motto on the Swansea armorial plate shown earlier in Fig. 1.9. Reproduced with permission from Gwyn Jones, Director of the John Andrews Charitable Trust Collection, Oriel Plas Glyn-y-Weddw, Llanbedrog, Abersoch, Gwynedd

"The Conspirator" for his part in the "Gunpowder Plot" (*aka* the Catesby Conspiracy) to blow up Parliament. The old villa probably stood on or near the site of the older part of the existing house, which was built in the late 16th or the early seventeenth century. In the 1780s, Farnley Old Hall was owned by Francis Fawkes, described as a "*the blunt squire of many acres*", a wealthy widower who died without surviving issue (Farnley Hall, Otley, Yorkshire, House Brochure). He bequeathed the Hall and his estate to a kinsman, Walter Beaumont Hawksworth of Hawksworth Hall, Guiseley, who inherited it in 1786 and thereby also acquired the additional surname and arms of Fawkes in the terms of the will (Burke, *A Genealogical and Heraldic List of the Landed Gentry of Great Britain and Ireland*, 1879), becoming High Sheriff of Yorkshire in 1789. His motto was *A Deo et Rege*, translating as *From God and the King*. His arms are given as: Quarterly: 1st and 4th *sable three hawks close proper*; 2nd and 3rd *a fess between in chief three mullets and in base as many chevrons interlaced sable* (Burke, *A Genealogical and Heraldic List of the Landed Gentry of Great Britain and Ireland*, 1879). His crest is cited as a *falcon ppr.* and is depicted in Fairbairn (Fairbairn, *Crests of the Families of Great Britain and Ireland and Their Kindred in Other Lands,* Volumes I and II, 1905; the crest is shown in Volume II: Plate 85/2).

Walter Hawksworth Fawkes died in 1792 but not before he had started to alter the existing Farnley Old Hall to make what is essentially two houses under the direction of the architect John Carr of York. His successor, Walter Ramsden Hawksworth Fawkes (1769–1825), duly completed the transformation of Farnley Old Hall into two distinct but connected buildings, the older wing at the north end and a newer four-square Georgian addition at the south end with several large bays. Walter Ramsden Hawksworth Fawkes married Maria, daughter of Robert Grimston of Neswick Hall

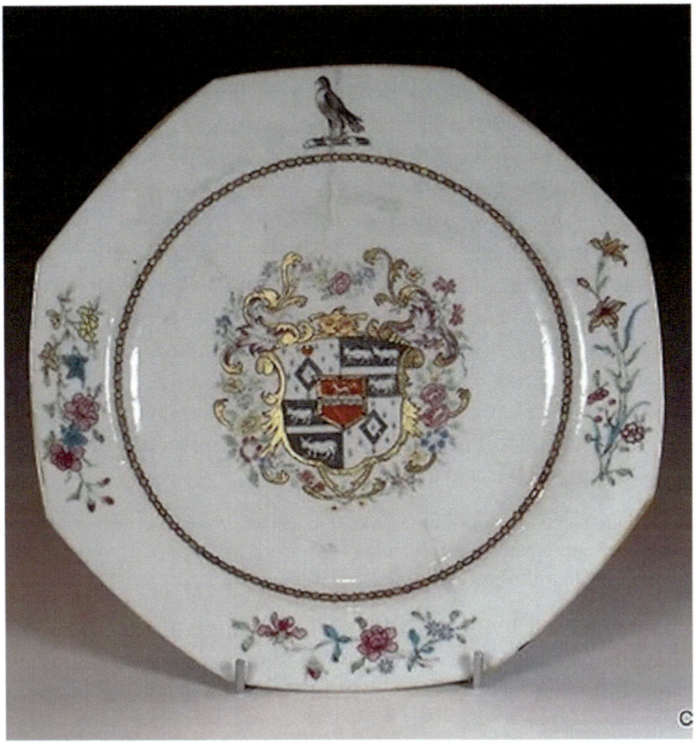

Fig. 1.11 Qianlong period Chinese armorial porcelain plate with the coat-of-arms of Fawkes impaling Wilkinson in a baroque escutcheon. Crest of a falcon on a torse is displayed in the verge. *Courtesy of Kingschina Antiques, Mayfair, London*

(who died in 1813), in 1794 and they had four daughters and three sons. Maria Fawkes died in 1813 and Walter then married Maria Sophia in 1816, daughter of John Vernon of Clontarf Castle, Dublin, widow of the Hon. Pierce Butler, third son of the Earl Carrick, Mount Juliet, Karryk MacGryffin, Kilkenny, who survived him. There were no children of the second marriage.

Walter Ramsden Hawksworth Fawkes, who became MP for Yorkshire in 1806 and High Sheriff in 1823, then filled his house with a remarkable collection of artworks, which included 250 watercolours and six large oil paintings by Joseph Mallord William Turner, which were purchased from Turner for the sum of £3,500, equivalent to £400,000 at present day values (Blayney-Brown, *J.M.W. Turner, Sketchbooks, Drawings and Watercolours,* 2012). Turner first visited Farnley Hall in 1796 whilst visiting Lord Lascelles in Harewood House with his friend and watercolourist rival Thomas Girtin. Whilst staying and sketching at Farnley Hall, which Turner did for several months in the summer of each year between 1808 and 1824, he compiled his celebrated five-volume *Ornithological Collection* between 1820 and 1824 (E.

Jacklin, *Natural History Studies, ca. 1820–1824,* in *J.M.W. Turner: Sketches, Drawings and Watercolours,* 2017). In 1819, Walter Fawkes held an exhibition in London which included many of the works by Turner in his possession; his wife, Maria Sophia, also befriended Turner and when viewing Turner's picture of a *Snow Storm* in 1820 quoted Turner as saying "*That he could never have painted the Snow Storm had he not asked a sailor to lash him to the mast to observe it. In this position I stayed for four hours and never expected to escape. I felt bound to record it if I did*". The acknowledged importance of Farnley Hall and its art collection was surely due largely to its tenure by Walter Ramsden Hawksworth Fawkes and his friendship with Turner and merited its inclusion in Neale's engravings of gentlemen's seats (Neale, *Gentlemen's Seats: Views of the Seats of Noblemen and Gentlemen in England, Wales, Scotland and Ireland,* 1830), shown here in Fig. 1.12.

Walter Ramsden Hawksworth Fawkes' art collection impressed John Ruskin (1819–1910), the first Slade Professor of Art at the University of Oxford in 1869, when he visited Farnley Hall and commented:

> Farnley is a unique place, there is nothing like it in the world – a place where a great genius has been loved and appreciated, who did all his best work for that place, where it is treasured up like a monument in a shrine.

It is certainly a credible hypothesis that such an astute collector of fine works of art such as Walter Ramsden Hawksworth Fawkes, with his access to fellow politicians, dignitaries and influential London society in the second decade in the nineteenth

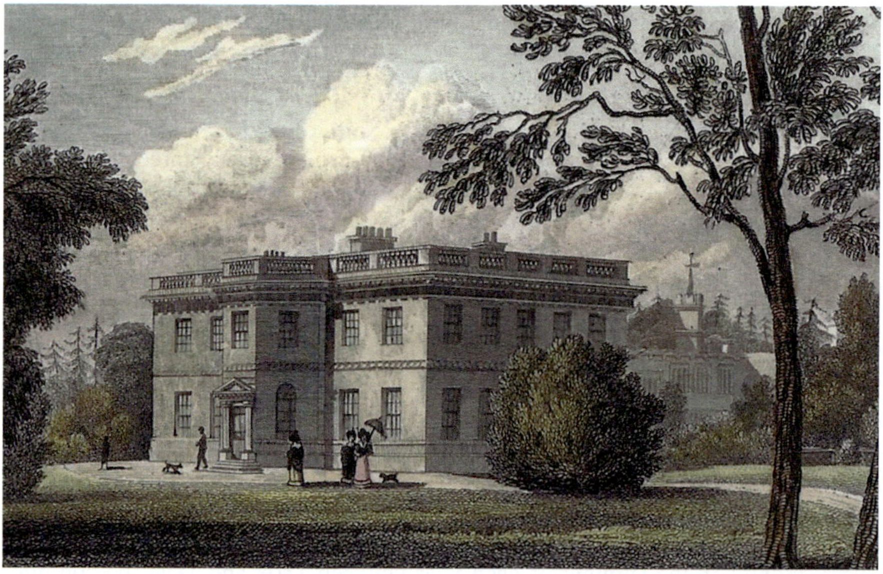

Fig. 1.12 An etching of *Farnley Hall, ca.,* 1830, by J.P. Neale for his book entitled *Gentlemen's Seats: Views of the Seats of Noblemen and Gentlemen in England, Wales, Scotland and Ireland,* H.W. Bond for Sherwood, Jones & Co., Paternoster Row, London, London, 1830. *Public domain*

century, would have been in a prime position to acquire a service of armorial porcelain which was so much in desirability for its elegance: however, the detailed genealogical research chronology is not compatible with the service illustrated in Fig. 1 being assigned to Walter Ramsden Hawksworth Fawkes as Qianlong reigned as the fourth emperor of the Qing Dynasty between 1735 and 1796, when he abdicated in favour of his son Jiaqing. Hence, it is not realistic to assume that Walter Fawkes commissioned this service perhaps in the last year or so of Qianlong's reign and we have to inspect the arms more closely to better define this commission. Furthermore, the coat-of-arms of Walter Ramsden Hawksworth Fawkes do not match those on the Qianlong period plate shown in Fig. 2.11: an image of the coat-of-arms of Walter Fawkes as given on the cover of Finberg's book (Finberg, *Turner's Watercolours at Farnley Hall*, 1912) is shown in Fig. 1.13 where the arms of Fawkes are impaling those of Vernon. It seems then that we should look to Francis Fawkes for the potential commission of the Chinese Qianlong armorial porcelain service; born in Otley in 1707, he died in July 1786, which nicely spans the Qianlong period—the plate also bears the Qianlong seal mark in underglaze blue (Fig. 1.14). Francis Fawkes married in April 1733 Christiana Wilkinson, daughter of William Wilkinson of Newhall, Kirkbrigg, who died in 1767, and as a widower with no surviving issue (a son born in 1734 who died unmarried in 1763) he bequeathed his Farnley estate to a distant relative, Walter Beaumont of Hawksworth, Guiseley, father of Walter Ramsden Hawksworth Fawkes. The arms of Wilkinson of Kirkbrigg are given in Burke's Peerage (Burke, *A Genealogical and Heraldic List of the Landed Gentry of Great Britain and Ireland*, 1879) as: *Escutcheon: azure, a fess erminois between three unicorns passant argent, a canton or;* Crest: *out of a mural coronet gules, a demi-unicorn rampant erased of the first,* which were first granted in 1615 to Lawrence Wilkinson of Ferryhill. It seems quite positive therefore that the arms depicted on the Qianlong plate are those of Fawkes impaling Wilkinson: the presence of the red crescent in the dexter part could be an indicator that Francis was the third son of Francis Fawkes of Farnley. A cursory consultation of some historical records indicates that Francis Fawkes never married, which is incorrect and therefore could have well sent a researcher down the wrong path in assigning the service: it would have been very tempting then to have selected his descendant, Walter Ramsden Hawksworth Fawkes as the originator of this commission, a totally incorrect assumption despite his historically proven support for the fine arts.

A further piece of interest in this context has surfaced from the workshop of James Giles and dated to around 1765–1768 (Hanscombe, 2008; Coke, 1983), being a trio of coffee cup, teacup and saucer from the Worcester factory exhibiting the arms of Beaumont-Ayscough (Fig. 1.15), but bearing a faux Meissen underglaze blue mark (Fig. 1.16), showing *the three unicorns argent with a canton argent* as in the Fawkes Qianlong service discussed above. However, the crest is now *a horned bull couped at the neck and quartered in gules and argent,* and the escutcheon again is of a derived baroque-rococo form but different stylistically from the Chinese Qianlong period service which possibly predates it. It is strongly implied that this Worcester trio could also be part of a similar service bearing the technically incorrect heraldic tinctures appearing with the three unicorns, namely, *a canton argent* in place of *a*

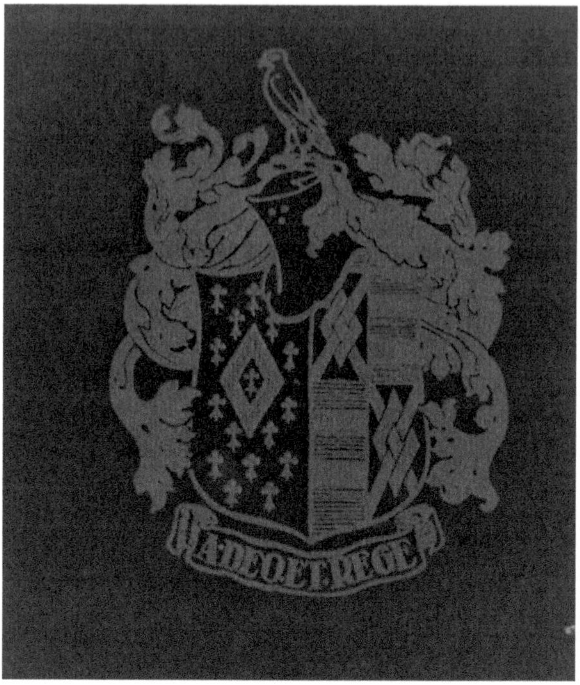

Fig. 1.13 Detail of the coat of arms of Walter Ramsden Hawksworth Fawkes as it appears on the cover of Finsberg's book, *Turner's Watercolours at Farnley Hall, 1912*

Fig. 1.14 The six-character seal mark of Qianlong (1735–1796) which appears on the Chinese porcelain plate shown in Fig. 1.13. *Courtesy of Kingschina Antiques, Mayfair, London*

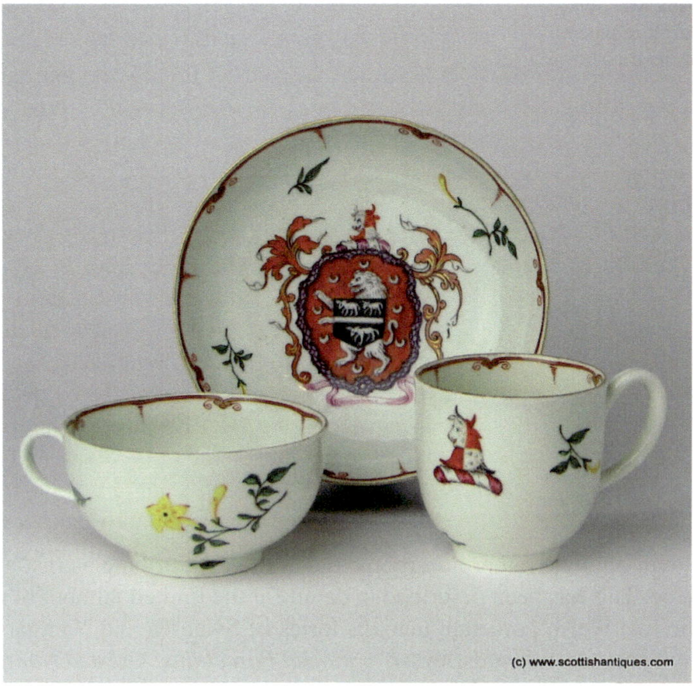

Fig. 1.15 Trio of Worcester porcelain, *ca.* 1765–1768, decorated by James Giles in his London workshop, bearing the arms of Beaumont -Ayscough

canton or, perhaps destined to be used together. It is certainly not uncommon to find early Chinese armorial services bearing the incorrect heraldic symbols and tinctures as the Chinese local decorators had little or no experience of Western heraldry and in the decipherment of the script instructions regarding the tinctures to be adopted on the drawings that were provided for their copying.

1.7 The Assignment of a Crest Only

The experienced researcher in armorial ceramics would undoubtedly agree that the most difficult armorial bearing to identify is the solitary crest and it is critical that a most exhaustive and critical assessment be made of the details portrayed and nuances contained in its depiction. We have already seen an example of this in the small differences that occur in the crested stork depicted for Sir Thomas Pitt and some other armigerous members of the wider Pitt family earlier in this Chapter, but at least in this case the researcher also had the escutcheon heraldic details to reinforce and support the correct assignment—although the associated motto is still being incorrectly ascribed by some historians. Where the escutcheon arms representation

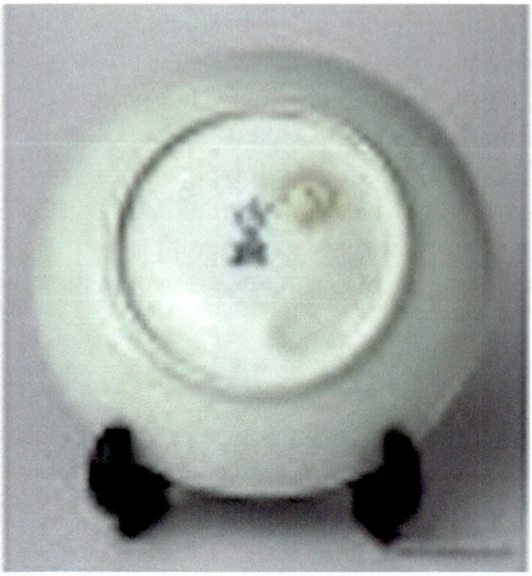

Fig. 1.16 Faux Meissen mark in underglaze blue on the saucer of the trio shown in Fig. 1.15

is absent it can often be rather hazardous to definitively attribute a crest alone on a ceramic piece. This has been described in detail for the limited number of armorials borne in the two Welsh porcelain manufactories of Swansea and Nantgarw by one of the current authors (Edwards, *Welsh Armorial Porcelains: Crested Nantgarw and Swansea China*, 2022): fewer than twenty distinctly different armorials are known for these factories, which operated broadly between 1817 and 1822, making them rather rare, and most of these were unattributed hitherto, the reason being that the majority depicted a crest alone. The precise details amount to 11 armorials for Swansea and 7 for Nantgarw of which 82% were crest alone for Swansea and 42% for Nantgarw. Only one armorial from both manufactories displays a full coat-of-arms and that is the *Viscount Weymouth* service, shown in Fig. 1.8., and even that has been incorrectly attributed hitherto. A typical example of crested china is shown in Fig. 1.17, which displays a simple gilt crest at the verge of a Nantgarw dinner plate, which is otherwise undecorated and has now been assigned to Baron Constantine Phipps of Normanby, through the combined application of detailed genealogical and armorial research. Occasionally, an armigerous family might have two crests granted for their heraldic use to highlight different ancestral forebears but only one would normally be used in the portrayal of arms on any particular armorial ceramic piece: such an example is that of Wyndham Lewis MP of Greenmeadow, Llantrisant, Glamorgan, whose crest on a Swansea duck-egg porcelain plate (*ca.* 1815–1817) is shown in Fig. 1.18, depicting a Paschal lamb bearing a pennant with the cross of St George. This crest was adopted by Wyndham Lewis to demonstrate his ancestry from Price/Pryce, and he additionally used the motto "*Duw ar fy Rhan*" meaning "*God for my Protection*". The alternative crest that he could have used would have been for Lewis and that carried the different motto "*Patriae fidus*" meaning "*Faithful to my Country*".

Sometimes, even the rendering of a simple crest is open to some minor inaccuracies as exemplified by that of the polychrome service ordered by Edward Harrison in the

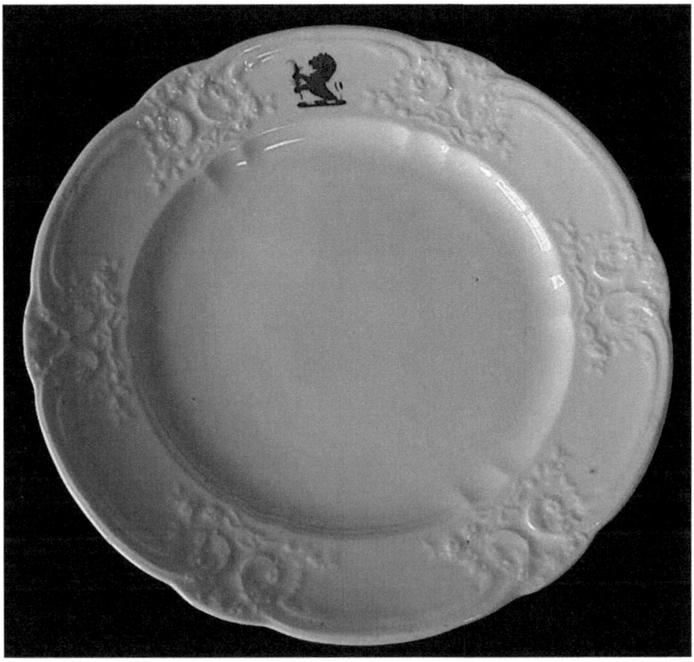

Fig. 1.17 Nantgarw dinner plate, *Baron Phipps*, Viscount Normanby, service, *ca.* 1817–1819, impressed mark NANT-GARW C.W., demonstrating the beauty of Nantgarw porcelain translucency at its very best with a simple gilt armorial crest decoration at the verge between moulded scrolls of foliage, flowers and stars and otherwise undecorated. Only two Nantgarw services are known of this type displaying just a crest and being otherwise completely undecorated with no enamelling, the other exemplar being the plain, unmoulded *Homfray* tea and coffee service of Penllyne Castle (Crest: an otter, pierced by an arrow in gold). Henry Phipps was Lord Chief Justice of Ireland, created Baron Mulgrave of York in 1794 and Viscount Normanby, Earl of Mulgrave in 1812. Marked impressed NANT-GARW C.W. His crest comprises a demi-lion rampant or holding in both paws a palm branch vert. *Private Collection*

famille verte style around 1720, whose arms are shown here in Fig. 1.19. The crest of Harrison as described in Fairbairn (Fairbairn, *Crests of the Families of Great Britain and Ireland and Their Kindred in Other Lands,* Volume I, 1905) is: *a cubit arm habited azure, cuffed argent and holding in the hand an arrow in bend sinister proper, headed or.* This is depicted on the armorial porcelain as shown in Fig. 1.19 and compared with that in Fairbairn (Volume 2, Plate 300.3) which is reproduced in Fig. 1.20: there are two potential points at issue here when the comparison is made between the heraldic crest and that depicted on the armorial porcelain, namely, the cuff on the cubit arm is not rendered as argent or white, the arrow is missing its stabilising fletchings and the whole arrow has been coloured in or (gold) and not just the arrowhead as indicated in Fairbairn.

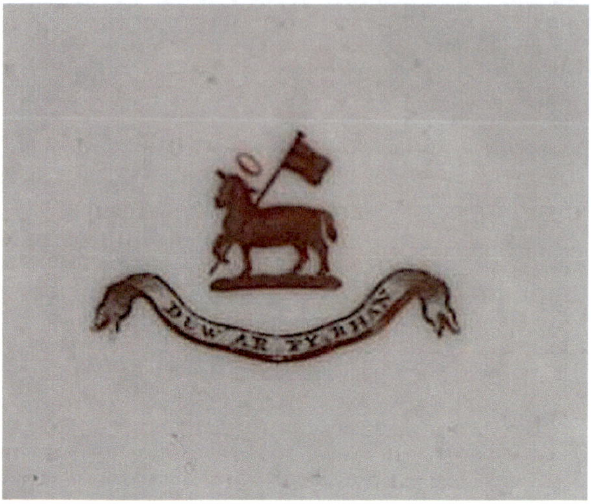

Fig. 1.18 The crest of a Paschal lamb in gilt facing dexter and holding in its right foreleg a pole with the cross of Saint George on a pennant and the motto in Welsh *"Duw ar fy rhan"*, translating as *"God for my Protection"*, is recorded on this C-scroll moulded Swansea dish in duck-egg porcelain, *ca.* 1817–1820. This is one of two crests recorded for Wyndham Lewis of Greenmeadow, Llantrisant, who was a descendant of Price (Pryce) and is one of only two crests and coats-of-arms on Welsh porcelain to bear an inscription in Welsh, the other being that of Thomas Lloyd of Bronwydd. Reproduced with permission from Gwyn Jones, Director of the John Andrews Charitable Trust Collection, Oriel Plas Glyn-y-Weddw, Llanbedrog, Abersoch, Gwynedd

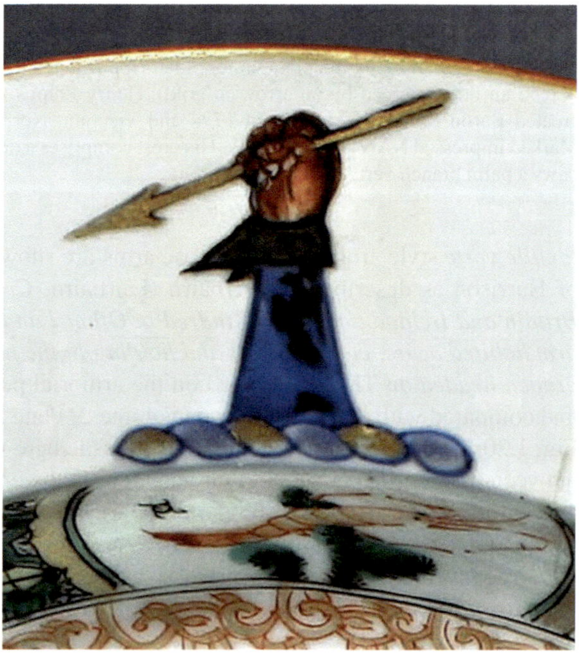

Fig. 1.19 The crest of Harrison on a plate from his Chinese armorial porcelain service, ca. 1720, as described in Fairbairn (Fairbairn, *Crests of the Families of Great Britain and Ireland and Their Kindred in Other Lands,* Volume I, 1905) is: a cubit arm habited azure, cuffed argent and holding in the hand an arrow in bend sinister proper, headed or. This is depicted on the armorial porcelain as shown here and should be compared with that in Fairbairn (Volume 2, Plate 300.3) which is reproduced in Fig. 1.20

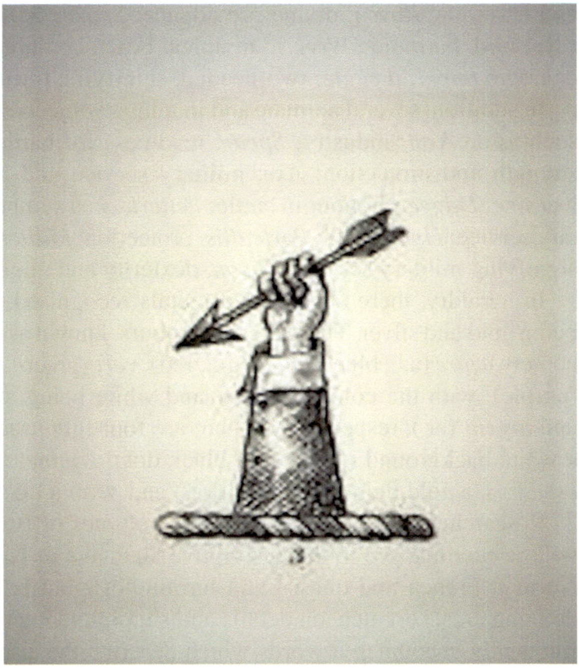

Fig. 1.20 The crest of Edward Harrison as depicted in Fairbairn (Fairbairn, *Crests of the Families of Great Britain and Ireland and Their Kindred in Other Lands,* Volumes 1 and 2, 1905; Volume 2, Plate 300.3)

1.8 The Meaning of Devices Used in Heraldry and Their Tinctures

The use of real and mythical beasts as supporters in the coat of arms can be traced back in English Heraldry to King Edward III and King Richard II, who used a combination of lions and stags as their supporters. Animals and birds are also used in crests and in the panels of the escutcheon, each one having particular significance in heraldic lore: these may be accompanied by objects of a particular heraldic significance. A summary of these is provided below:

Boar (sanglier): fiercely combative, courageous; *Otter*: denotes one who is intelligent and industrious; *Lion*: the noblest of the wild beasts, depicts strength and valour; *Leopard*: signifies one who is valiant and who braves danger with courage; *Dolphin*: charity; *Eagle*: fortitude and magnanimity; *Stag*: will fight if provoked; *Sprouting tree stump*, rebirth and growth; *Bird*: eloquence; *Escallop shell*: a successful leader and a protector of providence; *Paschal lamb*: possesses faith, innocence and is resolute in spirit; *Martlet*: continuous effort and virtue; *Dove*: exemplifying gentleness and peace; *Lamb:* symbol of Christianity; *Hound*: leadership; *Pelican*: self-sacrifice and piety; *Phoenix*: resurrection; *Falcon/Hawk*: signifies someone of action; *Frog:* harmony and peace; *Griffin*: mythical beast combination of an eagle and a lion, valour; *Cock*: virility and bravery; *Chough*: death; *Wildcat*: vigilance and courage; *Butterfly:* peace; *Bull*: valour; *Bee:* industrious; *Bat*: watchfulness; *Antelope:* peace and harmony: *Raven*: divine providence; *Snake:* wisdom and medicine; *Swan*: the royal bird, harmony; *Wyvern*: mythical beast, combination of a dragon and serpent, courage; *Imperial eagle:* two-headed, signifying fortitude and magnanimity.

In addition, several animate and inanimate objects signify a particular symbolism, such as an *Arm:* industry; *Spear*: readiness for battle, a knight in service; *Tower*: strength and protection; *Axe*: military service and duty; *Bezant*: banking; *Book*: learning; *Dagger*; honour in battle; *Fetterlock*: a symbol of authority; *Gauntlet*: military service; *Hand*: faith; *Portcullis*: protection; *Mullet:* a star with 5,6,8 or 12 points, signifying military service; *Pheon*: dexterity and nimbleness of wit.

In heraldry, there are but two metals recognised, namely *or* and *argent*, these being gold and silver. There are five colours, known as tinctures in heraldry, depicted, namely *azure* (az., blue), *gules* (gu., red), *vert* (green), *sable* (sa., black) and *purpure* (purple), with the colours yellow and white being additionally represented by *or* and *argent* (ar.), respectively. There are four furs used in heraldry: namely, *ermine*, a white background (field) with black dots, *ermines*, a black field with white dots, *erminois*, a gold field with black dots, and *pean* a black field with gold dots.

Researchers used to consulting the volumes of Burke and Fairbairn cited above will be acclimatised to the descriptive language of heraldry which originated using Norman French and quasi-Latin terminology but is nevertheless very accurate in defining crests escutcheon details and supporters in coats of arms. A list of the most frequently encountered words which describe the attitude of the animals, birds and reptiles and the geometric designations of the escutcheon composition is given in Table 1.1 and a familiarisation with the important differences described therein is a commendable starting point for the novice armorial ceramics researcher. In this task, reference to the classic heraldic work of Boutell (*The Handbook of English Heraldry,* 1914) is invaluable.

The listing of words and phrases used in heraldry in Table 1.1 accentuates their derivation from Norman French and mediaeval Latin and includes a summary of the different types of geometry used on the escutcheon to display the arms and also some of the different types of cross that have been used in heraldry. Although comprehensive, it is by no means exhaustive and readers are referred to the specialist textbooks on heraldry for terms that do not appear there (Fox-Davies, *Armorial Families: A Complete Peerage, Baronetage and Knightage, and a Directory of Some Gentlemen Coat -Armour and being the First Attempt to Show Which Arms are in Use at the Moment are Borne by Some Legal Authority*, Volumes I and II, 1895; Fox-Davies, *A Complete Guide to Heraldry: with Illustrations by G. Johnston, Herald Painter to the Lyon Court,* 1909; Neubecker, Brooke-Little and Tobler, *Heraldry: Sources, Symbols and Meaning* 1989; Scott-Giles, *Boutell's Heraldry*, 1954; Stanford London, "Crosses Paty, Patonce and Formy", *Coat of Arms, The Heraldry Society*, 1958; Woodcock, *The Oxford Guide to Heraldry*, 1988).

In particular, the work of Boutell (*The Handbook of English Heraldry*, 1914) gives a comprehensive analysis of heraldic devices, achievements and coats of arms and has chapters or sections on blazoning, heraldic language, tinctures, heraldry of the cross, ordinaries, heraldic classes, differencing, cadency, augmentation, marshalling, crests, badges, mottoes, supporters, crowns and coronets and insignia.

Table 1.1 Heraldic linguistic terms, their abbreviations and meaning

Heraldic term	Abbreviation	Meaning
abasement		Lowering of dignity or station
accollee		Side by side
accompanied		Between
accorne		Horned
accrued		Fully grown
achievement		Complete display of arms: shield, helmet, mantling, torse, crest, motto
addorsed		Turned back-to-back
affrontee		Full-face to the front
allerion		Eagle without beak or feet
allusive		Arms based on name
ambulant		Walking
anche		Curved
annulet		Small ring, circlet
appaumy		Open handed, showing palm
argent	ar	Silver, white
armed		Beast exhibiting claws, teeth, talons
aspectant		Animals facing each other
assurgent		Rising out of the sea
armed		Horn, beak, talon, usually in colour
at gaze		Animal looking directly at observer
azure	az	Blue
banner		Flag of knight
barbed		Bearded
barrulet		Half width of a bar
barry		Transverse bars of contrasting colours on shield
bars gemelles		Bars arranged in narrow couples
base		Lowest part of shield
basilisk		Cockatrice with dragon's tail
beaver		Chinguard of helmet
bend dexter		A diagonal from top dexter to bottom sinister on the shield
bend sinister		A diagonal from top sinister to bottom dexter on the shield
bird-bolt		Blunt arrowhead
blazon		Shield and coat-of-arms in colours
bordure		Tinctured border around shield distinguishing branches of the family

(continued)

Table 1.1 (continued)

Heraldic term	Abbreviation	Meaning
brased		Interlaced
brelesse		Embattlemented
brouchant		Overlying
bouche		An annulet set into the top dexter part of the shield to accommodate the lance during a jousting tournament
caboched		Head of beast, facing, without neck
cadency		Device to differentiate members of same family
cannelee		Inverted
canton		First quarter corner of shield
charge		Figures in escutcheon
chapeau		Cap of dignity used in place of torse
chaplet		Garland, wreath of leaves, flowers
checky		Alternate squares of different colours
chevron		An inverted vee, ^
chevronet		A small chevron
chimera		Mythical beast: griffin, dragon, gryphon, cockatrice, harpy, wyvern
closet		A half-bar
collared		Wearing a collar
combatant		Fighting
compony		Squares of alternate colours on shield
contournee		Facing sinister
couchant		Lying down with head raised
counterflory		Fleur-de-lys arranged around border
couped		Head, limbs cut off at neck
courant		Running
coward		Animal with tail between its legs
crenelle		Embattled
decrescent		Crescent rotated right by 90^0
demi		Top half: e.g. demi-lion, demi -leopard
denchee		Indented
dexter		Right side of, as viewed by bearer of shield
diadem		Pointed circle of gold
difference		Added to escutcheon to differentiate families
dimidiation		An alliance by marriage: shield divided per pale and quartered
displayed		Bird with wings expanded and legs spread
dormant		Sleeping

(continued)

Table 1.1 (continued)

Heraldic term	Abbreviation	Meaning
eaglet		Several eagles in an escutcheon
embrued		Bloodied, especially weapons
en arriere		Shown from the back
engrailment		Two concave regions on the top of the shield meeting at a central point
engreslee		Engrailed
ensigned		Ornamented
erased		Cut off, leaving jagged edge
erect		Horizontal parts, displayed vertically
ermine		White fur with black spots
ermines		Black spots on white fur
erne		Eagle
escallop		Seashell
estoile		Star with wavy rays
fess		Third part of shield between centre and top
fess point		Centre of escutcheon
fetterlock		Shackle and padlock
field		Surface of escutcheon
fitchee		Pointed at base
flanches		Curved lines each side of shield
flection		Modification of an edge
fleur-de-lys		Lily
fly		Side of flag furthest from staff
fructed		Bearing fruit
fusil		An elongated lozenge
gambe		Leg
garb		Wheatsheaf
gaze		Intent look
gauntlet		Gloved hand armour
gemel		Double, twinned
gorged		Collar around the neck of an animal
gules	gu	Red
guttee de sang		Drops of blood
habited		Clothed
hatchment		Arms in lozenge frame of deceased person
haurient		Fish, dolphin, vertical with head pointing upwards
helm		Visored helmet

(continued)

Table 1.1 (continued)

Heraldic term	Abbreviation	Meaning
herison		Hedgehog
hirondelle		Swallow
imbued		Spotted with blood
impaled		Two arms on one shield
increscent		Crescent rotated left by 90^0
inescutcheon		Small escutcheon in larger shield
in her piety		Pelican feeding her young
issuant		Rays
lambrequin		Mantling, cloth covering for armour
langued		Tincture of animal's tongue
luce		Pike (fish)
marshalling		Two or more coats of arms on one shield
martlet		A swallow
mullet		Five-pointed star, roundel of spur
ondee		Wavy
or		Gold, yellow
orle		False escutcheon
overt		Expanded wings
pale		Vertical band on a shield
pale tierced		Two vertical lines forming three divisions on shield
paly		Vertical bars in two tinctures
panache		Plume of ostrich feathers
pheon		Broad arrow, tang with converging blades and serrated inner edges
pineapple		Pinecone
popinjay		Parrot
proper	ppr	In standard tincture but stylised
quarter		Dexter quarter of the shield (top left)
rampant		Rearing beast with paws in attack position
rampu		Broken
sable	sa	Black
salient /salient		Leaping or springing beast, hind legs on the ground
saltire		St Andrew's cross, diagonal cross
segreant		Swimming
sejant		Sitting down on haunches with forepaws on the ground
sinister		Left side of, as viewed by bearer of shield
tennee		Tawny tincture

(continued)

Table 1.1 (continued)

Heraldic term	Abbreviation	Meaning
tierce		A third division
tincture		One of the heraldic colours
torse		Twisted silk cloth wreath of two colours between helmet and crest
tressure		Arranged around border
tun		Cask
vert		Green
voided		Central area removed
vulned		Ensanglante, animal wounded and bleeding
**Types of Shield Design*		
per fess		Shield divided in two horizontally
per pale		Shield divided in two vertically
per bend		Shield divided in two diagonally
per chevron		Shield divided by an inverted vee (chevron)
per pile		Shield divided by an inverted triangle whose base merges with the top of the shield
per saltire		Shield divided by a cross quarterly diagonally
per quarterly		Shield divided by a cross horizontally and vertically
per gyronny		Shield divided into eighths radiating from centre
***Types of Cross (Arms of Equal Length)*		
cross ancre		See cross moline
cross alaisee		See cross hummelty
cross avellane		Arms comprise four filbert nuts joined into a cross
cross barby		Arms terminate in arrowheads
cross bottony		See cross trefly
cross cercelee		A type of cross moline
cross charge		See cross remplie
cross cramponce		See cross gauche
cross crosslet		Arms terminate in T-bars
cross faussee		A cross which has the same tincture as the field but with a border tincture outlining the limbs
cross fitchy		Lower limb pointed to form a stake
cross fleury		Arms terminate in fleur-de-lys
cross fourchee		Arms terminate in forks
cross gauche		Arms end in truncated right-angled arms, a form of swastika or tetragammadion (Greek, four gammas)
cross hummelty		Greek cross, with four equal unadorned limbs

(continued)

Table 1.1 (continued)

Heraldic term	Abbreviation	Meaning
cross Maltese		Evolution of cross moline. arms narrow towards centre, indented at ends forming eight points
cross moline		Arms are bifurcated, split and curved backwards
cross patonce		The arms terminate in leaf shapes intermediate between a cross pattee and a cross fleury
cross pattee		Terminating in curved arms narrowing towards centre but with flat ends
cross pattee fitchy		Cross pattee with lower limb mounted on a stake
cross pommee		Arms terminate in narrow kinks
cross portate		Cross angled from vertical
cross potent		Limbs terminate in cross bars at end
cross remplie		Superposition of one tinctured cross upon another
cross trefly		Arms terminate in three knobs, trefoils
****Types of Cross (Arms of Unequal Length)*		
Several heraldic variations exist of crosses which possess arms of unequal length and these often have special names associated with their users such as; Latin Cross (elongated lower limb),St Peter's Cross (inverted Latin cross), St Philip's Cross (rotated through 90 ^{0}), Calvary Cross (mounted on steps), Patriarchal Cross (double Byzantine cross), Orthodox Cross, Cross of Lorraine		

References

Burke, Sir J. B. (1884). *Burke's General Armory—The General Armory of England, Scotland, Ireland and Wales: Comprising a Registry of Armorial Bearings from the Earliest to the Present Time*. Harrison & Sons.

Boutell, C. (1914). *The Handbook of English Heraldry* (10th Edition revised by A.C. Fox-Davies). Reeves & Turner.

Cooke, N. (2016, May 13). *Bookplates and porcelain: new research from the RCP archives*. https://history.rcplondon.ac.uk/blog/bookplates-and-porcelain-new-research-rcp-archives

Curtis' Botanical Magazine, or *The Flower-Garden Displayed*, Volume XLVII, ed, J. Sims, published by Sherwood, Nealey and Jones, Paternoster Row, London, Plate 282, 1819/1820.

Defoe, D. (1726). *The Complete English Tradesman*, Chapter 22.

Denyer, R., Denyer M., & Edwards, H. G. M. (2023). *The Pendock-Barry Derby Porcelain Service : A Forensic Reappraisal*. Springer-Nature.

Edwards, H. G. M. (2021). *Porcelain Analysis and Its Role in the Forensic Attribution of Cultural Heritage Specimens*. Springer-Nature.

Edwards, H. G. M. (2022). *Welsh Armorial Porcelain: Crested Nantgarw and Swansea China*. Springer-Nature.

Elvin, C. N. (1971). *A Handbook of Mottoes Borne by the Nobility, Gentry, Cities, Public Companies Etc.*, Bell & Daldy, Fleet Street, London,1860. Reprinted with Supplement and an Index by R. Pinches, Heraldry Today, Beauchamp Place, London and Genealogical Publishers, Baltimore, USA.

Fairbairn, J. (1986). *Crests of the Families of Great Britain and Ireland and Their Kindred in Other Lands,* Volumes I and II, T.C. and E.C. Jack, Edinburgh and London, 1905 and New Orchard Publishing, London.
Fiennes, C. (1888). *Through England on a Side Saddle in the Time of William and Mary—Being the Diary of Celia Fiennes.* Field & Tuer Publishers.
Fox-Davies, A. C. (1895). *Armorial Families: A Complete Peerage, Baronetage and Knightage, and a Directory of Some Gentlemen Coat -Armour and being the First Attempt to Show Which Arms are in Use at the Moment are Borne by Some Legal Authority* (Volumes I and II, T.C. and E.C. Jack). Grange Publishing Works.
Fox-Davies, A. C. (1909). *A Complete Guide to Heraldry: with Illustrations by G. Johnston, Herald Painter to the Lyon Court.* T.C. & E.C. Jack.
Gambier Howe, E. R. J. (1903). *Catalogue of British and American Bookplates Bequested to the Trustees of the British Museum by Sir Augustus Wollaston Franks, 1826-1897,* E. R. J. Gambier Howe (Ed.). Wm. Clowes & Sons Ltd.
Graca, J. (1977). The Portuguese Porcelain Trade with China. *Arts in Asia, 7,* 45–47.
Hesketh, H. E., & Phillpotts, E. (1920). *Bookplates by Frank Brangwyn RA.* The Morland Press Ltd.
Hobhouse, H. (1994). *Blackwall Yard: Development to c. 1819,* in Survey of London: Volumes 43 and 44, Poplar, Blackwall and the Isle of Dogs, ed. H. Hobhouse, London pp. 553–565.
Hopkinson, M. (2011). *Ex Libris: The Art of Bookplates.* British Museum Press.
Howard, A. (2013). *From Canton to the City: Export Porcelain for London* (pp. 104–114). Art Antiques.
Howard, A. (2023). personal communication.
Howard, A. (2023). Four centuries of blue and white. In B. Macguire (Ed.), *The Freylinghuysen Collection of Chinese and Japanese Porcelains.* Ad Illisum/Paul Holberton Publishers.
Howard, D. S. (1974). *Chinese Armorial Porcelain* (Vol. 1). Faber & Faber.
Howard, D. S., & Ayers, J. (1977). *China for the West: Chinese Porcelain and Decorative Art for Export Over Three Hundred Years.* Sotheby Parke Bernet.
Howard, D. S. (March 1984). Bookplate Designs and Chinese Armorial Porcelains. *The Bookplate Journal, 2,* 1.
Howard, D. S. (2003). *Chinese Armorial Porcelain* (Vol. II). Heirloom & Howard.
Howard, D. S. (1997). *A Tale of Three Cities—Canton, Shanghai and Hong Kong: Three Centuries of Sino-British Trade in the Decorative Arts.* Sotheby's Publishing.
Jewitt, L. (1878). *The Ceramic Art of Great Britain from the Prehistoric Times Down to the Present Day* (Vols. I and II). Virtue & Co. Ltd.
Latcham, P. (September 2008). Bookplates and Armorial Porcelain. *The Bookplate Journal, 6,* 2.
Le Corbellier, C., & Frelinghuysen, A.C. (2003). Chinese Export Porcelain. *The Metropolitan Museum of Art Bulletin, 60,* Winter issue, 1–60.
McGuire, B. (2023). *The Freylinghuysen Collection of Chinese and Japanese Porcelains.* Ad Illisum /Paul Holberton Publishers.
Mueller, S. M. (2001). Surface Silver Decoration on Chinese Export Porcelain: A Survey. *Oriental Art, 47*(3), 56–64.
Mueller, S. M. (2002). Surface Silver Decoration on Chinese Export Porcelain: An Analytic Approach. *Oriental Art, 48*(4), 43–46.
Mueller, S. M., & Bunney, M. (2016). A Chinese Export Porcelain Mystery Solved Using Intensive Surface Analysis. *Ceramics in America, 16***,** ed. R. Hunter, Chipstone Foundation, 118–121.
Neubecker, O., Brooke-Little, J. P., & Tobler, R. (1989). *Heraldry: Sources, Symbols and Meaning.* Tiger Books International.
Pepys, S. (1893). *The Diary of Samuel Pepys MA FRS, Clerk of the Acts and Secretary to the Admiralty,* Transcribed from the Shorthand Manuscript in the Pepysian Library, Magdalene College, Cambridge, by the Rev, M. Bright, George Bell & Sons, Covent Garden, London and Deighton Bell & Co., Cambridge.
Rui, L. (1990). Portugal Embusca de China: Imagens e Miragens (1498–1514). *Ler Historia, 9,* 33.

Scott-Giles, C., & Brooke-Little, J. P. (1954). *Revision of Boutell's Heraldry*. Frederick Warne & Co., Ltd.
Stanford London, H. (1958). Crosses Paty, Patonce and Formy. *Coat of Arms, The Heraldry Society, 33/34*, January and April issues.
Woodcock, T. (1988). *The Oxford Guide to Heraldry*. Oxford University Press.

Chapter 2
The Essence and Meaning of Armorials on Ceramics

Abstract The correct and incorrect usage of armorial bearings on porcelains is illustrated using specific examples of named armorial services such as the *Wittelsbach*, *Pendock-Barry* and *Viscount Weymouth* services. The appropriate usage of genealogy to exclude possible mis-attributions of an armigerous person in the identification of heraldic bearings is illustrated with considerations of the meaning of the heraldic devices. The use of monograms and ciphers in association with a crown or a coronet affords one of the simplest portrayals of an armorial ceramic artefact next to a simple unadorned crest. The difficulties in the interpretation of a crest without an escutcheon or motto is highlighted. Some analogies in modern miliary classification of heraldic bearings are reviewed.

Keywords *Wittelsbach* service · Genealogy · *Pendock-Barry* service · *Viscount Weymouth* service · Monograms and ciphers · Military heraldry

All truths are easy to understand once they are discovered. The point is to discover them.
Galileo Galilei (1564–1642)

2.1 The Meaning of Armorial Devices and Their Usage

A survey of the composition of coats-of-arms and the types of armorial device used in their construction has been carried out in Chap. 1. Historically, they served as recognisable identifiers of armigerous individuals and as rallying points for forces in battle and later became unique descriptors of one's ancestral line of descent. There were strict rules of heraldry imposed which meant that one could not adopt armorials to which one was not entitled to as adjudicated by the College of Arms for English heraldry, the Lyon Office of Arms for Scotland and the Ulster King of Arms for Ireland, and only specified members of an armigerous family were granted leave to use armorial bearings. Contrary to popular belief, others who bear that same family name do not have the right to assume the coat-of-arms or crest and terms such as

R. L. Denyer et al., *Armorial Porcelain*, https://doi.org/10.1007/978-3-031-63745-2_2

"family crest" need to be considered carefully by the researcher. Despite this, there are numerous occasions where arms have been adopted unofficially for use by those who are not entitled to use them, and these have been displayed illicitly at several vantage points on household effects or in an estate.

A classic example of this practice is the crest that was placed in the late eighteenth century by Eleanor Coade who was the influential creator of Coade artificial stone that was much admired and used in the eighteenth and early nineteenth centuries by Georgian architects such as Robert Adam and Sir John Soane (Kelly, *Mrs Coade's Stone*, 1990; van Lemmen, *Coade Stone*, 2006). The gateposts at Eleanor Coade's villa, Belmont House, in Lyme Regis, Dorset, which she inherited from her uncle Samuel Coade in 1784, bear the crest of a coot on a torse in Coade stone which embellish the limestone pillars, as seen in Fig. 2.1.

Research has indicated that Eleanor Coade did not have the right to bear arms and indeed, neither did any of her immediate family or ancestors: it has been suggested that the depiction of a coot is derived from the pronunciation in the local dialect of her family name, which is often represented there as Coode or Cood and the softening of the letter d makes this sound like "Coot". Hence, one could reasonably infer that the use of a coot on a crest is more akin to a rebus, which is a pictorial image of an object, animal or a bird that closely resembles one's surname and therefore offers an illustrative allusion to the ownership or possession of a nomenclature. It is true that several people who bear the name of Coade, Cood or Coad do feature in Fairbairn's authoritative text on crests (Fairbairn, *Crests of the Families of Great Britain and*

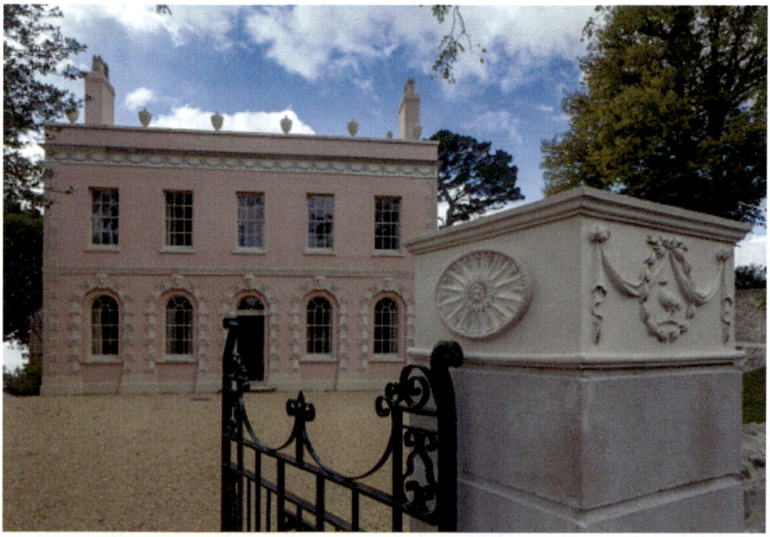

Fig. 2.1 Belmont House, Lyme Regis, Dorset, home of Eleanor Coade, who inherited the house from her uncle Samuel Coade in 1784 and thereon installed artificial Coade stone artefacts and embellishments. The moulded Coade stone frontage to the right-hand gatepost shows the heraldic crest of a coot and a torse. Belmont House is now in the possession of the National Trust

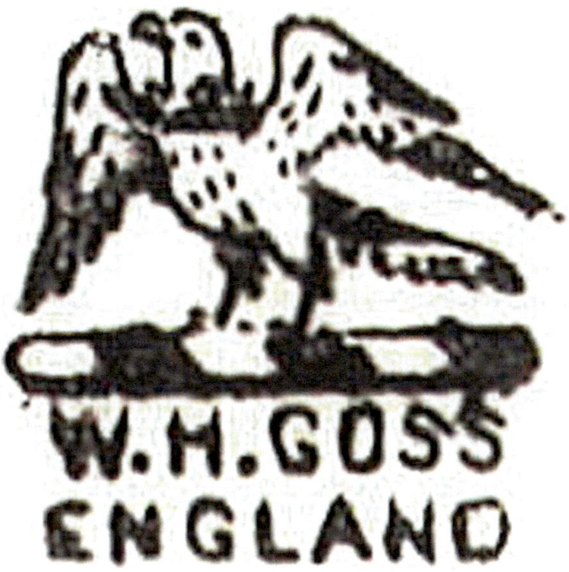

Fig. 2.2 Stencil mark of armorial Goss china with the rebus of a goshawk on a torse

Ireland and Their Kindred in Other Lands, Volumes I and II, 1905) but these do not have any familial connections with Eleanor Coade and it is also revealing that none of these display a coot on their crest, rather showing a Cornish chough for this purpose. Other examples of the use of a rebus on ceramics to indicate a particular association that is related to one's name are the adoption of a goshawk crest by the Goss family of armorial china manufacturers in mid-Victorian times (Fig. 2.2) and the anchor mark used by Richard Holdship on his Worcester porcelain (Edwards, *Welsh Armorial Porcelain: Crested Nantgarw and Swansea China*, 2022). The case of the *Frog* service of the Empress Catherine the Great in Wedgwood's *Queen's ware* that has been discussed in Chap. 1 also could be included in this category of being a rebus but this is perhaps a little further removed from the these other examples cited in that it depicts an object that is associated with a place of use, namely the Frog Marsh on which the Empress Catherine's Chesma Palace was constructed, rather than being a linguistic pun on the surname of the person involved.

2.2 Incorrect Armorial Usage

The essential requirement for the correctness of the depiction of an armorial bearing has already been emphasised and an example of the failure to execute this properly by the Chinese enamellers has been exemplified in the case of the armorial service of Sir Thomas Pitt in 1705 discussed in Chap. 1. The reason for this can probably be ascribed to the misconception of the Chinese enamellers who performed the armorial decoration in Jingdezhen or perhaps to a contributory mistake in the instructions provided to the HEIC supercargo who was charged with overseeing the production and delivery of the service in Canton. Yet, it is curious that Sir Thomas Pitt actually accepted the delivery of his service with its questionably deficient armorial decoration and perhaps this was treated as a regrettable but acceptable consequence of the necessary indirectness of the commissioning process: later in the eighteenth century, Josiah Wedgwood commented that he did not wish to undertake the production of armorial services for his British clients because of the associated problems in the achievement of an absolutely correct rendition of the coat-of-arms, which typically resulted in the refusal of the appropriate payment by the commissioning agent or arms bearer. He later relented this polarised stand and instead obtained the verified and correct drawings of the coat-of-arms, escutcheon and crest that were possessed by his armigerous client from the College of Arms, for which he received invoices for payment that are still retained in the archive of the Wedgwood Museum in Etruria. By this means, Wedgwood reasoned that, despite the additional cost incurred to himself, the client would never be able to argue that their arms had been portrayed incorrectly on the ceramic order and default in payment, as had been apparently the case with other manufacturers. So, it can be inferred that by this time the armigerous clients were more demanding that their armorial commissions were executed correctly by the manufacturers with whom they were dealing directly or through a sole agent. We should seek to broaden the discussion, therefore, into the reasons that armigerous families still accepted incorrect versions of their armorial bearings on china. Some specific examples of apparently straightforward armorial ceramics which in fact prove to be otherwise are highlighted below.

2.2.1 The Wittelsbach *Service*

This is an example of Wanli Chinese porcelain (late sixteenth to early seventeenth century) decorated in the *kraak* compartmented style in blue and white colours which contains an escutcheon enamelled centrally on the piece. At first, the escutcheon seems to be of an armorially acceptable standard but closer examination and comparison with the Wittelsbach arms given in the literature reveals that several things are not correctly portrayed here. For the purposes of direct comparison, Fig. 2.3 is placed alongside that of the Wittelsbachs of Zweibrucken, Counts of the Rhineland Palatinate, official coat-of-arms, as shown in the armorial represented on the porcelain plate in Fig. 2.16.

In juxtaposition, the discrepancies between the two armorials are now clearly evident: the heraldic arms are cited in the armories as quartered 1st and 3rd *sable a lion, armed, langued and crowned gules*, 2nd and 4th *fusilly bendiance azure and argent.* Perhaps the major difference between the two renditions in Fig. 2.3 lies in the representation of the two lions on the armorial plate, which certainly look more akin to a cross between a leopard and a horse, each of which bears a crown and a semblance of a three-forked tail. However, the animals here also occupy the 2nd and 4th quadrants of the escutcheon and not the 1st and 3rd as they should do. Also,

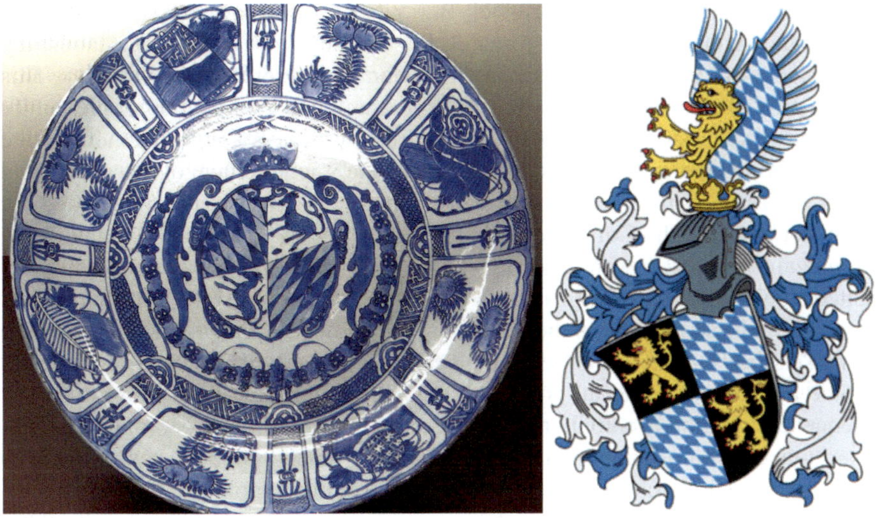

Fig. 2.3 Dinner plate from the *Wittelsbach* armorial service, Bavaria Rhineland Palatinate, in *kraak* porcelain. Wanli period (1573–1620), Chinese porcelain. For comparison, the actual coat-of-arms of Wittelsbach, showing the rampant lion of Wittelsbach quartered with the blue and white lozenges of their ancestral Bogen. *Munchener Residenz Muzeum, Munich. CC by SA 3.0*

the fusilly of blue and white lozenges, which relate to the Bogen family ancestry, are rather haphazardly and asymmetrically arranged, the 3rd quadrant being rotated through 90 ^{0} with respect to its congener occupying the 1st quadrant. The number of lozenges that occupy each quadrant are also strangely different. The "supporters", which should have been armed rampant lions bearing red crowns appear now as indifferent curlicues, as does the semblance of the bejewelled crown located atop the escutcheon. In summary therefore, these arms represented on the armorial plate are incorrectly rendered in several respects and yet were accepted in their commission; perhaps the sheer difficulty in arranging for a replacement with the necessity of something approaching a three-years' time delay would have been too negative to contemplate as armorial services were generally commissioned to commemorate some particular event of importance to the family concerned.

2.2.2 The **Pendock-Barry** *Service*

The *Pendock-Barry* Service of Derby porcelain, *ca.* 1805, is shown in Fig. 1.7 and has been the subject of much detailed genealogical research by Rachel Denyer (Denyer, Denyer and Edwards, *The Pendock-Barry Service: A Forensic Evaluation*, 2023). The crux of the matter regarding its evaluation chronologically lies in the identity of the armorial bearings used in the service decoration, which essentially appears to be not that of Pendock Barry Neale, but that of Barry-Barry, which Pendock Barry

Neale was not entitled to use. Pendock Barry Neale had requested the College of Heralds and Arms in London to research his ancestry with a view to establishing his lineage from the ancient Norman Barry family some centuries hitherto. This was accomplished satisfactorily after much research and consultation of the family records at Roclaveston Manor but still required by protocol the agreement of the monarch (King George III) to Pendock Barry Neale's acquirement of these arms and he had not actually received that formally when he commissioned the service .... the upshot was that the College of Heralds and Arms was then forced to take legal action against him to prevent his usage of the Barry arms prematurely and this was granted legally, Pendock Barry Neale being castigated publicly for his presumptuous behaviour in the acquisition of these arms and being fined in the process. So, in effect this makes the *Pendock-Barry* service a very special armorial porcelain service—a genuine and correct armorial porcelain service but chronologically portrayed out of sequence with the eventual grant of the arms. The correct chronological dating, decoration and commissioning, therefore, required much assiduous genealogical as well as armorial research for the proper assignment of this important and documentary armorial porcelain service.

2.2.3 The Viscount Weymouth *Service*

The *Viscount Weymouth* Service of Nantgarw porcelain, *ca.* 1817–1820. Technically, this should be a relatively straightforward armorial service to identify absolutely and unambiguously as it portrays a full coat-of-arms including the motto. However, detailed research has again revealed that all is not as clear as it seemed hitherto, and the established literature is as a result incorrect in its assignment of this service to its armigerous holder. Firstly, despite the presence of the full coat-of-arms, the service has been mis-attributed and recent genealogical, and family historical research has revealed that the incorrect armigerous assignment has actually been made to the commission, from which a complex tale of family intrigue has been forthcoming from genealogical and historical documentary sources. Dr W. D. John, G. J. Coombes and Katherine Coombes in their book (*Nantgarw Porcelain Album*, 1975, Plate 56) illustrate a very fine armorial Nantgarw plate (diameter 9″), impressed NANT-GARW C.W., *ca.* 1817–1820, depicting a full coat-of-arms with crest, escutcheon, supporters and motto (illustrated here in Fig. 2.4) which they duly assigned to Viscount Weymouth, Lord Henry Frederick Thynne (1797–1837), who became the 3rd Marquess of Bath in 1837. His father, Thomas Thynne, was created the 2nd Marquess of Bath in 1796. Lord Henry was the second son but his elder brother, Thomas Thynne, who was Viscount Weymouth, predeceased him in January 1837, so Henry then inherited the title, firstly of Viscount Weymouth in January 1837 and then becoming the 3rd Marquess of Bath in March 1837, just four months before he died. The Thynne/Weymouth escutcheon is described in the armories as quarterly, 1st and 4th, *a barry of ten or and sable*; 2nd and 3rd *or a lion rampant tail nowed gules and erected.* Crest *on a wreath of gules and or, a reindeer statant or collared sable.* Supporters: *dexter, a reindeer or gorged with a plain collar sable; sinister, a lion tail nowed and erect, gules.* Motto: "*J'ay bonne cause*" which translates as "*I have good reason*". Family seat, Longleat House, Wiltshire.

There is extensive gilding applied to this armorial plate with its 12-lobed moulded border, which with its very white glaze clearly belongs to the Billingsley/Walker period of production at the Nantgarw China Works, and would probably have been commissioned in London through John Mortlock's china agency in 47, Oxford Street, and decorated at either Robins & Randall's or John Sims' ateliers. This is apparently an unequivocal attribution of a coat-of-arms, yet all is not straightforward on a closer examination and there are several very curious heraldic anomalies being portrayed here: firstly, the escutcheon is not quartered as expected for the Bath/Weymouth arms but comprises the ten bars component indicated solely on the shield and these are coloured white and green (argent and vert) rather than black and gold (sable and or) as were specified in the heraldic armory literature. The tincture description given in the armories is confirmed in the monochrome bookplates of Viscount Thynne shown in Figs. 2.5a and b, where the "tricking" (an heraldic term to indicate the required tincture represented in a monochrome engraving) is clearly indicative of

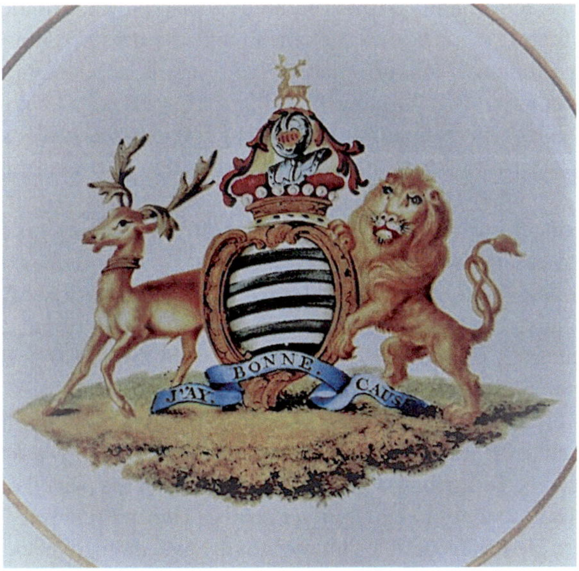

Fig. 2.4 Dinner plate from the Nantgarw China Works, *ca.* 1817–1820, impressed NANT-GARW C.W., bearing the full cost-of-arms of Viscount Weymouth, Lord Henry Frederick Thynne (1797–1837), who became the 3rd Marquess of Bath in 1837 just months before he died. The escutcheon is quartered: 1st and 4th, *a barry of ten or and sable;* 2nd and 3rd, *or a lion rampant tail nowed gu.* Crest: *on a wreath of gu. and or, a reindeer statant or collared sable.* Supporters: dexter, *a reindeer or gorged with a plain collar sable*; sinister, *a lion, tail nowed and erect, gules.* Motto: "*J'ay bonne cause*" which translates as "*I have good reason*". Family seat, Longleat House, Wiltshire. Illustrated in Plate 56 John et al., *Nantgarw Porcelain Album*, 1975. *Private Collection*

the barry being *or* (dotted background) and *sable* (hatched background) rather than *argent* (clear) and *vert* (diagonal lines, top left to bottom right) as portrayed on the Nantgarw plate in Burke's Armory (Burke, *Burke's General Armory—The General Armory of England, Scotland, Ireland and Wales: Comprising a Registry of Armorial Bearings from the Earliest to the Present Time*, 1884) does not differentiate in the arms borne between Viscount Weymouth and the Marquess of Bath, who appears under the family name of Thynne. This raises the question as to whether there is some special significance or hidden message in this amendment of the barry heraldic tinctures—green signifies hope and loyalty in love, white signifies sincerity and peace, whereas yellow signifies generosity and black, knowledge and constancy?

Secondly, the shape of the escutcheon itself is not the standard heater shield as would be expected but is shown as a distorted and asymmetric ellipse which is contained within a gadrooned baroque shield type frame that is referred to heraldically as a "*baroque cardidoid*". In contrast, the escutcheons shown in the two bookplates in Figs. 2.5a and b are known as "*square-eared, indented top and lobed base*" in type. The construction of this coat-of-arms on the armorial porcelain plate is also rather strange in several respects; firstly, the helm is shown facing dexter and not frontal as is depicted on the Viscount Thynne arms from his bookplate (Figs. 2.5a and b) and it also exhibits additionally a full armoured breastplate and shoulder guards. The coronet upon which it sits also has a chapeau and the dexter reindeer supporter is facing away from the arms and not placed in a properly supporting role. Likewise, the lion in sinister support is looking frontally and not dexter as it should be. The reindeer also does not have the correct coloured collar *(sable)* as is specified in the heraldic armory. The number of inconsistencies in this portrayal of these coat-of-arms require an explanation which is not immediately forthcoming and would surely give cause for concern to the noble client as patently displaying his incorrect arms, unless a hidden message is contained therein.

A more serious genealogical anomaly arises in the assertion of Dr John et al. that this plate is attributed to Lord Henry Frederick Thynne, because as the second son of the Marquess of Bath he was not entitled to call himself Viscount Weymouth until after his elder brother died childless in 1836; this was, of course, the title assumed by his elder brother Thomas Thynne in 1796 when his father succeeded to the Marquessate of Bath. Hence, the Nantgarw *Viscount Weymouth* service must surely be more correctly attributed to Thomas Thynne for the period of its production between 1817–1820 and not to his younger brother Henry Frederick Thynne. It is to be noted that Lord Henry Frederick Thynne would have attained the age of 21 in 1818 and this would quite possibly have merited a reasonable suggestion for the purpose of the original commissioning of this service in Nantgarw porcelain had he been truly Viscount Weymouth at that time, which apparently, he was not! This service does not seem to be referred to elsewhere in the literature and only a single plate is exhibited in the work of John et al. (*Nantgarw Porcelain Album*, 1975); it is not clear if other examples of this armorial creation are known to exist at this time. It is quite possible therefore that this plate could perhaps have been an experimental specimen or perhaps a cabinet sample for a service that was never fully commissioned later. The production of several specimen plates to demonstrate the use of alternative types

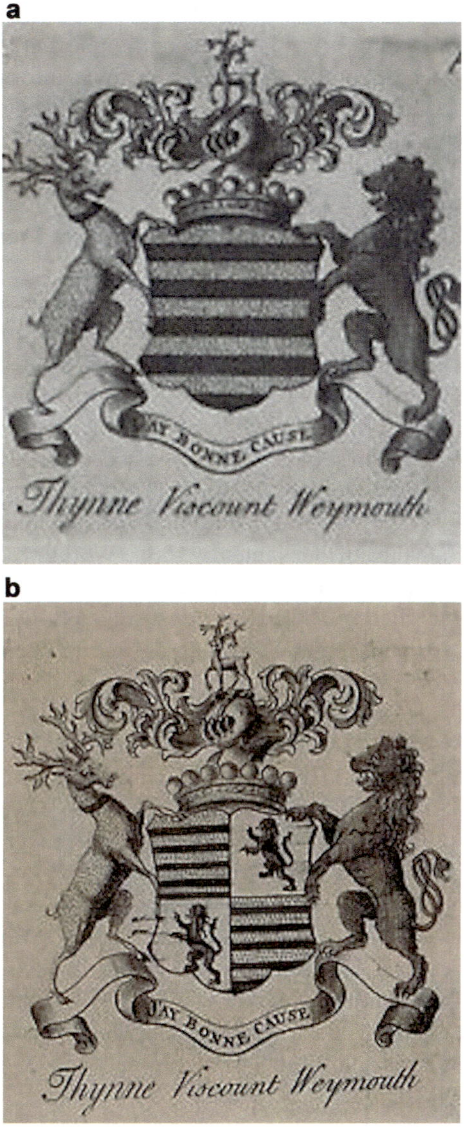

Fig. 2.5 **a** Bookplate of Thomas Thynne, Viscount Weymouth, showing his arms comprising a *ten-fold barry of or and sable*, yellow and black, which are represented by dots and cross-hatching, respectively, in what is termed heraldic "tricking". **b** Bookplate of Thomas Thynne, Viscount Weymouth, showing his fuller escutcheon comprising the quartered arms of the Marquess of Bath

of decoration and gilding patterns for the final selection by the host commissioner was quite common practice in porcelain manufactories and a classic example of this is manifest in the twelve or so specimen armorial plates that were created for King William IV and Queen Adelaide for their Royal Service in the 1830s by John Brameld at the Rockingham Porcelain Manufactory at Swinton in Yorkshire, one of which was selected by Queen Adelaide for its eventual manufacture and delivery in 1837 (Cox and Cox, *Rockingham Porcelain, 1745–1842*, 2001). It is revealing to explore further the genealogy of the Thynne family at Longleat House, as this uncovers some interesting information which is relevant in the correct attribution of this armorial plate and will prove to be especially informative regarding the use of the incorrect heraldic tinctures that were used in the composition of the barry of the escutcheon, which instead of being gold and black is actually white and green.

2.2.3.1 Marquess of Bath/Viscount Weymouth Genealogy

The Marquessate of Bath was created on the 18th August 1789 and was held first by Thomas Thynne, the 3rd Viscount Weymouth. The first Viscount Weymouth was Thomas Thynne (1640–1714) created in 1682, followed by his son, also Thomas Thynne (1710–1751) as the 2nd Viscount, and then by his son, Thomas Thynne (1734–1796), Baron Thynne of Warminster, as the 3rd Viscount, who then became the 1st Marquess of Bath. He was succeeded by his son, Thomas Thynne (1765–1837), as the 2nd Marquess of Bath in 1796, having previously succeeded his father as the 4th Viscount Weymouth until this time. Thomas Thynne was hence the 2nd Marquess of Bath during the Nantgarw China Works production period from 1817–1820. The 2nd Marquess of Bath had 11 children by his wife, Isabella, daughter of Viscount Byng, who died in 1830. The eldest son, also Thomas Thynne (1796–1837), held the title of 5th Viscount Weymouth during the second decade of the nineteenth century and would therefore be the name that would naturally be associated chronologically with the armorial *Viscount Weymouth* Nantgarw service. He attended Eton College and St Johns College, Cambridge between 1814 and 1816, graduating with an Honorary MA, and then became a diplomat under the Earl of Clancarty, who was "*not impressed with his ability*". Then in 1818, he represented Weobley in Parliament as Tory MP, a rotten borough, which his father had previously represented before him (Lord Henry Cavendish-Bentinck was the other MP for Weobley along with Thomas Thynne)—but in 1820 he did not stand in the General Election that arose because of the death of King George III. Thomas, Viscount Weymouth, clashed with his father over the behaviour of his friends and the heavy debts incurred in his lifestyle. On the 11th May 1820, Thomas, the 5th Viscount Weymouth, eloped with Harriet Matilda Robbins, a tollgate keeper's daughter, and they married in Marylebone; afterwards they resided in Paris and Florence, having been cut off by his father, the 2nd Marquess of Bath, who disapproved of his son's lifestyle and socially unacceptable marriage and who attempted legally to disinherit him from succession to the title, but failed to do so in the courts under the law of primogeniture. Thomas returned to England in the late 1820s and reappeared on the social scene once more, as noted in columns such as *Fashionable Movements* and *Fashionable Intelligence* in society newspapers, which is indicative of his acceptance on the social circuit, although the Viscountess is rarely mentioned except that she travelled from Paris occasionally. Thomas Thynne is recorded as selling objects of art and china that he had collected to raise money. His father meanwhile maintained that his only recourse was to outlive his son to prevent his inheritance. There was no reconciliation with his father in the years that followed but his mother did visit Thomas and Harriet in Paris before she died in

1830. Thomas died as tenant of Shanks House, Dorset, on the 16th January 1837, aged 40. Some 5 weeks later his father Thomas, the Marquess of Bath, died then the younger brother Henry Frederick, now Viscount Weymouth, became the Marquess of Bath until he too died on 24th June 1837. Henry Frederick served in the Royal Navy, reaching the rank of Captain, and was also MP for Weobley between 1824—1826 and 1828–1832; Henry married a Harriet Baring on the 10th April 1830. John, Viscount Lonsdale of Lowther, a good friend of Thomas Thynne, Viscount Weymouth, died in 1818 and Viscount Weymouth was an executor of his estate, who was bequeathed the sum of £100 "*....to be bestowed in a piece of plate in remembrance of me*". A piece of plate could relate to silver but maybe this gift enabled Viscount Weymouth to commission the Nantgarw armorial porcelain service—his relationship with his father being so strained at this time it is difficult to see otherwise how an impecunious Thomas would have had the wherewithal to commission a rather expensive Nantgarw service? Henry Frederick's son, John Alexander Thynne (1831–1896), then became the 4th Marquis of Bath and the 7th Viscount Weymouth in 1837 at the tender age of six. Thomas Thynne's widow, Harriet (1800–1873), remarried an Italian Count Inghirami but never had any children.

The coat-of-arms of the Marquess of Bath, which are shared with the Viscount Weymouth, are: Quarterly—1st and 4th, *a barry of ten or and sable* (Botteville), 2nd and 3rd *argent a lion rampant with tail nowed and erect gules* (Thynne). Crest: *a reindeer statant or.* Supporters: *Dexter—a reindeer or gorged with a plain collar sable; Sinister—a lion with tail nowed and erect gules.* Motto: "*Jay bonne cause*", meaning "*I have good reason*". A portrayal of these arms is given in the bookplates of the Viscount Weymouth in Figs. 2.5a and b, from which it can be seen there is a match with the armorial records but certainly not with the armorial porcelain plate as depicted in Fig. 2.4!

It is apparent that the current attribution of the Nantgarw armorial plate described here is incorrect as Lord Henry Frederick Thynne did not hold the title of Viscount Weymouth in the second decade of the nineteenth century; this belonged to his elder brother, Thomas, who was living overseas and in disgrace for some years from 1820. The *Viscount Weymouth* service must, therefore, relate to Thomas Thynne. It is also realistic to postulate that it could have been ordered by the 2nd Marquess of Bath, Thomas Thynne, possibly to celebrate the coming of age of his second son, Lord Henry Frederick, in 1818 and perhaps the anomalies in the coat-of-arms may reflect this. An alternative explanation could be that it was ordered by the elder brother Thomas Thynne, Viscount Weymouth, before he eloped with Harriet in 1820. Thereafter, the family differences and enforced separation were maintained but Isabella, Marchioness of Bath, did try to keep in touch with Thomas and Harriet by visiting them in Paris in the 1820s. What is still unexplained is the clear misplacement of the heraldic colours or tinctures in the escutcheon—instead of *a barry of ten or and sable* (gold and black) we actually have *a barry of ten white and green* (argent and vert).

What is clear is that this apparently indisputable assignment historically still retains some unanswered questions—but at least it can be amended to exclude Lord

Henry Frederick Thynne, who was not the Viscount Weymouth that had been previously assigned to this service; even if we accept the premiss that it could have been manufactured to celebrate his coming of age in 1818, it should not then be referred to as the *Viscount Weymouth* service but preferably as the *Lord Henry Frederick Thynne* service and, of course, he would not have been entitled to bear these arms before he properly acceded to the Viscountcy some twenty years later! Alternatively, the assignment to Thomas Thynne, who was Viscount Weymouth at that time before his disgrace, does then carry with it the complexity of the incorrect heraldic tinctures! In this case it must now be called the *Thomas Thynne, Viscount Weymouth*, service. As a final speculative thought, the service could have been commissioned by him in celebration of his impending marriage to his socially unacceptable Harriet and declaring his sincerity, hope and loyalty in love with the unusual heraldic tinctures, whilst symbolically adopting the odd stance of the reindeer supporter in his arms as his family turning against him? The motto of "*I have good reason*" would also strike a very poignant note in association with these heraldic clues referring to his family estrangement.

2.2.4 Coat-of-Arms on Cambrian Pottery Earthenware: Foreign Nobility?

An armorial plate in Cambrian Pottery, Swansea, earthenware and now in the collection of the Victoria and Albert Museum, South Kensington, London, has occasioned much comment because of its rather unusual armorial design which has been assumed hitherto to belong to foreign nobility (Morton Nance, *The Pottery and Porcelain of Swansea and Nantgarw*, 1942, Plate XXX): the two supporters are lions but one is *rampant regardant* whilst the other is lying down and *couchant* with its head raised in front of the escutcheon—a rather strange combination of heraldic "supporters" when it is apparent that one of them is clearly not actually supporting the escutcheon and arms (Fig. 2.6a) ! A further example of this unusual heraldic depiction is shown in the oval dish in Fig. 2.6b. The original function of supporters heraldically was to guard and protect the armour and arms of the noble involved in battle or in jousting tournaments, hence they were usually animals depicted in rampant, segreant or aggressive poses, and it would not be possible to accomplish their designated role effectively if the supporter was lying down (couchant) beneath the arms they were meant to protect and support. A search of other national armories reveals that the design of the coronet shown on this armorial Cambrian Pottery, Swansea, earthenware plate immediately rules out the Danish, Swedish and Spanish nobility, but the five pearls visible could match the designation of a Vicomte (France) or a Visconti/Nobile (Italy)—who were ranked equivalent to a Baron in English nobility. Further than that, the only clues to a potential assignment would be the closed visor helm and lambrequins, but the armorial bearings are minus a crest and torse, and the winged figure of a rampant Pegasus below it which is surmounted by three *estoiles* (*etoiles rayonnant*) all enclosed in an oval escutcheon. It is interesting that the emblem of Pegasus is extremely rarely found in an escutcheon, being usually featured as a crest heraldically, and the heraldic symbolism of the escutcheon is also intriguing—the *estoiles* are emblems of God's goodness and the eminence of the bearer, with the rampant Pegasus representing fame through brilliance of the mind, deep friendship and faithfulness. Ernest Morton Nance's (*The Pottery and Porcelain of Swansea and Nantgarw*, 1942) attribution of the flower painting to Thomas Pardoe's work on the basis of the rather limited associated painting and artwork, comprising scattered single floral sprays, would apparently date this intriguing armorial piece to before the end of the first decade of the nineteenth century as Pardoe had by then left Swansea in about 1809/10 and set up a ceramics decorating business in Bristol. This plate is now in the Victoria and Albert Museum (No. 604, acquired in 1906; donated by J. H. Fitzhenry), where it was examined by Bernard Rackham, at one time the Curator of Ceramics there, who believed it to have been decorated in London for an Italian noble family at the Cambrian Pottery Warehouse prior to its closure in 1807—in which case it would debar Thomas Pardoe from its enamelling since he had never worked in London. Also, this would definitively have dated the commissioning of the armorial service to before 1807. The plate has a spade device impressed on its reverse, a characteristic mark of the Cambrian Pottery. Morton Nance (*The Pottery and Porcelain of Swansea and Nantgarw*, 1942, pp. 64, 83) also quotes an extract from the *Cambrian* newspaper, dated December 18th, 1807, which advertises the closure of the Cambrian Pottery Warehouse of Haynes, Dillwyn & Co. in 64, Fleet Street, London, and of the sale of its stock of "*an elegant assortment of every kind of service and article in earthenware ... to be sold off at a 10% discount for ready money*".

Ernest Morton Nance has found a record that John Mortlock of 47, Oxford Street, London, a prime agent for china in London, then took over the agency for the Cambrian Pottery, Swansea, earthenwares and creamwares in London and were fully operating as its agents on this basis by 1808—which itself must be a tribute to the quality and desirability of the earthenwares and creamwares then being offered for sale by the Cambrian Pottery in the capital. Morton Nance also noted that there were several titled people interested in acquiring some of these ceramics from Mortlock's, including a Mrs Hester Lynch Salusbury Piozzi, a noted connoisseur of ceramics at that time. It is intriguing to speculate upon the assigned foreign nobility source of this armorial as this period was at the height of the Napoleonic Wars and the commissioning of this service by foreign nobility at either Swansea or London would suggest strongly that the noble family concerned would have to have been possible émigrés who had escaped the French "Terror" in 1792 or equally the invasion of Italy by Napoleon, as at this time the French, Italian and Spanish ports were being blockaded effectively by the Royal Navy, which would have interfered with the delivery of such a service when completed if the noble family were still based in France or Italy. Napoleon, between 1796 and 1799, had acquired and occupied in the name of the French Revolution most of the Italian states, such as Lombardy, Sardinia, Emilia Romagna, Tuscany, Venice, Piedmont, Genoa, Rome and Milan, as satellites of France, whilst appointing himself Emperor of France and King of Italy by 1805. Perhaps realistically it would be appreciated that such an émigré family would have

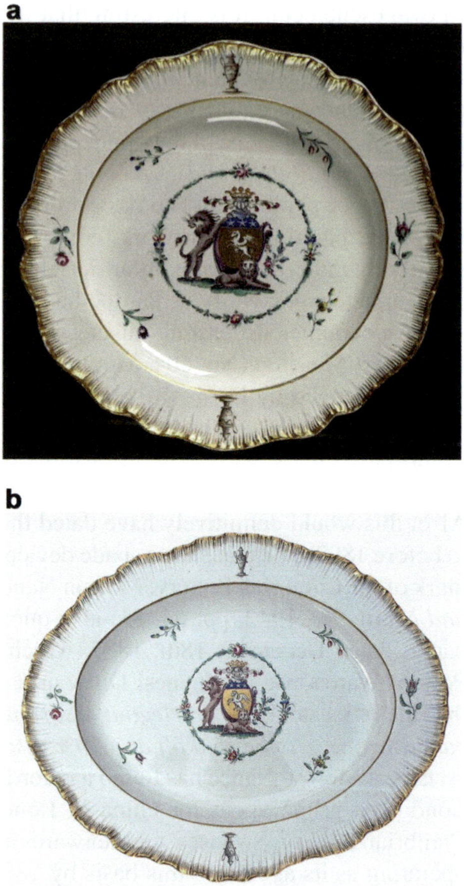

Fig. 2.6 **a** Cambrian Pottery earthenware armorial plate, *ca.* 1810, with a rather unusual heraldic depiction of an escutcheon, coronet, helm with visor and supporters which has been the cause of some controversy in the Welsh armorial literature relating to the Cambrian Pottery. It was formerly assigned to foreign nobility, probably Italian, but here a hypothesis is advanced for it belonging to Mrs Hester Thrale Piozzi Salusbury and that it contains some anecdotal commentary on her family lineage. Now in the Victoria and Albert Museum, South Kensington, London, the gift of J. H. Fitzhenry, Accession number 604–1906. *Reproduced by courtesy of the Victoria and Albert Museum, South Kensington, London.* **b** Cambrian Pottery armorial oval dish *en suite* with the plate shown in **a**. *Courtesy of Kingschina Antiques, Mayfair*

been based in London and this hypothesis would then reinforce Bernard Rackham's conclusion, but still keep open the potentially French sourcing of the coat-of-arms. Alternatively, the mention of the name Piozzi, clearly being of an Italian origin, generated the examination of *Burke's Armory* in this context, which gives the following citation for Mrs Hester Lynch Piozzi:

> Piozzi Salusbury: Bachergraig, co. Denbigh, descended from Sir John Salusbury of Lleweny Hall, Denbighshire. Hester Lynch, daughter and heiress of Thomas Salusbury, of Brynbella, co. Flint, married first Henry Thrale, then Signor Grambattista Piozzi of Brescia in Lombardy, who assumed by royal licence in 1813 the surname of Salusbury and became high sheriff of Flint in 1816, receiving a knighthood from the Prince Regent in 1816. Arms: gules a lion rampant or ducally crowned between three crescents of the last a canton erm. Crest, a demilion rampant couped ar. collared gules and ducally crowned holding in its dexter paw a crescent gold. Motto: "Sat est prostrasse leoni", which means "It is enough to have laid the lion prostrate".

This last statement is very interesting as it surely provides a graphic description of the strange configuration of the two lion supporters of the coat-of-arms depicted here, where one rampant lion is visibly triumphant over another prostrate lion, heraldically termed *couchant*, at the front—and could this not therefore be an example perhaps of a "pictorial motto"? Further research into the genealogical background of Hester Lynch Salusbury Piozzi is very revealing and demonstrates how fascinating and intriguing this background can be for the correct assignment of armorial ceramic artefacts. Hester Lynch Salusbury Piozzi (1741–1821) was born at Bodfel Hall in Caernarfonshire, the daughter of Hester Lynch and Sir John Salusbury, one of the most influential and wealthiest Welsh landowners of his time, whose seat was at Lleweni Hall, Llysmarchweithian, Denbighshire. Sir John Salusbury founded Halifax in Nova Scotia, Canada between 1749 and 1753, and named it in honour of his patron, Lord Halifax, then President of the Board of Trade, but he then lost most of his fortune in several failed speculative Canadian venture investments. After the death of her father in 1761, Hester and her mother then moved to live with her uncle, Sir Thomas Salusbury, at Offley Park in Herefordshire. In 1763, aged 22, she married a wealthy London brewer, John Thrale, who in 1765 became MP for Southwark. The same year she was befriended by Dr Samuel Johnson, who often enjoyed the company of his well-educated hostess and her family and friends at their London home of Streatham Park. In this context, she came into contact with Susana Dawson Dobson, whose husband was Mrs Thrale's physician, an historian and writer on the origins of heraldry and chivalry (Dawson Dobson, *Historical Anecdotes of Heraldry and Chivalry: Tending to Shew the Origin of Many English and Foreign Coats of Arms, Circumstances and Customs*, 1795). Mrs Dobson had strong ambitions to join Mrs Thrale's eminent literary group, which included James Boswell and Oliver Goldsmith as well as Dr Johnson. However, Hester was not keen at all on her joining the group and said, "*Mrs Dobson persecutes me strongly as if with a violent and undesired friendship. she is jealous*". John Thrale died in 1781, upon which his widow Hester sold the Anchor Brewery, Southwark, and thereby increased her wealth; it was widely speculated that Hester would then marry Dr Samuel Johnson but instead she married Gabriel Mario Piozzi, an impecunious musician and Italian music teacher to her 12 children (of whom only four daughters survived infancy), in July 1784. Her marriage to Piozzi was not favoured at all by London society, who viewed him as a gold-digger, and she complained that the newspapers gave her some bad press on his account:

> The English newspapers are full of gross insolence towards me by raising an obscure and penniless Italian Fiddler into sudden Wealth.

She became estranged from Dr Samuel Johnson and other *literati* upon her marriage to Piozzi and with Gabriel Piozzi she travelled widely on the Continent, after which they both then retired to her country home, Brynbella, on her Bach-y-graig (mis-spelled in *Burke's Armory*) estate at Tremeirchion in North Wales. Her marriage to Piozzi was strained and she managed eventually to strike up a renewed correspondence with Dr Samuel Johnson just before he died in December 1784. In 1786 she wrote her "*Anecdotes of the Late Samuel Johnson LLD During the Last Twenty Years of His Life*" (Hester Lynch (Thrale) Piozzi, 1786), which was well received and is now acclaimed to be the best personal account written of Samuel Johnson in a new anecdotal style and termed "*a new fashioned biography*", which excited much literary comment as this had never been experienced before in literary circles (Clifford, *Hester Lynch Piozzi (Mrs Thrale)*, 1987; Birkbeck-Hill, *Hester Thrale*, 1897; D'Ezio, *Hester Lynch Thrale Piozzi: A Taste for Eccentricity*, 2010). The anecdotal nature of her writing must have been a major reason for Susana Dawson Dobson wanting to join her literary circle as Mrs Dobson's survey of heraldry and chivalry was itself written in a similar anecdotal form!

The entry for Hester Lynch Thrale Piozzi Salusbury in *Burke's Armory* is incorrect in several aspects: in the spelling of her country estate (Bach-y-Graig), in the spelling of her husband's name, which is given as Grambattista Piozzi, instead of Gabriel Mario Piozzi, and in his appointment as High Sheriff of Flintshire which was dated seven years after he had died (!), but an important message is that she was the one who was entitled to bear arms under the name Hester Lynch Piozzi Salusbury. Grambattista (*sic.* Giambattista, Giovanni Battista) was actually Gabriel Piozzi's brother and in 1798, Hester and Gabriel Piozzi adopted the orphaned 5-year-old son of Giambattista and renamed him John Salusbury Piozzi. Gabriel died of gout in 1809 and in 1814 Hester gifted her estate to her adopted son John upon his marriage to Harriet Maria Pemberton of Ryton Grove; John then acquired the additional family surname, with an accompanying knighthood, and became Sir John Salusbury Piozzi Salusbury, becoming the High Sheriff of Flintshire in 1816 (which now agrees with *Burke's Armory*)! Hester then went to live in Bristol in 1814 and died there in Sion Row, Clifton, in 1821 (Franklin, *Piozzi Hester Lynch (1741–1821)*, 2004). This is a most interesting discovery, as Thomas Pardoe was then in business in Bristol from 1812 up to 1821 decorating ceramics, when thereafter he left to work with William Weston Young at Nantgarw. It would hence be a reasonable supposition that Hester Lynch Piozzi Salusbury would have been in an ideal situation to engage Thomas Pardoe to locally decorate Cambrian Pottery ceramics in Bristol (perhaps directly shipped over the Bristol Channel from Swansea), where they were then both resident. This scenario would have created an armorial Cambrian Pottery service with her arms and even incorporating into them an "anecdotal" pun, for which she was renowned in her literary circles (!), based upon her motto in having the most unusual situation of one of her lion supporters lying prostrate before the other. This situation also means of course that both Bernard Rackham and Morton Nance were both partially correct in their supposition that the coat-of-arms were possibly of an Italian family in origin and that it would be perfectly possible for Thomas Pardoe to have decorated the service, as alleged by Morton Nance, but this would have occurred hypothetically in Bristol and not in Swansea nor in London. This reasoning is supported by Pountney (Pountney, *Old Bristol Potteries*, pp. 115–116, 1920), who stated that Pardoe definitely decorated Swansea "china" (earthenwares and creamwares) in Bristol (see Appendix II) up to 1821—these having been shipped directly in small coasters from Swansea into Bristol docks. The historical research evidence then is clearly supportive of Mrs Hester Lynch (Thrale) Piozzi Salusbury as being the commissioner and originator of the Cambrian Pottery armorial plate under investigation, probably around the middle of the second decade of the nineteenth century, for which a celebratory date would surely be the award of a knighthood to her adopted son John in 1814/15 upon his accession to her estate? The rather unusual form of the armorial depicted would surely have appealed to someone who has been cast as the doyenne and avowed mistress of the literary anecdote in Georgian England. Does this then perhaps constitute a "pseudo-armorial" according to our definition in that it contains elements of a true armorial but also some very unusual attributes, although of course the lady concerned had every right to bear arms? Other small heraldic clues place this particular armorial in a distinctly different category to others: the escutcheon is oval (as befits a European continental influence), whereas an armigerous lady's escutcheon in English heraldry is lozenge shaped, the helm itself is more in the style of Italian armour with its gilt visor, and the coronet is accompanied by a wreath or torse in the helm, which is unusual in English heraldry. Crests are rarely used in Italian heraldry as are supporters and these are therefore not regulated (Von Volborth, *Heraldry of the World*, 1973). The coronet depicts five strawberry leaves which in Italian heraldry signifies a Ducal status of non-regal birth, which Hester Piozzi certainly was not. The presence of a two-handled ewer above the arms in the verge is also strange as this could be assumed to be a crest, which exists only rarely in Italian heraldry. The consensus must be therefore that this is an enigmatic armorial ceramic piece which displays elements of English and Italian heraldic symbols but is clearly anecdotal in its construction and contains a hidden message.

2.3 The Interpretation of Armorial Bearings

The interpretation of the various component heraldic devices in armorial bearings is essential for the correct attribution of an armorial ceramic service or armorial piece to its armigerous owner. Experience in armorial research has shown that the most difficult armorial bearing to define and interpret correctly is the crest in isolation, whereas in contrast a full suit of armorial bearings or achievements as displayed in a coat-of-arms with escutcheon, supporters, helm, mantlings, coronet or chapeau and motto affords the best possible scenario for its correct assignment. Even so, as observed in several of the case studies discussed in Sect. 2.2 above, sometimes the researcher has to be aware the incorporation of extraordinary objects, designs or strange and unusual styles which can create problems but which in themselves also can provide a source of enlightenment as to the reasons and thoughts behind the origins of the armorial commissioning. It is clear that the application of the appropriate genealogical research in parallel with the heraldic interpretation of the partial or full coat-of-arms is then essential to arrive at a satisfactory conclusion.

It is important to realise that the information retained in a coat-of-arms is critical for familial attribution and there are several heraldically dictated protocols that must be adopted for the proper succession of arms that are extremely useful for the armorial ceramics researcher to recognise. Firstly, when the marriage occurs between entitled armigerous man and a similarly entitled woman then the arms are quartered with elements of both, the man retaining the dexter side and the woman the sinister side. However, when another escutcheon is enclosed within the main shield, known heraldically as an "*inescutcheon of pretence*", this signifies heraldically that the male arms bearer has married an heiress who is the sole daughter of an armigerous father, who herself has no male siblings. Hence, rather than impaling his wife's paternal arms as would be normally expected upon such a marriage with an entitled bride, the paternal arms of his wife are now placed in the centre of his own shield as a pretence (i.e. a claim) of headship of his wife's family, which will become his or will do when it has become extinct heraldically on his wife's father's side. Any son of the marriage would then be expected to quarter his arms accordingly in the normal fashion to absorb his maternal arms. An interesting exemplar of this practice can be recounted here. Jane Thelwall (1665–1759) was the daughter and heiress of Edward Thelwall and Lady Sydney Wynn of Plas-y-Ward, Denbighshire, and she married Sir William Williams, 2nd Baronet, of Llanforda, Oswestry, in 1687 at the age of 21. Their son, Watkin, born in 1692, graduated from Jesus College, Oxford, and became MP for Denbighshire in 1716. In 1719, Watkin Williams became Williams-Wynn and acquired the large estate of Wynnstay, Ruabon, Denbighshire, through his mother's inheritance from her relative, Sir John Wynn of Gwydir, thereby assuming the surname of Wynn as well, in addition to his succession to his father's title and lands upon his death in 1740. The inescutcheon of Thelwall that is represented on the Jesus College punchbowl donated by Williams-Wynn in 1732 (Fig. 2.7) contains quartered arms as follows: 1st and 3rd, *a chevron with three boar's heads erect and erased* (Thelwall); 2nd and 4th, *a stag trippant facing sinister* (Powers). It is revealing that the official crest of Williams-Wynn, *an eagle displayed with wings spread and beaked and talons drawn,* does not appear in the arms on the punchbowl but it is represented in the 1st and 3rd quartering of the arms. It seems apparent, therefore, that the arms in the Jesus College punchbowl are those of the Williams-Wynn family and of his father, the second Baronet, as in 1732, Watkin Williams-Wynn had not yet succeeded to the baronetcy and claimed the title.

Another manifestation of heraldic discipline is the methodology of representing colours in a monochrome or monotone rendition of a coat-of-arms: the tincture or heraldic colour used in heraldic arms is most critical for the maintenance of their accuracy and there are strict rules for indicating its source, especially for monotone bookplates and engraved arms upon silverwares. The earliest representation of a description of a heraldic coat-of-arms is believed to be that of Chretien of Troyes in the late twelfth century (*ca.* 1170), which used a form of words which thereafter became known as a *blazon*. This has remained essentially unchanged in English heraldry from the reign of King Edward I (fourteenth century) to this day, which explains its predominant usage of Norman French linguistic terminology. The derivation of *blazon* is from the German *blasen* meaning "to blow a horn" and its representation of the correct heraldic description of a coat-of-arms is also intriguing historically in that it originally announced the arrival of a knight, his arms and supporters at a tournament. Then, with the growth of printed books it was realised that another means of describing coat-of-arms and heraldic emblems was required, especially for armigerous library bookplates, and the concepts of *tricking* and *hatching* were devised. Primarily, tricking started with the adoption of a code

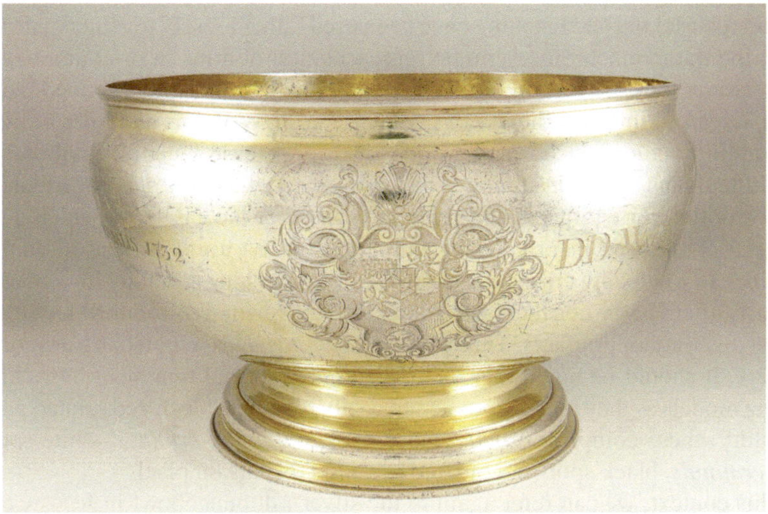

Fig. 2.7 The very large silver gilt punch bowl presented to Jesus College, Oxford, by an alumnus, Sir Watkin Williams—Wynn of Wynnstay, Denbighshire, in 1732, on the occasion of his receiving the Honorary Doctor of Civil Law, D.C.L., degree of the University of Oxford. The arms of Sir Watkin Williams-Wynn are shown engraved on the side of the bowl comprising an escutcheon but with no helm, crest, supporters or motto. *Reproduced with the permission of Jesus College, Oxford*

of single letters which indicated the tincture or base colour that would be used on a heraldic armorial diagram: this code was expanded and modified somewhat with usage in the fifteenth and sixteenth centuries, but a version of colour indication based upon single text letters representing the main heraldic tinctures is shown below, along with their colour equivalents:

Heraldic Colour	Diagrammatic Abbreviation	Origin
Or	A	aureus /yellow
Argent	a	argentum/white
Azure	c	caeruleum/blue
Gules	r	rubeus/red
Vert	v	viridium/green
Sable	–	sable/black (using a black colour)

An Italian monk called Borghino, among others, devised an alternative list in the sixteenth century, which then became adopted by some, which included the letters N for sable (nigrum), B for white (bianco), R for red (rosso), A for blue (azzuro) and G for yellow (gialbo) and omitting green altogether. Clearly there was scope for some confusion here with the depiction of heraldic arms and several tricking lists running at the same time.

Although tricking and hatching have been used rather loosely together historically to describe the monochrome heraldic representation of arms in print and engraved on silverware, they are sensibly different in approach as hatching (derived from the French, *hachure*) involves the use of geometric designs of dots and lines to indicate the heraldic tincture instead of the upper and-lower-case letters described above. More difficult and time-consuming to endorse and use, hatching was nevertheless the favoured option for engravers and printers because of the ease with which a single letter in tricking used incorrectly or ambiguously was found to lead to confusion and thence to a palpably erroneous interpretation. Hatching certainly increased in its popularity after 1630 when there was a noted increase in the requirement for engraved arms on a variety of artefacts. The major hatching designs adopted are described in Table 2.1. It should be noted that furs do not have special treatment in tricking or hatching models as they already possessed easily recognisable differentiated patterns in heraldry: these comprise *ermine*, black spots on white, *ermines*, white spots on black, *erminois*, black spots in gold and *pean*, gold spots on black.

In this context, we can refer again to the silver-gilt punchbowl in Jesus College, which illustrates the engraved arms of Watkin Williams-Wynn of Wynnstay, which uses hatching to describe the tinctures of the heraldic colours. On the other side of the punchbowl is the escutcheon of Jesus College, Oxford, described heraldically as "*an escutcheon vert with three stags trippant argent attired or*" as represented in Fig. 2.8: however, the hatching on the punchbowl arms comprises horizontal lines for the field of the escutcheon which clearly indicate an *azure* tincture is present and not a green tincture. It transpires that Watkin Williams-Wynn was an admirer and supporter of Bonnie Prince Charlie, Prince Charles Stuart, the Young Pretender to the Throne, whose standard was a saltire argent on a blue field and his clansmen supporters wore blue bonnets. His Jacobite supporters became associated with wearing the colour

Table 2.1 The geometric designs adopted for heraldic tinctures in hatching

Tincture/Colour	Hatching
Or/Gold/Yellow	Black dots on white background
Argent /Silver/White	Blank/white
Gules /Red	Vertical lines
Sable/Black	Squared grid
Azure /Blue	Horizontal lines
Vert/Green	Diagonal lines: top left to bottom right
Purpure/Purple	Diagonal lines: top right to bottom left
*Murrey /Mulberry	Diagonal grid: diagonal lines crossing each other
*Sanguine/Blood-red	Lozenge grid: vertical lines crossing vert
*Tennee /Tawny	Lozenge grid rotated through 90 ^{0}: vertical lines crossing purpure

*These represent additions to the seven accepted tinctures in the heraldic tincture list as they are formally stains that have been derived from organic plant extracts and used to dye armorial textiles and seem not to be represented on engraved silverwares, for example

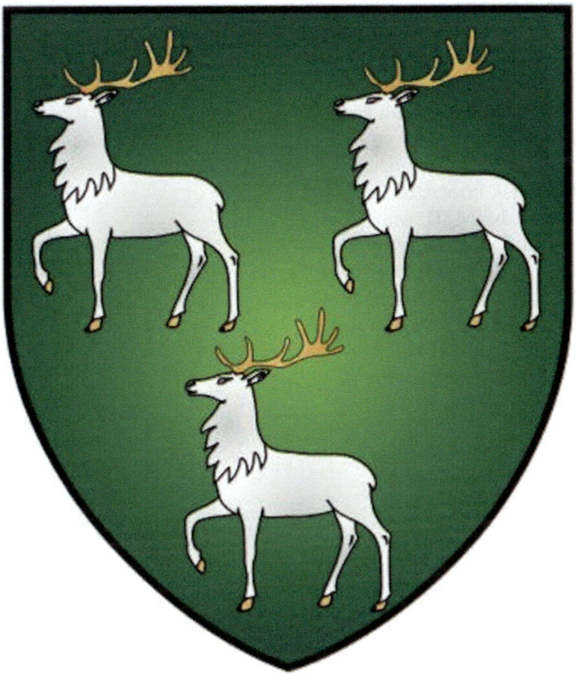

Fig. 2.8 The arms of Jesus College, Oxford, comprising an escutcheon *with three stags trippant argent, attired or, in a vert (green) field*, as they appear today

blue and it is quite conceivable that Watkin Williams-Wynn was here declaring his secret support for Charles Stuart and the Jacobite cause, rather than there being an unfortunate error in the engraving of the arms on the punch bowl.

The mineral pigments that were recommended for use in heraldic tinctures varied over the course of time as new pigments became available for artwork but some of these that are noted in heraldic lists are: *Azure:* French blue or Prussian blue, *White*: Chinese white, *Gules*: vermilion, *Yellow*: cadmium yellow, aureolin, *Scarlet:* carmine, *Vert*: emerald green, *Purpure*: French blue plus carmine, *Sable*: Indian ink, *Gold*: gold leaf plus the white of an egg, *Silver*: silver leaf, aluminium, *Grey*: French blue plus Indian ink.

The use of a *badge* heraldically is worthy of further description here. A *badge* is an armorial ensign which is complete in itself and can be borne without the necessary accompaniment of a shield or escutcheon: it has no connection whatsoever with a crest, coronet or a helm but has frequently been confused in the literature with a crest (Boutell, *The Handbook of English Heraldry*, 1914). The badges of the realms of England, Scotland and Ireland are the rose, thistle and shamrock, but perhaps the most famous badge in heraldic usage in English heraldry is that of the Prince of Wales' feathers, the badge of the Heir to the Throne of the United Kingdom with the motto "*Ich Dien*" which was apocryphally acquired by Edward the Black Prince when he vanquished King John of Bohemia at the Battle of Crecy in 1346. Prince Edward frequently referred to this emblem as his badge, but it has since been mistakenly called his crest. It comprises a plume of three ostrich feathers argent

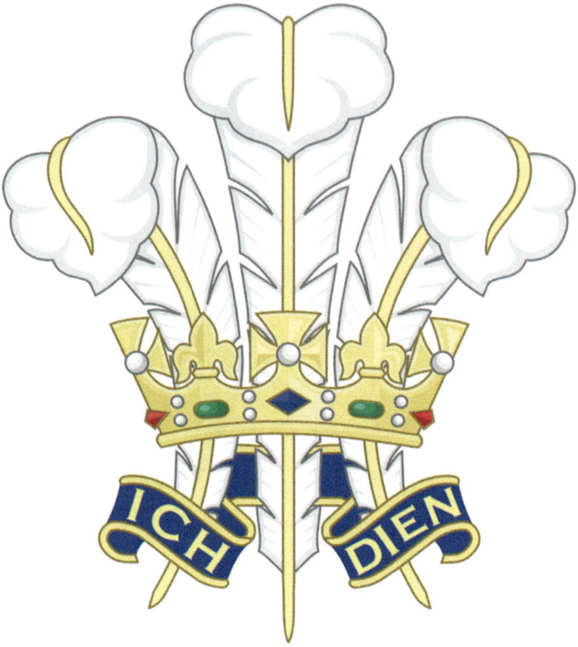

Fig. 2.9 The heraldic badge of the Prince of Wales, Heir to the Throne of the United Kingdom of Great Britain and Northern Ireland, which historically is believed to derive from its acquisition by the Black Prince, the son of King Edward III, who vanquished King John of Bohemia in the Battle of Crecy in 1346. This is not a heraldic crest although it is frequently described incorrectly as such. *Public domain CC by SA 3.0*

enfiled by a royal coronet of alternate crosses and fleur-de-lys or, with the motto "*Ich Dien*" meaning "*I Serve*" (Fig. 2.9). An alternative origin of the three feathers badge is ascribed to Queen Philippa of Hainault, Queen Consort of King Edward III and the Black Prince's mother.

Finally, mention must be made of hatchments: these are arms displayed in a lozenge framework which are raised upon the decease of an armigerous individual and placed at the entrance to a residence or estate and afterwards lodged in the local parish church or mausoleum where the entombment has taken place. The field of the arms of the deceased is represented by a black field and his widow retains her normal quartered arms. In the case of the decease of the wife, her arms are retained in a field of black whereas the husband retains his normal quartered arms. Much heraldic information can be derived from a study of hatchments in association with family genealogy.

2.3.1 The Frog *Service of Empress Catherine the Great of Russia*

A type of armorial service, which strictly cannot come under the standard classification of an armorial service as we have defined here as it does not portray arms, badges, a coronet or heraldic achievements but rather displays a device, emblem or a motif which nevertheless could perhaps be seen to be indicative of its individual armigerous family, is exemplified by the famous *Frog* service in Wedgwood's ceramic creamware. Although not strictly porcelain, this case study will be illustrative of several features which are shared with its porcelain analogues. On the 1st May 1759, Josiah Wedgwood founded his earthenware pottery at Burslem in Staffordshire and then successfully expanded his business to its new location at Etruria in 1766, which was built on the success of his fine glazed white earthenware production that had been achieved the year before. In 1767, Queen Charlotte, the Queen Consort of King George III, admired Wedgwood's creamware so much that she allowed him to call it *Queen's Ware* and she invested in several items of this especially admired lightweight and robust earthenware pottery. This creamware body was lighter than normal earthenwares and also much more robust and was considered by some at that time to be equivalent in quality to many of the porcelains then being made elsewhere: being lighter in weight, it also was not subjected to the punitive national trade tariffs then operating throughout Europe that were based upon the weight of the ceramic items being exported and imported. This made the European ceramics manufacturers rapidly sit up and take notice as these creamwares were then effectively rendered more economically viable than contemporary porcelains to purchase and afforded a fine quality and robust ceramic that was clearly financially competitive with their other contemporary fine china and heavier faience and majolica earthenware analogues. The French manufactories, in particular, immediately responded to this challenge with a lighter version of their own heavier creamwares called *faience fine*. The supreme accolade accorded to Wedgwood's creamware, however, was the commission from another Queen, the Empress Catherine the Great of Russia (Ekaterina II of the Romanov Dynasty), for a very large dinner and dessert service in *Queen's Ware*, which thereafter became known as the *Frog* service because of the green enamelled emblem that was painted on each piece. The Empress Catherine wished to celebrate the victory of her Russian naval forces against the Ottoman Turks in the Russo-Turkish War (1768–1774) at the Battle of Chesma Bay, so she embarked upon the construction of the Chesma Palace, which was situated midway between her Winter Palace in St Petersburg and her Summer Palace at Tsarskoe Selo, on marshy ground known locally as the Kekerekeksinsky, which has been translated simply as "Frog Marsh", and known as *La Grenouillere* in the French language which was then officially used in the Russian Court. The Empress Catherine modelled her new Chesma Palace on Longford Castle in Great Britain, the seat of the Earl of Radnor, whose ancestor Laurens de Bouverie purchased the estate in the early eighteenth century and landscaped it with the assistance of Lancelot "Capability" Brown. Empress Catherine the Great's order with Josiah Wedgwood, which was placed through her Russian Consul in London in 1773, was for a very large service of 944 pieces, in total some 50 place settings, comprising 680 pieces in a dinner service and 264 pieces in a separate dessert service. These services were to be hand painted in monochrome sepia enamel, depicting in all 1222 different views of English stately homes and landmarks that were much admired by the anglophilic Empress Catherine, and with each piece bearing a green coloured enamelled motif of a frog in an escutcheon, in deference to the site chosen for her new palace (and thereby so providing us also with an example of a *rebus*). At first, so Llewellyn Jewitt reports (Jewitt, *The Ceramic Art of Great Britain from the Prehistoric Times Down to the Present Day*, 1878), it seems that Josiah Wedgwood was extremely reluctant to "deface" his esteemed *Queen's Ware* with such a motif:

> Wedgwood was very unwilling to disfigure the service with this reptile but he was told that it was not to be dispensed with.

The landscape views were selected by Thomas Bentley, Josiah Wedgwood's business partner, and featured Wedgwood's own home, Etruria Hall, on a serving dish. This service was the most expensive ever made at Etruria and the Empress Catherine was invoiced for £2,700, equivalent to approximately £400,000 today, i.e. approximately $500,000 (Edwards, *Porcelain Analysis and Its Role in the Forensic Attribution of Ceramic Specimens*, 2021). The service was made in Wedgwood's manufactory in Etruria and then sent to London to Wedgwood's decorating atelier in Little Cheyne Row, Chelsea, where his team of 30 enamellers and painters decorated it with the requisite borders and scenes and then overglaze fired it. It is alleged that Wedgwood did not actually make a profit from this prestigious royal commission, but he shrewdly put the service on show in London after its decoration and prior to its being shipped out to Russia, and it is said that he charged an entrance fee to the public for their admission and inspection of his noteworthy ceramic creation. The assembled finished service must have made a wonderful sight and undoubtedly provided Wedgwood with new aristocratic clients for his esteemed creamwares. Most of the *Frog* service that has survived into modern times is now held in the Hermitage Museum, St Petersburg, which now comprises some 300 pieces only, although some pieces do still occasionally surface at auction—the Wedgwood Museum in the Staffordshire Potteries now has the largest collection of *Frog* service items outside of St Petersburg. Recently, a serving platter from this *Frog* service made the sum of $46,000 at auction in 2019, a price clearly rivalling that of its contemporary fine porcelains in support of its importance, historical provenancing and ceramic quality. An example of the *Frog* service in Wedgwood's *Queen's Ware* is shown in Fig. 2.10, which illustrates a serving platter depicting Ditchley Park in Oxfordshire, the seat of the Earl of Litchfield, and is hand painted in sepia enamel monochrome with a green enamelled frog motif at the verge. The border comprises scrolls with oak leaves and acorns in sepia, signifying that this platter is actually part of the Empress' dinner service. In Fig. 2.11 is shown the detail of the green enamelled frog in its engrailed "escutcheon" and in Fig. 2.12, a pair of covered vessels or porringers from the Empress Catherine's dessert service is shown with their characteristically distinctive and different border of intertwined ivy leaves in sepia. It should be noted that the enamelled frog motif which characterises these services is enclosed within an engrailed escutcheon or shield, known as a "wave" type escutcheon with the double curvature on its top line on the dexter and sinister halves (technically known heraldically as an *engrailment*), which gives the ceramic artefacts in the *Frog* service their "pseudo-armorial" context. A ceramics researcher could easily be mistaken for believing that the compound *Frog* dinner and dessert services were genuine armorial services, which they clearly are not! The Empress Catherine's genuine heraldic coat-of-arms, in contrast, comprises

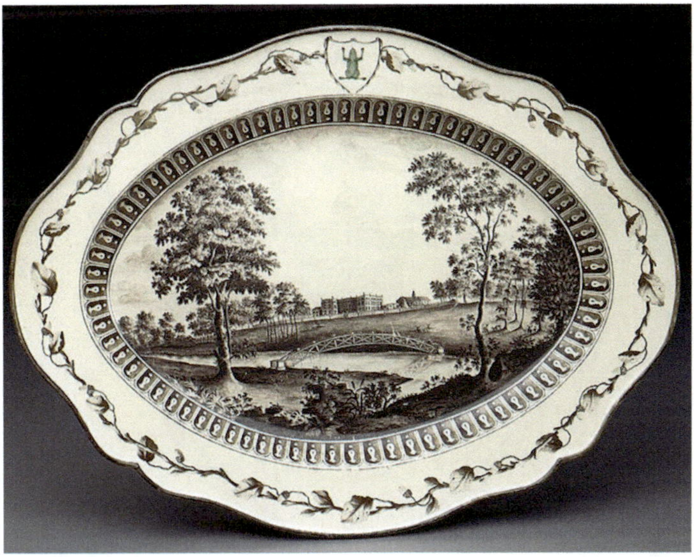

Fig. 2.10 Serving platter from Josiah Wedgwood's Queen's Ware *Frog* service made for Empress Catherine the Great of Russia (Ekaterina II) in 1773 for use in her Chesma Palace—a very large, combined dinner and dessert service originally comprising 944 pieces with hand painted illustrations in sepia monochrome of individual English houses and estates, each containing a green enamelled frog motif contained within an escutcheon. This piece shows Ditchley Park in Oxfordshire, seat of the Earl of Litchfield, and is from the dinner service, which had a distinctive border of oak leaves and acorns. *City of Birmingham Museum of Art. Public domain*

a Romanov double eagle bearing a central escutcheon of St George on horseback slaying a dragon, with no supporters, the eagle holding a sceptre and an orb in its talons and being surmounted by the Imperial Russian crown, as it appears on other armorial services in porcelain commissioned elsewhere by the Empress Catherine in the mid-to-late 1700s. Figure 2.13, in comparison, shows an example of the Empress Catherine's Imperial double-headed eagle on a Chinese porcelain Qianlong (1736–1795) service, known as the *Gatchina Palace* service (Howard, *Chinese Armorial Porcelain*, Volume 1, 1974; Howard, *Chinese Armorial Porcelain,* Volume II, p. 455, 2003).

2.3.2 *The Chamberlain Worcester* **Carnatic** *Service*

Here, it is appropriate to mention a special porcelain service which has been universally described in the ceramics literature as an "*unusual English armorial service*". An order for an armorial porcelain service was placed by His Highness Willajah Nabob Azam Jah with Chamberlain's Worcester manufactory through their agents in Madras, Griffiths, Cooke & Co., in 1820 for a large compound porcelain service

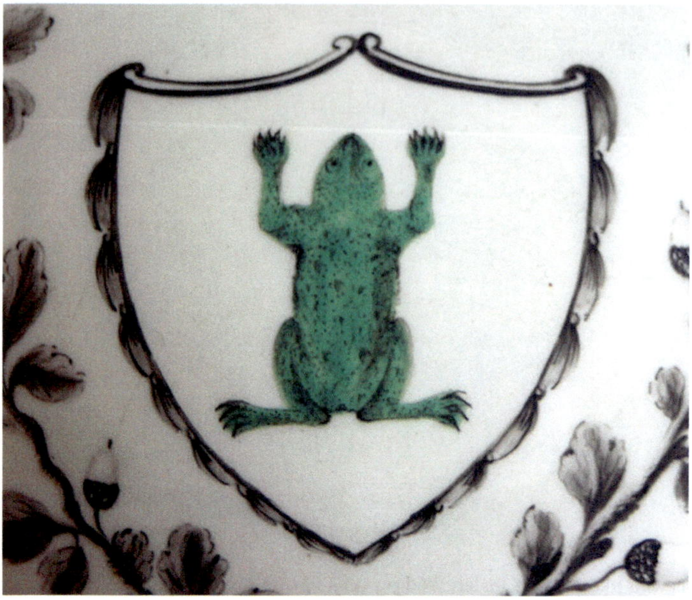

Fig. 2.11 Detail of the green enamelled frog motif in an escutcheon from the *Frog* service; demonstrating that one could easily be misled into believing that this was an armorial service!

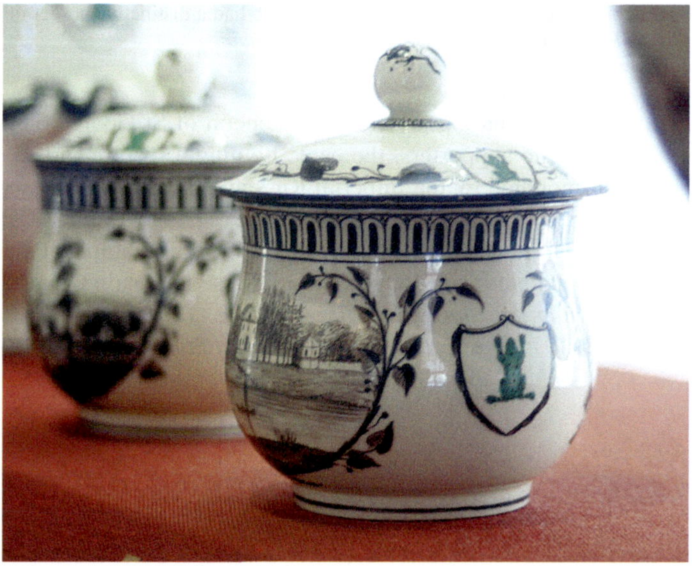

Fig. 2.12 Pair of porringers or covered bowls from the Queen's Ware *Frog* service of Empress Catherine the Great of Russia, showing the border of intertwined ivy leaves, indicating that these items were part of the dessert service component. *Hermitage Museum, St Petersburg, Russia. Public domain*

Fig. 2.13 Armorial bearing of Empress Catherine the Great, dinner plate, Qianlong period (1736–1795) Chinese porcelain, destined for the Gatchina (Khotchino) Palace, St Petersburg, occupied by Count Orlov, a favourite of the Empress Catherine. *Reproduced with the permission of Kingschina Ltd, Mayfair, London*

comprising integral breakfast and dinner-dessert services of over 1000 individual pieces in total; these services are unusual in that the "coat-of-arms" or heraldic device depicted thereon is not immediately recognisable as such but has nevertheless been designated by many writers to be armorial in its appearance. This dinner-dessert service comprised a cobalt blue ground border and central illustrations taken from *Curtis' Botanical Magazine*, with the Arabic inscription in the form of an armorial "escutcheon" included in a vignette in the reserve. Founded in 1787 at Kew Gardens and still extant today, this magazine is the world's longest running botanical periodical. An example of this particularly fine botanical service is shown in Fig. 2.14 where the "escutcheon" containing the Arabic script is also seen: the oval dish also bears the picture of a plant labelled in script on the reverse as "*Shewy Arctotes*", which is taken from Plate 282, Volume 47, of *Curtis' Botanical Magazine* (1819/1820). The breakfast service in contrast has a deep pink border and the same inscription is contained in a cartouche that is now located centrally and which is surrounded by a wreath of garden flowers. The "escutcheon" actually depicts the name and title of Azam Jah, Nawab of the Carnatic (reigned 1819–1825) in Arabic script and reads "*Amir al-hind nawab azam jah bahadur*" above a date of *1236 Hijri*, which equates to 1820 in the Western calendar (Fig. 2.15). Known in the ceramics literature as the *Carnatic* service, it took three years to complete at the Chamberlain's Worcester manufactory and was eventually shipped out to India in 1823 for the Nawab Azam Jah on the HEIC East Indiaman, *Windsor Castle*. It is indeed a moot point as to whether this constitutes an armorial service on the basis of our definition since it does not

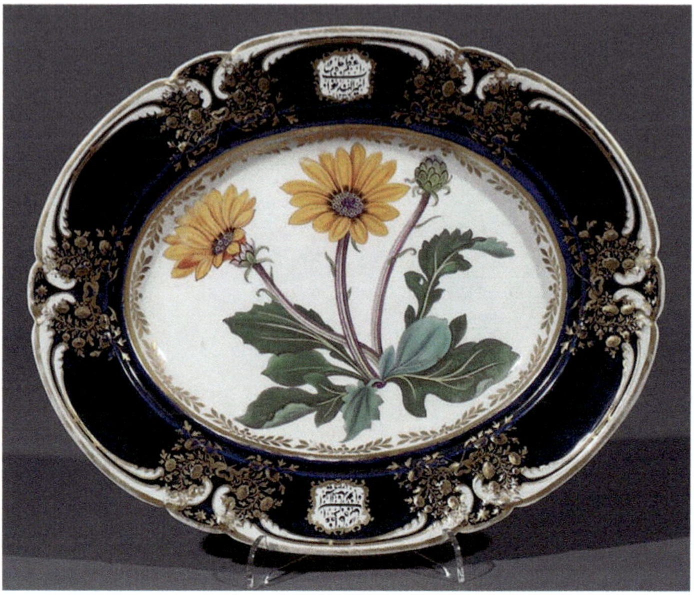

Fig. 2.14 Large oval dish from the dinner-dessert service ordered from Chamberlain's Worcester manufactory in 1820, for Azam Jah, Nawab of the Carnatic (reigned 1819–1825). This service has a dark cobalt blue ground border, and this particular dish displays a central botanical plant taken from *Curtis' Botanical Magazine* of 1819/1820 (Volume XLVII, Plate 282), inscribed on the reverse "*Shewy Arctotes*". Reproduced with permission from the Collection of Indar and Aruna Pasricha, Kensington, London

contain any heraldic bearings, devices or achievements that are recognisable as such and really is more akin to several of the personal inscriptions found on commemorative earthenwares, which are not considered to be armorial. Nevertheless, porcelain dealers, museum curators and collectors do still refer to this Chamberlain's Worcester *Carnatic* service as armorial porcelain even though it is devoid of any true heraldic devices or a coat-of-arms because the inscription sits in a cartouche which could be loosely regarded as an escutcheon but is strictly not so heraldically as it does not contain any symbols relating to familial descendancy.

2.3.3 The Wittelsbach *Armorial Service*

A special type of early armorial porcelain service is exemplified by the *Wittelsbach* service from the ancient noble family who ruled the Rhineland Palatinate and Bavaria and was founded in the eleventh century. It is an example of the blue and white Chinese porcelain known as *kraak* ware, which was imported into the Netherlands by the Dutch VOC and the Portuguese in the seventeenth century, particularly during the Wanli (1573–1620), Tianqi (1620–1627) and Chongzheng (1627–1664) periods

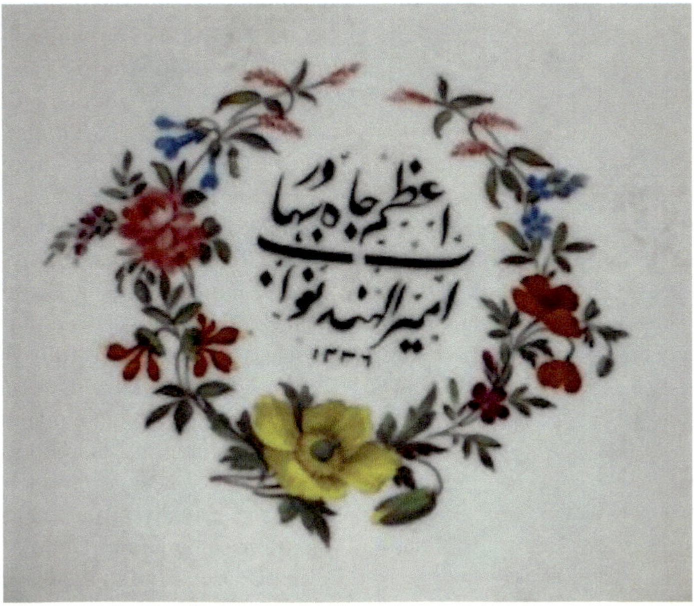

Fig. 2.15 Detail of the "armorial" feature resembling an escutcheon on the *Carnatic* breakfast service ordered from Chamberlain's Worcester manufactory in 1820, for Azam Jah, Nawab of the Carnatic (reigned 1819–1825) bearing the Arabic script, "*Amir al-hind nawab Azam Jah bahadur*" above a date of *1236 Hijri* (Hegira). Reproduced with permission from the Collection of Indar and Aruna Pasricha, Kensington, London

of the late Ming Dynasty (see the example illustrated in Fig. 1.4 bearing the VOC monogram). It is believed that it derived its name from the Portuguese carracks that were used to transport it to Europe from China. The decoration is exclusively cobalt blue and white and is characterised by the use of compartmentalised geometric designs which may further contain vignettes of scenes or patterns. After 1644, the Ming Dynasty collapsed and China was subjected to civil war for some twenty years until stability was once again restored with the formation of the Qing Dynasty, which finally accessed the Southern Ming Dynasty in 1662. In this unsettled period, the Japanese porcelain manufactories, aided by the Dutch VOC imported the Japanese porcelain variant of kraak ware into the Netherlands where it proved to be extremely popular with clients. In Fig. 2.16 an example of the Wittelsbach armorial service decorated in blue and white and displaying an escutcheon centrally in a circular format which contains in quartered arms (1st and 3rd) the blue and white lozenges of Ludwig of Bavaria and what should be the lion rampant of Bavaria in the 2nd and 4th quarters, although these seem to be rather poor representations of lions! There is some semblance too of a coronet above the escutcheon and what may be lambrequins on either side of the escutcheon in place of supporters. Otherwise, the plate shows a typical kraak decoration with geometric compartments depicting scenes and objects. The plate dates from the Wanli period and consultation of the records indicates that the

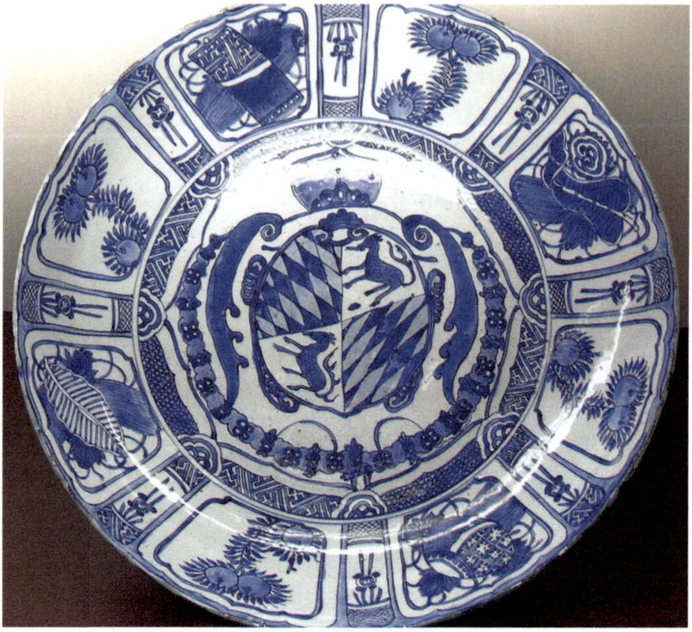

Fig. 2.16 Dinner plate from the *Wittelsbach* armorial service, Bavaria Rhineland Palatinate, in *kraak* porcelain. Wanli period (1573–1620), Chinese porcelain. *Munchener Residenz Muzeum, Munich. CC by SA 3.0*

most likely Wittelsbach commission came from either Prince Christian I or II (1598–1654, and 1637–1711, respectively), Count Palatinate of Zweibrucken, or from either Prince Frederick V or Prince Rupert of the Rhineland Palatinate (1596–1632, and 1619–1682, respectively).

2.4 Non-Specific Armorials

An interesting category of ceramic armorials are those that may be termed non-specific armorials: by this it is implied that although the ceramic artefact does bear an armorial device it alone is not specific enough to identify unambiguously the armigerous individual.

An armorial ceramic that bears a badge or insignia of an honour that has been conferred upon the armigerous holder. Often, the commissioning of the armorial artefact or service is a celebration or commemoration of that award: such is the exemplar shown in Fig. 2.17, which is a plate from the famous royal *Duke of Clarence* service which the Duke of Clarence commissioned from the Royal Worcester Porcelain Manufactory in 1789 to commemorate his elevation to a Royal Dukedom and the award of the Order of the Thistle. Prince William, the Duke of Clarence and St

Andrew's (1765–1837), was the third son of King George III and he later became King William IV in succession to his elder brother King George IV in 1830, the second son Prince Frederick, the Duke of York, had died in 1827: with reference to our earlier discussion of armorial cadency the escutcheon bears the label with the three points on the arms. The extreme points bear an anchor in azure and the central point a red cross of St George, and these can be discerned on the escutcheon in Fig. 2.17. This magnificently decorated plate is also a genuine armorial in that it displays the Duke's full royal coat-of-arms centrally, but also has the badge of the Order of the Thistle, shown below the escutcheon and also in the reserve where the badge of the Order of the Garter also appears. Clearly, this plate can be uniquely identified as part of the *Duke of Clarence* service because of the arms displayed, but this is not always the case, even with royal armorials. A second example is that shown in Fig. 2.18, where the garter of the Order of the Garter is displayed centrally as a badge with the legend "*Honi Soit Qui Mal Y Pense*", but now without any additional arms this becomes a non-specific armorial: undoubtedly an armorial representation, it nevertheless fails to subscribe to the basic tenet that the arms should facilitate the identification of their unique ownership. In this particular case, there is documentary evidence that the service was commissioned by Queen Victoria from the Coalport Porcelain Manufactory in 1860 through their agent William Mortlock of 18, Regent Street, London and their backstamp shown in Fig. 2.19 confirms this assignment. A further and related example, also ordered by Queen Victoria from the Coalport Porcelain Manufactory for her use in Windsor Castle, is shown in Fig. 2.20 which now contains the full Order of the Garter badge and the William Mortlock backstamp now shown in Fig. 2.21 indicates its designated usage there appropriately.

The presence of the *label* as a sign of cadency on a coat-of-arms is an example of a heraldic *brisure* or differentiation between armigerous individuals in a family: perhaps surprisingly, it is not a stipulation in English heraldry to display cadency and it is more regarded as voluntary. In normal usage, a first son displays a three pointed label on the escutcheon field, and the eldest son of the eldest son a five-pointed label: the royal arms discussed hitherto also charge the points with heraldic symbols such as the anchor and cross used for the Duke of Clarence, his elder brother, the Duke of York, had a three-pointed label in which the extreme points were left blank and the central one bore the red cross of St George. A second son normally bore a crescent on the escutcheon field, a third son a mullet, a fourth son a martlet, a fifth son an annulet, a sixth son a fleur-de-lys, a seventh son a rose, an eighth son a cross moline and a ninth son a double quatrefoil. Daughters of armigerous fathers were not differentiated by *brisures* or cadency but could bear their father's arms in a lozenge. Upon marriage to an armigerous husband the father's arms are impaled sinister in a new escutcheon, which bore the husband's arms dexter. As commented earlier, if, in addition the woman is an heraldic heiress then her husband can adopt the father's arms upon his own as an *inescutcheon of pretence*.

Another type of armorial bearing that very often defies unambiguous identification without the presence of supporting achievements such as an escutcheon, a crest or a motto, is the coronet. Although it is possible to narrow down the field to a particular rank in the peerage or nobility it sometimes is not practical to make any further

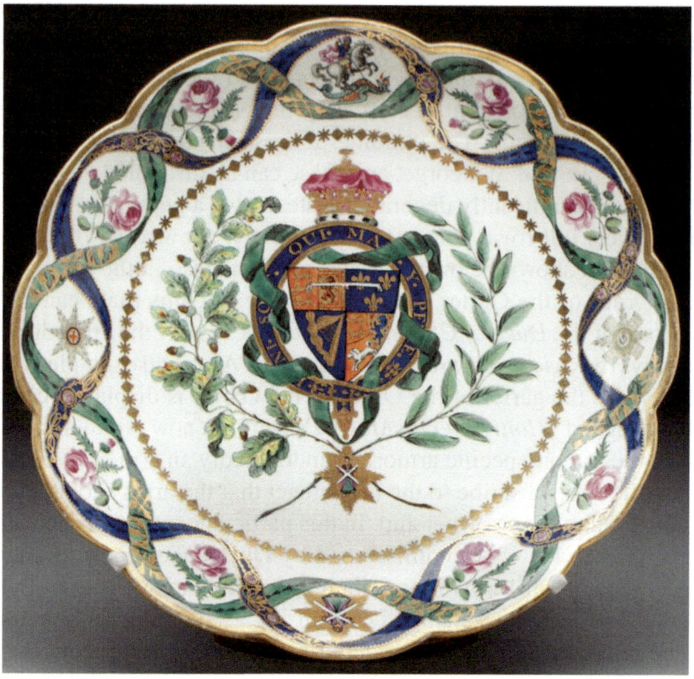

Fig. 2.17 Armorial plate from the *Duke of Clarence* dinner-dessert service in Royal Worcester porcelain, Flight period, ordered by His Royal Highness Prince William in 1789 (who later became King William IV in succession to King George IV in 1830) to celebrate his accession to the royal Dukedom. Sumptuously decorated with his coat-of-arms centrally located and enclosed in the Order of the Garter with its motto, *Honi Soit Qui Mal y Pense*, and displaying the heraldic badges of the Order of the Thistle and Order of the Garter in the verge, which also shows St George impaling a dragon with a lance. Reproduced from The Art Institute, Chicago (CCO; public domain): Joseph Maier and Arthur Lewis Liebman Memorial. Gift of Kenneth J. Maier MD, ref 1994.90

progress in identification of the armigerous person involved. However, the simple coronet shown on the Derby porcelain armorial dish shown in Fig. 2.22 has the additional advantage of bearing cursive entwined initials of the armigerous individual who commissioned the service and has facilitated its absolute identification, without which it would have been extremely difficult or impossible for an armorial researcher to assign the piece. In this case, the dish portrays the coronet of a Viscount identified from the number of pearls it bears. This armorial dessert comport of unusual and unique shape was commissioned from the Derby China Works by Lord and Lady Cremorne: Thomas Dawson (1725–1813), Baron Dartrey of Dawson's Grove, County Monaghan, Ireland, was raised to the peerage as Viscount Cremorne in June 1785. In 1770 he married Philadelphia Hannah Freame, granddaughter of William Penn, founder of Pennsylvania, who was born in Philadelphia. This service, which can be dated to 1789 from archival correspondence between Lady Cremorne and William Duesbury of the Derby Porcelain manufactory, bears a Viscount's coronet

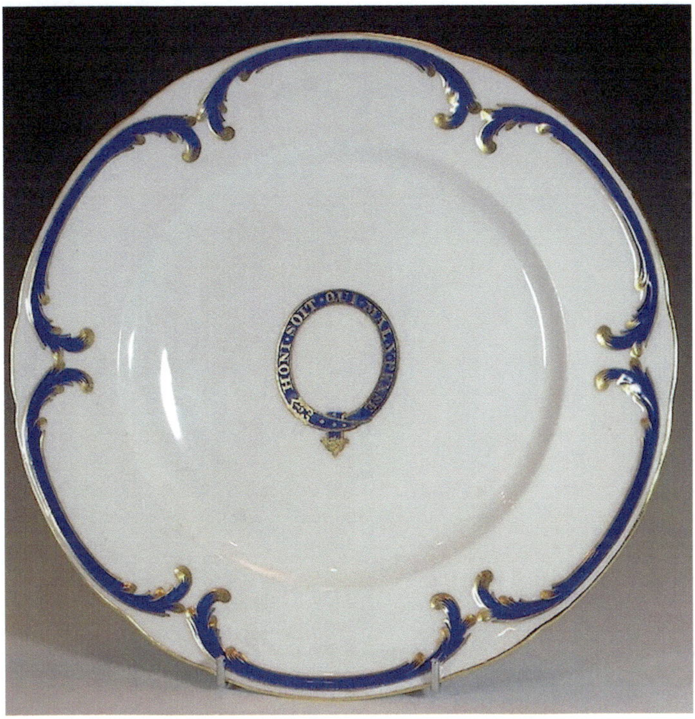

Fig. 2.18 Armorial service in Coalport porcelain, *ca.* 1860, depicting an Oder of the Garter band in blue with the motto "*Honi Soit Qui Mal y Pense*" located centrally, which was used in Buckingham Palace. *Reproduced with Courtesy of Kingschina Antiques Ltd., Mayfair, London*

below which are the entwined initials *PHC* (Philadelphia Hannah Cremorne) inside a circlet of gilt foliage. The decoration is a simple pattern of chantilly sprigs of blue cornflowers and gilded foliage. This service depicts perhaps the simplest kind of armorial on porcelain, namely a coronet and initials, but it nevertheless serves to attribute the commissioning of this service unequivocally. Without the initials and the associated documentary correspondence the task in researching and identifying this armorial service would have been otherwise immense and would probably have been irresolvable.

Another category which often causes confusion in armorial ceramics is that of monograms and ciphers: of course, these are not armorial in themselves but are often associated with an armorial device as we have seen for the *Cremorne* service above, where the entwined initials *PHC* alone would not have been sufficient to enable it to be called an armorial service. Many early earthenwares and porcelains contained a monogram or cipher, and a good example is the *kraak* porcelain service commissioned by the Dutch East India Company in the seventeenth century, an example of which is shown in Fig. 2.23, where the Japanese Arita porcelain contains the logo VOC as a cipher in underglaze blue within a six-panelled geometric border.

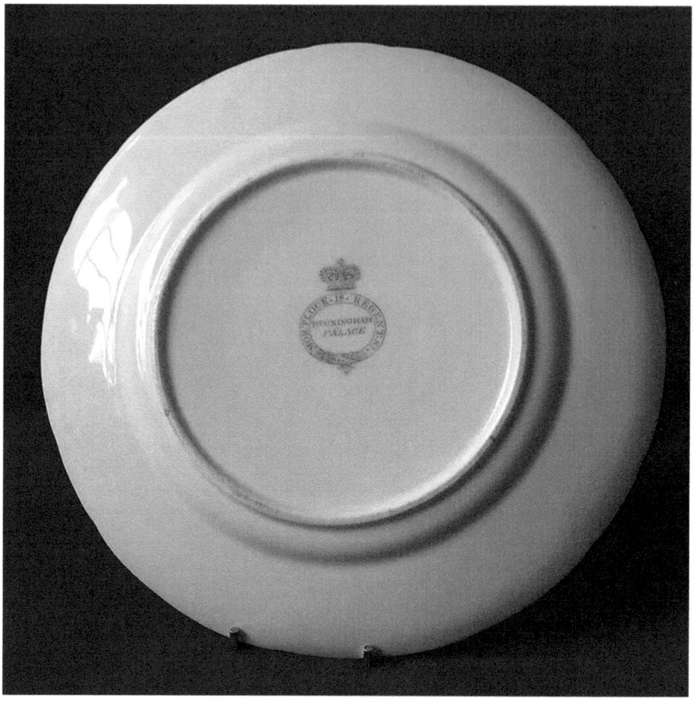

Fig. 2.19 Reverse of the armorial plate shown in Fig. 2.11, showing the stamp of William Mortlock, 18 Regent Street, London, and the place of use, Buckingham Palace. *Reproduced with Courtesy of Kingschina Antiques Ltd., Mayfair, London*

In contrast, the dessert comport dish shown in Fig. 2.24 is a true armorial in that it displays the coat-of-arms of the Honourable East India Company (HEIC) granted by King William III in 1698: this is a full coat-of-arms with escutcheon, crest, helmet, torse, motto and supporters. The arms are given as: *argent a cross gules,* in the dexter chief quarter an escutcheon of the arms of France and England quartering the shield, ornamented and regally crowned *or*. The supporters comprise *two lions rampant guardant or each supporting a banner erect argent charged with a cross gules* and the crest is *a lion rampant guardant or holding between the forepaws a regal crown proper.* The motto is "*Auspicio Regis et Senatus Angliae*", translating as "*By Command of the King and Parliament of England*" and the torse is gules and argent with an erminois and or mantling. This was part of a very large porcelain service commissioned from the Robert Chamberlain Worcester manufactory in 1817 for use in the Fort St George headquarters of the HEIC in Madras.

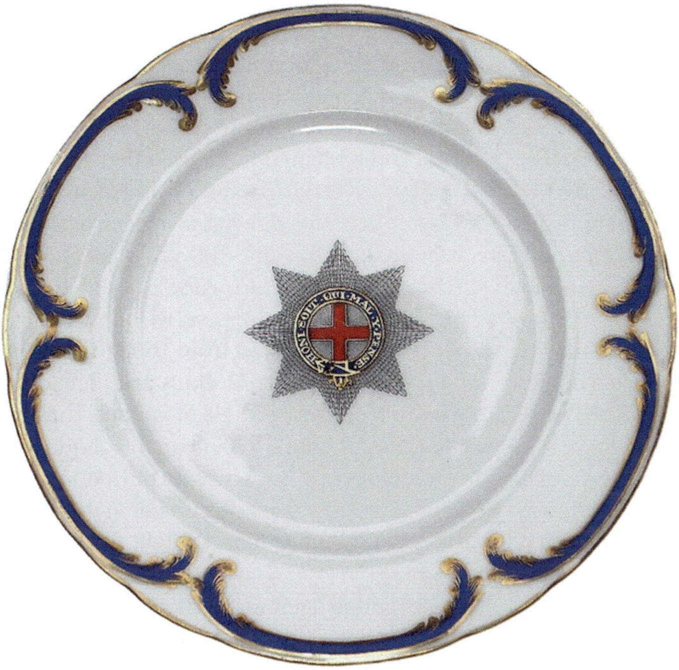

Fig. 2.20 Armorial service in Coalport porcelain, *ca.* 1860, similar to that shown in Fig. 2.19 but here depicting a full Order of the Garter badge bearing the cross of St George with the motto "*Honi Soit Qui Mal y Pense*" in a blue garter located centrally in the silver eight-pointed Garter Star., which was used in Windsor Castle. *Reproduced with Courtesy of Kingschina Antiques Ltd., Mayfair, London*

2.4.1 Distinction Between Monogram, Cipher and Initials

In several instances in the above text the phraseology of monogram, cipher and initials has been used: these are not synonymous terms, although all involve combinations of letters, and it is important at this point to explain the distinction between them as they have been used rather interchangeably and often fairly loosely to describe their presence on armorial china in the past. A *monogram* is a combination of two or more letters in which one letter forms part of another and cannot be separated from it: the monogram, therefore, represents an artistic design in which the letters are interdependent, the removal of one letter component then causes the design to fall apart. A *cipher,* on the other hand, is composed of an interlacing of the letters whereby the removal of one letter does not affect the others and they still exist in their original design form (an example is the *VOC* logo of the Dutch East India Company shown in Fig. 2.23). *Initials*, in contrast, are simply either a single letter or combinations of two or more letters which are separate spatially and these also may have a design component (an example is the *Cremorne* service in Fig. 2.22

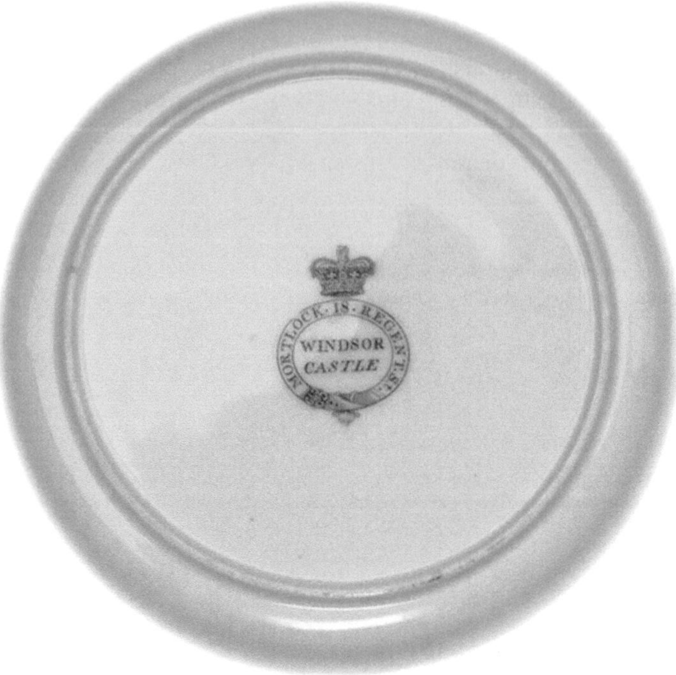

Fig. 2.21 Reverse of the armorial plate shown in Fig. 2.20, showing the stamp of William Mortlock, 18 Regent Street, London, and the place of use, Windsor castle. *Reproduced with Courtesy of Kingschina Antiques Ltd., Mayfair, London*

which displays the initials *PHC*). Historically, there has been much confusion in the use of these terms, which prompted Albert Turbayne to write his classic text (*Monograms and Ciphers: Designed and Drawn by Albert Angus Turbayne and the Other Members of the Carlton Studio*, 1905) in which he illustrates examples of artistically designed monograms and ciphers; some monograms and ciphers are so artistically drawn that it is sometimes difficult to extract from them the correct sequence of initials. It is important to consider the basis upon which a monogram, cipher or initials have been associated with armorial devices in the past as, without any armorial achievement at all, such as a crest, coronet, escutcheon and motto, they really cannot be fully implemented as armorial devices in their own right. In some cases, such as a royal cipher, for example EIIR or CIIIR, the identity of the armigerous person is abundantly clear, but this does not mean that this person actually commissioned such an item personally, although it is most likely that due permission or a royal warranty has been secured for its proper usage There are many pieces of early earthenware and porcelain that bear simple ciphers, monograms and initials without armorial bearings and it is necessary to exclude these from an armorial canon: such an example is given in Fig. 2.25 and this type of ceramic is better described perhaps as a commemorative. It is intriguing that the artefact shown in Fig. 2.25 has

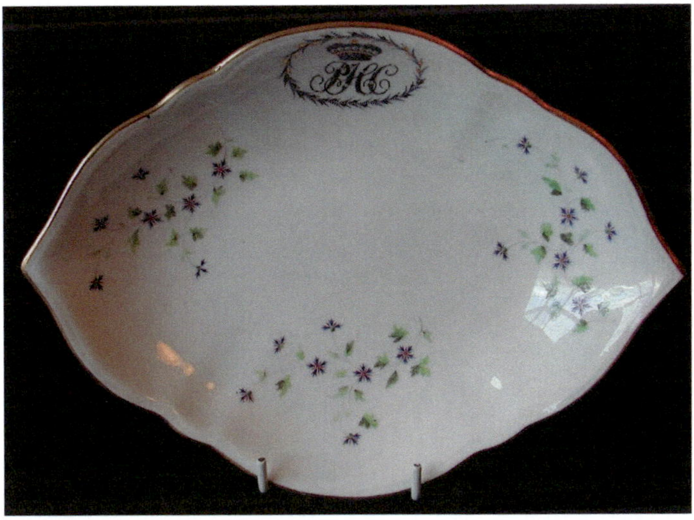

Fig. 2.22 Dessert comport of unusual shape commissioned from the Derby China Works by Lord and Lady Cremorne: Thomas Dawson (1725–1813), Baron Dartrey of Dawson's Grove, County Monaghan, Ireland, was raised to the peerage as Viscount Cremorne in June 1785. In 1770 he married Philadelphia Hannah Freame, granddaughter of William Penn, founder of Pennsylvania, who was born in Philadelphia. This service, which can be dated to 1789 from archival correspondence between Lady Cremorne and William Duesbury of eth Derby Porcelain manufactory, bears a Viscount's coronet below which are the entwined initials *PHC* inside a circlet of gilt foliage. The decoration is a simple pattern of chantilly sprigs of blue cornflowers and gilded foliage. This service depicts perhaps the simplest kind of armorial on porcelain, namely a coronet and initials, but it nevertheless serves to attribute the commissioning of this service unequivocally. *Private Collection*

been identified and classed as an armorial piece despite lacking any armorial heraldic bearings because the cursive initials *EG* shown centrally belong to the Gosling family of Bentfield House and Hassobury, Essex, and forms part of a service commissioned in 1795 from the Derby Porcelain Manufactory (Twitchett, *Derby Porcelain, 1748–1848: An Illustrated Guide*, 2002). Apparently, over 80 pieces of this attractively decorated service, marked with a Derby puce mark, survive. The Goslings were very wealthy bankers, partners in the Gosling & Sharpe Bank founded in 1742, which was acquired by Barclays Bank in 1896: Robert Gosling became High Sheriff of Essex and at that time was the wealthiest man to play international football for England! The crest associated with this family service (Fig. 2.26) is confirmed by Fairbairn (*Crests of the Families of Great Britain and Ireland and Their Kindred in Other Lands*, Volumes I and II, 1905) as the crest for Robert Gosling of Hassobury Manor, Farnham, Bishop's Stortford, which the Gosling family purchased as an armorial from the Derby Porcelain Manufactory in 1773 and given as *a bear's paw erased ppr. holding a fleur-de-lys and crescent in front or.*

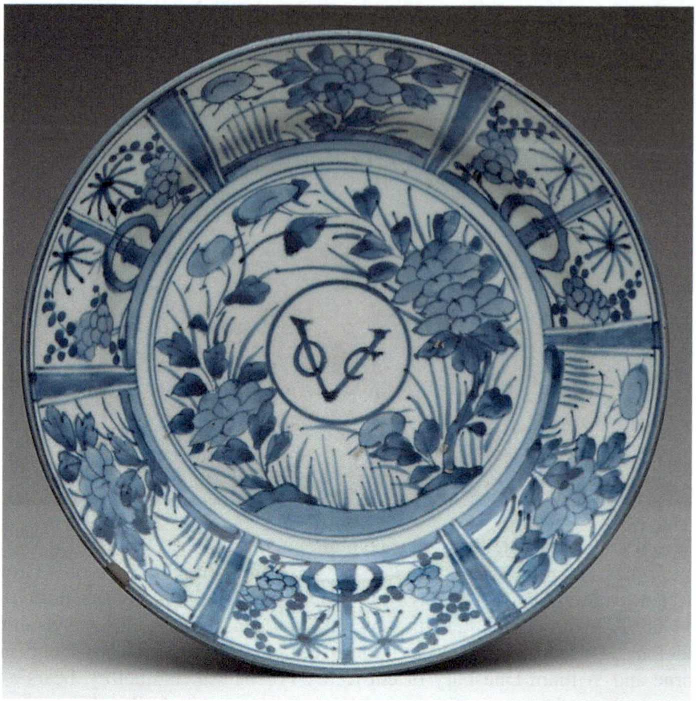

Fig. 2.23 Dutch East India Company *kraak* style large porcelain plate, diameter 12.5″ (31.3 cm), Japanese Arita ware, bearing the logo VOC as a monogram in underglaze blue with a six-panelled geometric border, Edo period, *ca.* 1660. Metropolitan Museum of Art, New York, the bequest of Dr and Mrs Roger G. Gerry, Roslyn, New York, 2000, Accession number 2002.447.40. *Public domain*

2.5 The Use of Bookplates in the Decoration of Armorial Porcelains

Another concept which is raised to the fore by this detailed consideration of the finer points and issues relating to the correct execution of the heraldic coat-of-arms and crest of the *Pitt* armorial service is one that relates to the actual mechanism of the undertaking of the armorial decoration in China: several historian observers have noted hitherto that sometimes some serious flaws exist between what the armorial bearings should indicate and their actual execution in enamels on the porcelain artefact, which has been ascribed quite naturally to the unfamiliarity of the Chinese artists and decorators with the niceties of Western heraldic designs. The enamelling decoration in the earliest days was accomplished by Chinese artists at Jingdezhen, before the Chinese Hongs were established as active brokers between the HEIC agents and the Jingdezhen Imperial manufactory on the waterfront in Canton to undertake the porcelain decoration under the watchful eyes of the East Indiamen supercargoes responsible, who may have been able to alert them immediately to any serious

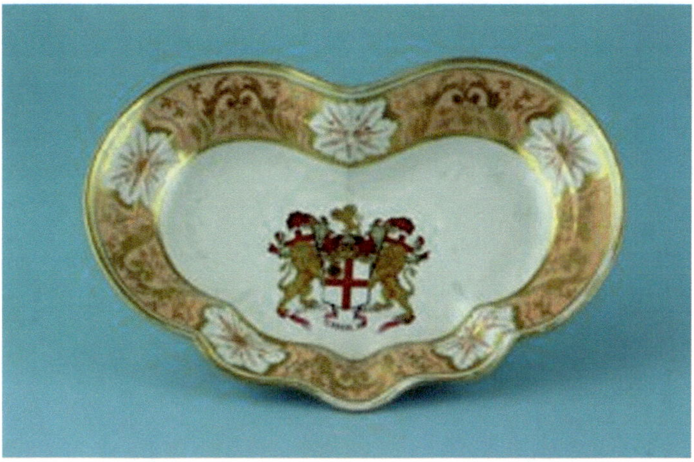

Fig. 2.24 Dessert comport in Chamberlain's Worcester porcelain displaying the coat-of-arms of the Honourable East India Company (HEIC) granted by King William III in 1698: this is a full coat-of-arms with escutcheon, crest, helmet, torse, motto and supporters. Arms: *Argent a cross gules, in* the dexter chief quarter an escutcheon of the arms of France and England quartering the shield, ornamented and regally crowned or. Supporters: *Two lions rampant guardant or each supporting a banner erect argent charged with a cross gules.* Crest: *A lion rampant guardant or holding between the forepaws a regal crown proper.* Motto: "*Auspicio Regis et Senatus Angliae*", translating as "*By Command of the King and Parliament of England*". Torse: *gules and argent.* Mantling: *erminois and or. Private Collection—CC by SA.3.0*

errors being compounded in the specific armorial decoration being undertaken there. Hence, it has been proposed that the adoption of copies of the printed bookplate that was already an established feature of gentlemen's libraries, and which itself saw a significant increase in usage from the late seventeenth century, would have been a major advance in demonstrating to the Chinese artists exactly what was required for the decorative emblazoning of an armorial porcelain artefact, hence contributing materially to a perceived diminution in the anticipated errors in this accomplishment.

An intriguing example of such a glaring error that should have been noted before the delivery of the porcelain artefact but was clearly not done is provided by an early sixteenth century Ming porcelain vase bearing the royal armorial of the King of Portugal: this is depicted in Fig. 2.27 as an argent escutcheon bearing five shields with azure ground each containing five silver bezants and bordered in gules with seven castellated towers in or, all surrounded by a golden armillary sphere as used in navigating. This Ming porcelain vase, which is now in the Metropolitan Museum of Art, New York, surely enough depicts such an armorial but it is enamelled upside down as shown in Fig. 2.28!

From Fig. 2.28 it can be seen that the five blue ground shields within the escutcheon do indeed contain five silver bezants (although these have not been located in the correct escutcheon shape of a heater shield) but the seven castellated towers in the border have now been reduced to six unidentifiable shapes and the encasing armillary

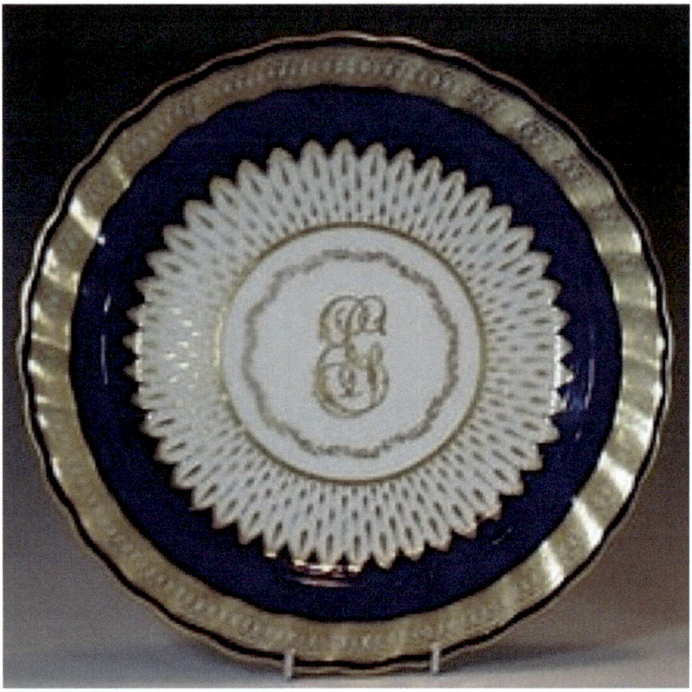

Fig. 2.25 Plate from a dinner service commissioned by the Gosling family of Bentfield House and Hassobury, Farnham, Essex, for the Derby Porcelain Manufactory in 1795 and marked with the Derby puce mark of crown, crossed batons and six dots. Decorated with a cobalt blue cavetto and profuse gilding patterns it centrally bears the script cursive initials *EG*. Described as an armorial piece it nevertheless does not bear any arms which facilitate the unambiguous identification of the owner EG. *Private collection and courtesy of Kingschina Ltd, Mayfair, London and Cambridge*

sphere on the coat-of-arms is missing altogether. Previous research into early European porcelains has found that in 1488, Bartolomeo Diaz rounded the Cape of Good Hope and entered the Indian Ocean, so starting the establishment of the Portuguese Oriental colonial empire, the Seculo de Oro, which was soon followed by Vasco da Gama sailing eastwards and reaching India in 1498 and Pedro Alvarez Cabral sailing westwards and reaching Brazil in 1500. The Portuguese initially discovered Chinese porcelain in India and the first pieces were brought back to Lisbon by Vasco da Gama in his flagship *Sao Gabriel* and presented to King Manuel I (reigned 1495–1521) on his return to Portugal in 1499 after navigating the Cape of Good Hope, crossing the Indian Ocean and making a landfall in Calicut. For his exploration exploits, Vasco da Gama was created "Admiral of the Seas of Arabia, Persia, India and All the Orient". King Manuel I was so pleased with his gift of porcelain that he ordered some more from the first Portuguese expedition that he sent to China in 1517 to be decorated with a Portuguese armillary sphere device as exhibited on his personal standard (Fig. 2.27) and the order of an ewer bearing his heraldic coat-of-arms, which probably represents the first specific European commission to be made for Chinese

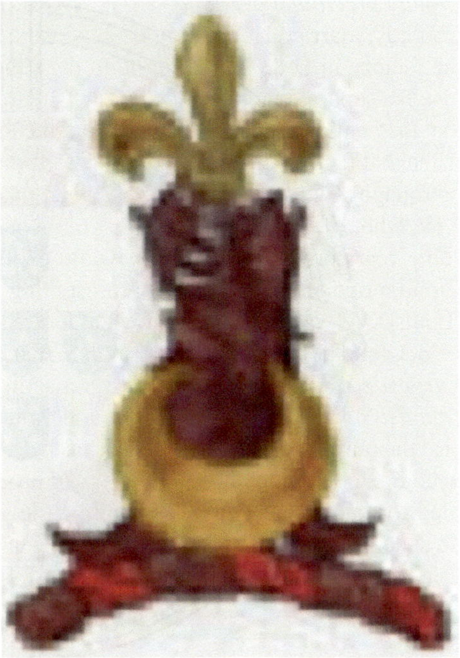

Fig. 2.26 Crest of Robert Gosling, Hassobury Manor, Farnham, Essex, described in *Fairbairn* (*Crests of the Families of Great Britain and Ireland and Their Kindred in Other Lands,* Volumes I and II, 1905) as a bear's paw erased ppr. holding a fleur-de-lys and in front a crescent or

armorial porcelain historically (Rui, 'Portugal Embusca de China: Imagens e Miragens (1498–1514)', *Ler Historia* 1990; Graca, 'The Portuguese Porcelain Trade with China', *Arts in Asia*, 1977). King Manuel I commissioned Fernao Pires de Andrade to establish trade with China in command of a fleet of seven caravels, lateen-rigged and highly manoeuvrable small vessels of approximately 150 tons burthen, which became the first European vessels to officially reach the Chinese coast, landing first in Macao and then navigating the Pearl River estuary to reach Canton on the 15th August, 1517. It is interesting that Andrade and his second-in-command, Tome Pires, were both apothecaries and this reinforces the suggestion that the main purpose of this Portuguese expedition was not primarily to seek porcelain but rather to evaluate the potential of imported Chinese herbal drugs and medicines for European adoption. The expedition delegation had to wait two years in Canton before they could secure an audience with the Emperor Zhengde (Zhu Houzhao) to advance the trading protocols of their mission. In the meantime, diplomatic relations between the local Chinese and the Portuguese expedition members in Canton deteriorated seriously, it is said due to the unacceptable behaviour of the visitors, and with the installation of the new Emperor Jiajing (Zhu Houcong) the visiting delegation was incarcerated punitively for several years. In view of this, it is reasonable to suggest that the armorial ewer bearing the royal coat-of-arms of the King of Portugal is perhaps the very one illustrated here in Fig. 2.28: however, King Manuel I died in 1521, whilst his delegation was still in China awaiting an audience and he was succeeded by King John III (reigned 1521–1557), which covers the date of production of this vase /

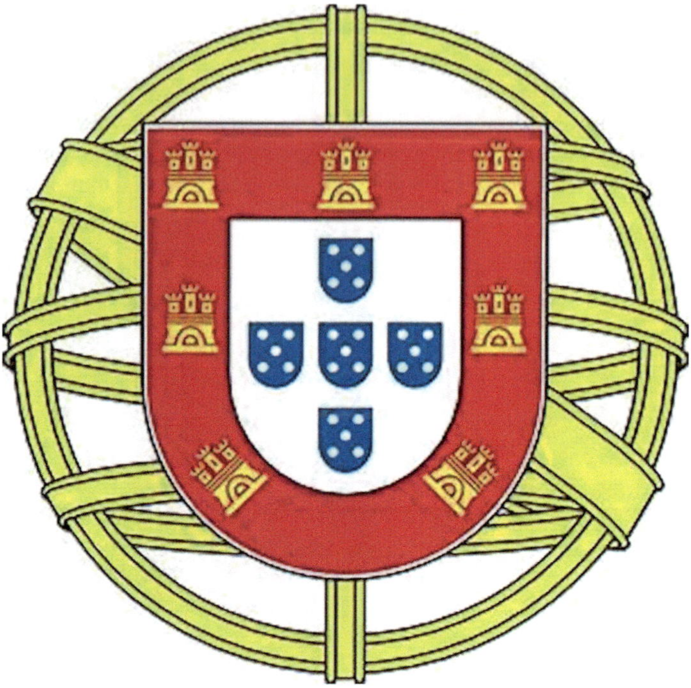

Fig. 2.27 Royal coat-of-arms of the Kingdom of Portugal from the 16th Century comprising five escutcheons contained within the shield and an armillary sphere adopted as the heraldic supporters. *Public domain*

ewer, namely 1520–1540, and therefore he was its probable eventual royal recipient. The design of the ewer is modelled on an Islamic silver jug, but the importance of this piece is in its being probably the very first piece of Chinese armorial porcelain to be commissioned by a European monarch, although it is not the first to be manufactured for Europeans, as this is generally designated as the Gaignieres-Fonthill vase (see Chaps. 3 and 4), which dates from about 1320. What is highlighted in this discussion, of course, is the incorrect display of the armorial bearings themselves, the execution of which would have been most strange for the Chinese enamellers that were assigned to this task.

The ineffective rendition of Western heraldic bearings by the Chinese decorators on their export porcelain in the late seventeenth and early eighteenth centuries from perhaps imperfectly drawn sketches resulted in the necessary provision of a detailed and correct facsimile of the crest, supporters and the escutcheon that comprise the desired coat-of-arms for copying onto the porcelain in China: in this context, the ready access to existing printed armorial bookplates designed for the libraries of armigerous gentry at this time afforded a means of sending exact copies of what was required for the reproduction of armorial bearings that were to be enamelled onto porcelain. As an example, the Royal College of Physicians of London, in conjunction

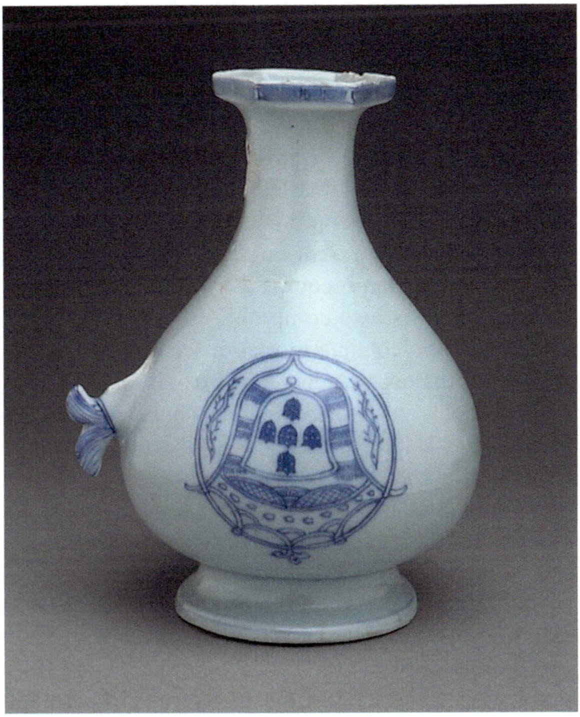

Fig. 2.28 Ming porcelain Chinese ewer/vase in an Islamic silverware style, ca. 1520–1540, decorated in underglaze cobalt blue with the arms of the King of Portugal (identified as King John III in this time frame). Metropolitan Museum of Art New York. *Public domain*

with the Bookplate Society, have demonstrated in an article written by Nigel Cooke (Cooke, *Bookplates and porcelain: new research from the RCP archives*, 2016) the depiction of a polychrome Chinese Chienlung period coffee cup showing the arms of Dr James Parsons MD FRS (1705–1770) being impaled by those of his wife, dated to around 1750, and bearing the crest of a peacock (Fig. 2.29). The excellence of the armorial depiction in this particular instance is claimed to verify the suggestion that bookplates, which it was estimated provided some 15–20% of the basis of all armorial decorations on Chinese porcelain artefacts (Howard, *Chinese Armorial Porcelains*, 1974), were the best vehicle for the generation of a correct armorial bearing however, the reality is somewhat rather different upon closer inspection of this particular piece (Latcham, *Bookplates and Armorial Porcelain*, 2008). Upon comparing the bookplate for James Parsons MD FRS (Fig. 2.30) with the armorial coffee cup derived from it, there are in fact apparently several minor incorrect renditions made heraldically which raise a serious issue in the decorative accomplishment resulting from this process. Firstly, because the bookplate is a monochrome print of an armorial bearing it necessarily requires the adoption of heraldic tricking or hatchings (see further details in Table 2.1 for a discussion of this heraldic aspect) to indicate the correct colour or tincture that is to be used in the coat-of-arms or crest. Here, in the James Parsons armorial bearing the vertical lines drawn on the sinister side of the escutcheon in the bookplate for his wife, Elizabeth Reynolds, are correctly reproduced as a gules, or red, background in the polychrome armorial bearing. However, the chevron that is displayed as a muddy brown pigment on the dexter side of the escutcheon (Parsons) is indicated correctly as vert, or green, properly in the bookplate upon an argent ground as are the three leaves above and below it (the hatching extending diagonally from top left to bottom right), which appear on the armorial coffee cup rendered as a dull grey, as does the leaf in the beak of the peacock crest. The lambrequins enclosing the escutcheon are also represented with a curious fish scale or shingle effect which is not present on the bookplate and the pheons that are present in the bookplate on the lambrequins are missing completely from the porcelain armorial. The torse itself is also compromised similarly, being rendered as a two-tone grey colour on the porcelain armorial instead of the alternating proper silver (argent) and green (vert) tinctures. A depiction error is also seen to apply on the plumage tinctures of the peacock crest where a rather haphazard and arbitrary number of eyes appear on the peacock's tail which bear no resemblance at all to the number indicated that should be present on the bookplate (there are eleven eyes shown on the peacock's tail on the bookplate versus only eight shown on the coffee cup). The source of several of these errors is easily explained when the Chinese artist is offered a design with a code of tinctures which may be unfamiliar to him for the production of a resultant polychrome decoration on the porcelain artefact. However, the relative dimensions of the various components of the heraldic design and their relative placement would nevertheless at least be correctly reproduced, unlike those deficiencies that we have commented on earlier for the *Pitt* service regarding the relative location and size of the crest and escutcheon, which we can now presume not to have involved a printed bookplate or similar design ! Finally, the shape of the escutcheon itself in the Parsons armorial is worthy of some comment as it is a rather earlier baroque version of the standard heater shield format used in the eighteenth century with ten points or crescented indentations. The motto, *Arte Salutari*, is surprisingly not recorded in the many thousand listed in Fairbairn (*Crests of the Families of Great Britain and Ireland and Their Kindred in Other Lands*, Volume I, 1905) but can be translated literally as *A Skill for Health,* which is certainly appropriate for adoption by a practising member of the medical profession!

Nevertheless, the sheer scale of production of armorial printed bookplates for armigerous book owners, which are known as *ex-libris* and were designed to be pasted onto the front cover of a book to indicate their ownership, can be gauged from the Sir Augustus Wollaston Franks Collection which he established in the nineteenth century and which he bequeathed to the British Museum after his death in 1897 (Gambier Howe, *Catalogue of British and American Bookplates Bequested to the Trustees of the British Museum by Sir Augustus Wollaston Franks*, 1903). In several volumes of bookplates catalogued in this comprehensive collection, dating from the earliest example of a British bookplate recorded for Sir Thomas Tresame in 1585, there are some 35,000 catalogued examples of British and American bookplates alone and many others additionally recorded for international armigerous families. Some of the most renowned artists and engravers were commissioned to make bookplates for the aristocracy and landed gentry, such as Albrecht Durer, Edward Burne-Jones, Aubrey Beardsley and Frank Brangwyn (Hopkinson, *Ex Libris: The Art of Bookplates*, 2011).

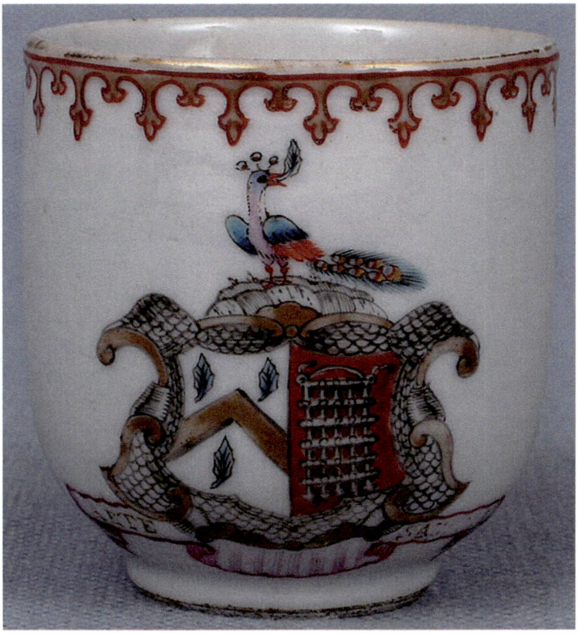

Fig. 2.29 Chienlung period, armorial Chinese coffee cup, ca. 1750, depicting the arms of Dr James Parsons MD FRS impaled by those of his wife, Elizabeth Reynolds, with a gilt enamelled spearhead border. Modelled upon the printed bookplate of Dr James Parsons MD FRS. David Sanctuary Howard Collection of Coffee Cups at the Reeves Centre, Washington, and Lee University, Virginia. Reproduced by courtesy of Angela Howard, Heirloom & Howard Ltd., West Yatton, Chippenham, Wiltshire, UK

Frank Brangwyn designed 130 bookplates for his friends and clients in the early 1900s, which were themselves works of art and were not necessarily armorial in concept but often bearing an image that was a pun on the owner's name (*Bookplates by Frank Brangwyn RA*, Hesketh Hubbard & Phillpotts, 1920). Despite the fact that bookplates had been in existence for over 120 years before they were considered for use as templates or models for the reproduction of the armorial bearings on porcelain, the earliest dateable bookplate that was used in this context was that of Mungo Haldane, the 15th Chief of Gleneagles, whose bookplate exemplar dated to 1707 appears on his armorial service of Chinese porcelain in the mid-1720s, compete with its transcribed errors including the surround itself (Figs. 2.31 and 2.32). A second armorial service was commissioned in 1730 using a monochrome *grisaille* technique (French: shades of grey) which facilitated the better and more accurate reproduction of the hatching lines on the porcelain from the printed bookplates, and this persisted into the 1740s (A. Howard, personal communication).

A quite interesting and relevant exemplar of how even the use of a monochrome printed model provided to the Chinese decorator to ensure the correct display of an enamelled armorial on a piece of porcelain could still result in an incorrect portrayal of the armorial bearings is given by the Chinese dish (Kangxi, Chinese porcelain, ca. 1715) now in the collection of the Metropolitan Museum of Art, New York (Helena Woolworth McCann Collection, Winfield Foundation Gift, 1974.195) which shows the armorial bearings of Edward Harrison (1674–1732) and his wife Frances Bray (1674–1752), whom he married in 1708 and who was the daughter of Reginald Bray of Great Barrington, Gloucestershire. Edward Harrison was the son of Richard

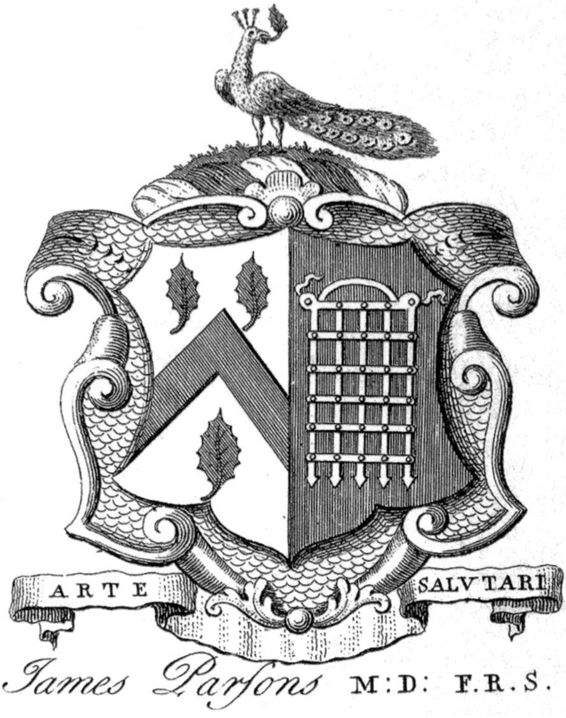

Fig. 2.30 Printed bookplate of Dr James Parsons from the database of Paul Latcham, originally photographed from the print in the Franks Collection at the Department of Prints and Drawings, British Museum, ref F22822, and reproduced with permission from Anthony Pincott, Hon. Treasurer of The Bookplate Society and also Angela Howard, Heirloom & Howard Ltd., West Yatton, Chippenham, Wiltshire, UK. James Parsons (1705–1770) was born at Barnstaple and educated in Dublin. Tutor to Lord Kingston. Studied medicine in Paris; MD Rheims 1736, then practised in London. Physician to the Public Infirmary of St. Giles in 1738; FRS 1741. Married 1739 Elizabeth Reynolds. Licentiate of the College of Physicians 1751. Died 4th April 1770

Harrison and Audrey Villiers, daughter of George Villiers, 4th Viscount Grandison. He captained ships of the East India Company trading with China in the first decade of the eighteenth century, then in 1711 he went out as Governor of Madras in Fort St George, India. He established the postal service for the EIC in India and on his return home in 1718 became a Director of the EIC and MP for Weymouth and Melcombe Regis, living in Balls Park, Hertfordshire (Tallboys Wheeler, *Madras in the Olden Time*, 1861). For Harrison's first service, the Chinese painter copied the printed picture exactly in blue and white, even down to the reproduction of the script information about the heraldic colours that were to be used thereon by heraldic tricking, in this case *sa,* meaning sable (black) and *arg*, meaning argent (silver) (Fig. 2.33)! Technically, it is believed that Harrison, who was Governor of Madras at the time of his armorial commission in 1717, actually used an alternative diagram to a bookplate for this commission: it is believed that he would have used a sketch of his

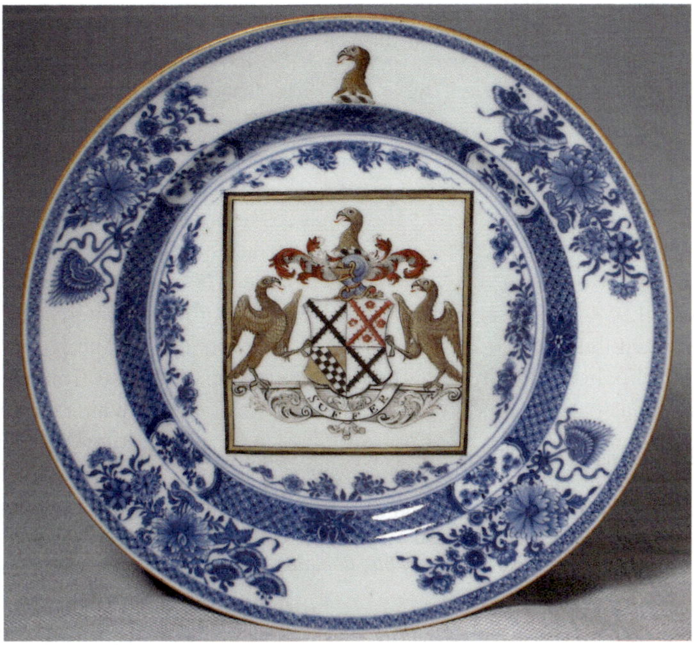

Fig. 2.31 Armorial plate from a service commissioned by Lord Haldane of Gleneagles, ca. 1720, in which the bookplate surround has also been reproduced on the porcelain. Reproduced by courtesy of Angela Howard, Heirloom & Howard Ltd., West Yatton, Chippenham, Wiltshire, UK

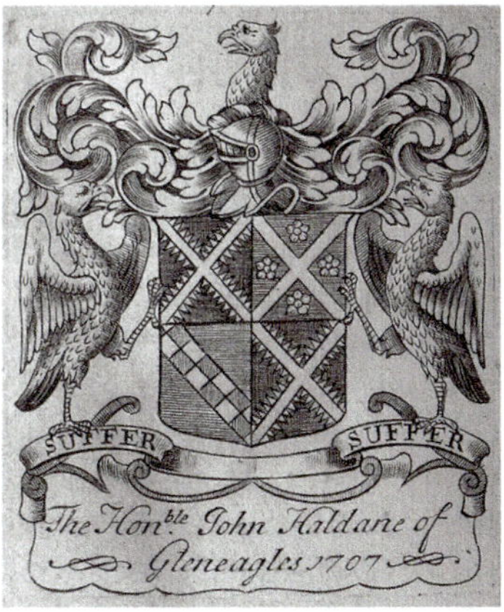

Fig. 2.32 Haldane of Gleneagles bookplate dated 1707, the earliest known example for use in enamelling armorial Chinese porcelain. Reproduced by courtesy of Angela Howard, Heirloom & Howard, West Yatton, Chippenham, Wiltshire, UK

heraldic coach panel (Fig. 2.34) for this purpose and that some 200 or so examples of these that were adapted for armorial porcelain decoration can be identified today. Edward Harrison cannot have been impressed by this inadequate rendering of his armorial bearings as, on his return from India, he later commissioned a further armorial service of Chinese porcelain, specifying that it must be decorated in polychrome

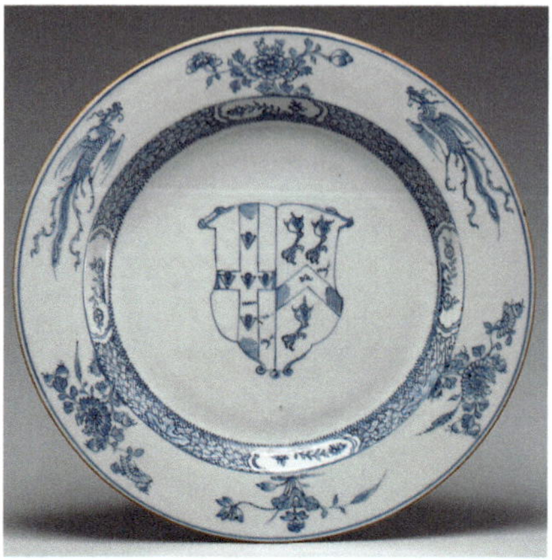

Fig. 2.33 Edward Harrison and Frances Bray blue and white armorial Chinese dish from a service commissioned in 1717 when he was Governor of Madras, India. The script details of the heraldic tricking have also inadvertently been reproduced in the armorial bearings. From the Helena Woolworth McCann Collection, Winfield Foundation Gift, 1974, accession number 1974.195, Metropolitan Museum of Art, New York. *Public domain*

enamels in the *famille verte* palette with the correct heraldic blazoning (Le Corbellier & Freylinghuysen, *Chinese Export Porcelain*, 2003)—an example of his properly emblazoned escutcheon on this polychrome service is shown in Fig. 2.35. Even then, some variants of the bearings on this armorial service have caused a problem in interpretation for connoisseurs and armorial historians because of the apparent absence of any gold or silver decoration, which was attributed initially to errors in transcription or perhaps to an economy in the decorating. However, further research by Shirley Mueller using intensive surface analysis does in fact reveal that the armorial bearings do indicate a vestigial presence of gold and silver in their proper placements on the escutcheon (Mueller, 'Surface Silver Decoration on Chinese Export Porcelain: A Survey', *Oriental Art*, 2001; Mueller, 'Surface Silver Decoration on Chinese Export Porcelain: An Analytic Approach,' *Oriental Art*, 2002). An interesting teapot from the collection of Angela Howard displays the armorial bearings of Harrison impaled by Bray (Fig. 2.36) on Chinese porcelain in which the silver (argent) field tincture has tarnished to silver sulfide and has become blackened upon atmospheric exposure. The use of metallic silver as a background field for armorial bearings indicates a pre-1725 date for this service as afterwards it was discontinued and the argent tincture was represented either as a white enamel or by merely leaving the white porcelain substrate undecorated (Angela Howard, personal communication, 2023).

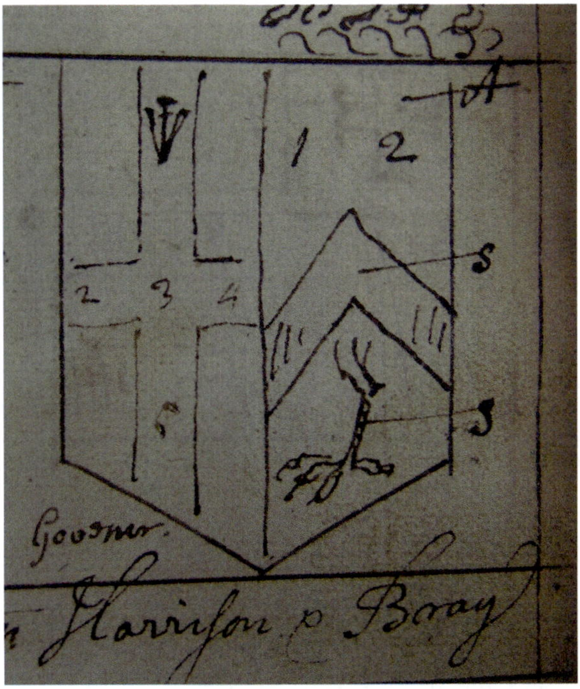

Fig. 2.34 Edward Harrison and Frances Bray drawn carriage print (coach panel) with heraldic details and labelled "Governor" in script. From collection of Sir William Betham. Reproduced by courtesy of Angela Howard, Heirloom & Howard, West Yatton, Chippenham, Wiltshire, UK

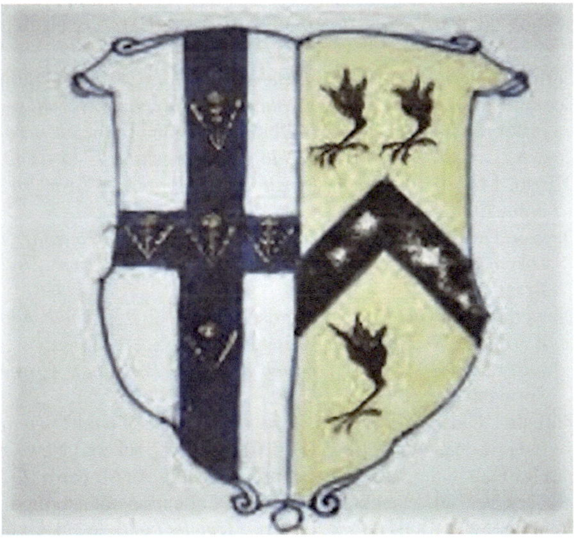

Fig. 2.35 Shows the arms of Edward Harrison impaled by those of Frances Bray on a polychrome *famille vert* style service. Metropolitan Museum of Art, New York. Public domain

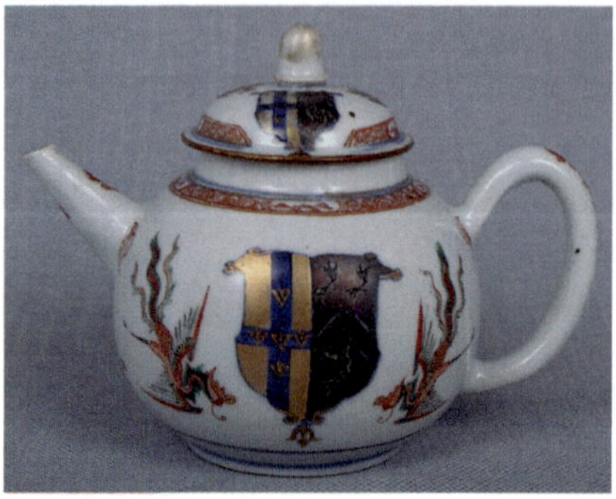

Fig. 2.36 Teapot from armorial service prepared for Edward Harrison and Frances Bray, ca. 1720, in polychrome enamels with or and argent gilding: the silver has tarnished and is blackened with the formation of silver sulfide from atmospheric pollution. Reproduced by courtesy of Angela Howard, Heirloom & Howard, West Yatton, Chippenham, Wiltshire, UK

References

Bernard Burke, Sir J. (1884). *Burke's General Armory—The General Armory of England, Scotland, Ireland and Wales: Comprising a Registry of Armorial Bearings from the Earliest to the Present Time.* Harrison & Sons, Pall Mall.

Birkbeck-Hill, G. (1897). Hester Thrale: Anecdotes of the Late Samuel Johnson LLD During the Last Twenty Years of His Life. In *Johnson Miscellanies* (Vol. 1, pp. 141–351). Clarendon Press.

Boutell, C. (1914). *The Handbook to English Heraldry* (10th ed., Revised by A. C. Fox-Davies). Reeves & Turner.

Clifford, J. L. (1987). *Hester Lynch Piozzi (Mrs Thrale).* Columbia University Press.

Cooke, N. (2016, May 13). *Bookplates and Porcelain: New Research from the RCP Archives.* https://history.rcplondon.ac.uk/blog/bookplates-and-porcelain-new-research-rcp-archives

Cox, A., & Cox, A. (2001). *Rockingham Porcelain, 1745–1842.* The Antique Collectors Club.

D'Ezio, M. (2010). *Hester Lynch Thrale Piozzi: A Taste for Eccentricity.* Cambridge Scholars Publishing.

Dawson Dobson, S. (1795). *Historical Anecdotes of Heraldry and Chivalry: Tending to Shew the Origin of Many English and Foreign Coats of Arms, Circumstances and Customs.* Holl and Brandish.

Denyer, R., Denyer, M., & Edwards, H. G. M. (2023). *The Pendock-Barry Derby Porcelain Service: A Forensic Reappraisal.* Springer-Nature.

Edwards, H. G. M. (2022). *Welsh Armorial Porcelain: Crested Nantgarw and Swansea China.* Springer-Nature.

Fairbairn, J. (1986/1905). *Crests of the Families of Great Britain and Ireland and Their Kindred in Other Lands* (Vols. I and II). T.C. and E.C. Jack and New Orchard Publishing.

Farnley Hall, Otley, House Brochure, The Home of Mr & Mrs G.N. le G. Horton-Fawkes, Engraving Services Ltd, Manchester, Date of Publication not supplied.

Franklin, M. J. (2004). *Piozzi, Hester Lynch (1741–1821).* Oxford University Press.

Graca, J. (1977). The Portuguese Porcelain Trade with China. *Arts in Asia, 7*, 45–47.

Howard, D. S. (1974). *Chinese Armorial Porcelain* (Vol. 1).

Howard, D. S. (2003). *Chinese Armorial Porcelain* (Vol. II).

John, W. D., Coombes, G. J., & Coombes, K. (1975). *The Nantgarw Porcelain Album.* The Ceramic Book Company.

Kelly, A. (1990). *Mrs Coade's Stone.* Self-Publishing Association.

Morton Nance, E. (1942). *The Pottery and Porcelain of Swansea and Nantgarw*. Batsford.
Mueller, S. M. (2001). Surface Silver Decoration on Chinese Export Porcelain: A Survey. *Oriental Art, 47*, 56–64.
Mueller, S. M. (2002). Surface Silver Decoration on Chinese Export Porcelain: An Analytic Approach. *Oriental Art, 48*, 43–46.
Neale, J. P. (1830). *Gentlemen's Seats: Views of the Seats of Noblemen and Gentlemen in England, Wales*. H.W. Bond for Sherwood, Jones & Co., Paternoster Row.
Pountney, W. J. (1920). *Old Bristol Potteries* (pp. 115–116). J.W. Arrowsmith.
Rui, L. (1990). Portugal Embusca de China: Imagens e Miragens (1498–1514). *Ler Historia, 9*, 33.
Thrale Piozzi, H. L. (1786). *Anecdotes of the Late Samuel Johnson LLD During the Last Twenty Years of His Life*. T. Cadell, The Strand.
van Lemmen, H. (2006/1897). *Coade Stone* (pp. 141–351). Shire Publications Ltd.
von Volborth, C. A. (1973). *Heraldry of the World* (pp. 126–133 and 213–216). Batsford Books.

Chapter 3
Genesis of Chinese Porcelain and the Western Trade

Abstract The genesis of Chinese porcelain manufacture is discussed. The establishment of the trade routes and early contact with Western Europe is traced. Development in decorative techniques is considered. The early trade with the Portuguese, and the first armorial commission by Manuel I of Portugal is described. The Dutch East India Company and the establishment of the English East India Company is considered. The role of King William and Queen Mary in setting the fashion for blue and white porcelains is discussed, as is the role of the imported drinks punch, coffee, chocolate and tea in creating demand for ceramics.

Keywords Genesis of Chinese porcelain · Kiln development · Trade routes · Mongol Empire · First blue and white porcelains · The Ming Gap · Ming Empire · Imperial kilns · Cobalt · Portuguese trade · Manuel I of Portugal · First contact with Japan · Nanban trade · Dutch East India Company (VOC) · Philip of Spain · Salt trade · Kraak porcelain · Queen Mary & King William · Garniture · Punch · Coffee · Chocolate · Tea

3.1 The Genesis of Chinese Porcelain

To understand the genesis of armorial ceramics it is useful to understand the importance, value and history of ceramics, particularly porcelain and its genesis. Porcelains have considerable advantages over earthenwares and stonewares, specifically being impervious to water and being resistant to fracturing under heat stress. The first water impervious proto-porcelains consisting of a hard kaolinic stoneware fired at temperatures of between 1100 and 1200 °C (Demandt, 'Tracking Proto-Porcelain Production and Consumption in the Dongjiang Valley of Bronze Age Lingnan', *Cambridge Archaeological Journal*, 2019, 2017). These proto-porcelains were initially believed to have been produced in China during the Shang period (Kerr, Needham, Wood, *Science and Civilisation in China: Volume 5, Chemistry and Chemical Technology, Part 12, Ceramic Technology*, 2004). However, recent evidence associated with the discovery of proto-porcelain sherds in Northern China from the Erlitou Site in Yanshi,

R. L. Denyer et al., *Armorial Porcelain*, https://doi.org/10.1007/978-3-031-63745-2_3

Henan province indicates that the first proto-porcelains probably date from the almost mythical Xia dynasty of around 1800–1600 BCE, (Liu & Chen, *The Archaeology of China: From the Late Palaeolithic to the Early Bronze Age*, 2012). No early kilns have been identified in the Erlitou site, but sherds of similarly decorated pottery and proto-porcelain have been found together at the earliest known proto-porcelain kiln (the Piaoshan Kiln) located at Huzhou in the more Southerly, Eastern province of Zhejiang. The proto-porcelains were less absorptive than other early ceramics but were more absorptive than, and less fine in terms of thickness of body, translucency and glaze than true porcelains of later periods.

The proto-porcelains seemed to undergo slow and steady improvements as evidenced by comparative proton-induced X-ray analysis (PIXE) of shards of proto-porcelain from the early Xia period, the Xia–Shang period, the Early Shang period, the Late Western Zhou/early Eastern Zhou Spring and Autumn period, middle Spring and Autumn period, late Spring and Autumn period and Early Warring States period (Li, Zhang, Cheng, Zheng, 'The Earliest Chinese Proto-Porcelain Excavated from Kiln Sites: An Elemental Analysis', *PLoS One*, 2015) In the late Xia period, the body of the proto-porcelain sherds consisted of mostly aluminium (III) oxide and silicon dioxide (around 92 wt%). This remained relatively constant for around 1500 years from the early Xia to the Early Warring States periods. However, the relative proportions aluminium (III) oxide and silicon dioxide did slowly change as did the content of iron (III) oxide. In the late Xia period, the aluminium (III) oxide and silicon dioxide ratio of the body was around 1:4 and the iron (III) oxide wt4% was around 4.29 wt%. The ratios of aluminium (III) and silicon dioxide seemed to fluctuate around the 1:4 ration throughout the Late Xia through to the late Western Zhou/early Eastern Spring and autumn and then steadily increased during the middle and late Spring and Autumn period to the early warring States period to around a ratio of a ratio of 1:5. Between the Xia period and the early Warring States period iron (III) oxide content steadily decreased to around 2–2.6 wt%, which would have contributed to the development of a whiter body.

The glazes of the shards also underwent a similar slow but steady development. In the late Xia/Xia–Shang periods the glaze, like the body mostly consisted of aluminium (III) oxide and silicon dioxide (around 81 wt%), at a ratio of 1: 4.3 to 1:4.5 with a potassium oxide content of around 3.4 Wt%, a low calcium oxide component of around 6–7.5 wt%, a low magnesium oxide component of around 1.4 wt%, a low phosphorus pentoxide of 0.37 wt% and a high iron (III) oxide content of around 5.2–4.5 wt%. By the Early Shang period the glaze had become more calcified and phosphatic, with the calcium oxide content increasing to 10.7 wt% and the phosphorus pentoxide content increasing to 0.69 wt%, nearly double that in the late Xia period. To accommodate the increased calcium and oxide and phosphorus pentoxide content the aluminium (III) oxide and silicon dioxide content had dropped to around 77% at a slightly lower ratio of 1.4. Like the body of proto-porcelains, the iron (III) oxide content of the glaze also decreased by the Early Shang period to around 3.7 wt%. By the early Warring States period further changes in the glaze had occurred. The aluminium (III) oxide/silicon oxide content had decreased further to around 71–71 wt% at a ratio of aluminium (III) oxide to silicon oxide of around 1:5. The calcareous and phosphatic content continued to increase through the late Western Zhou / early Eastern Spring and Autumn, the middle Spring and Autumn and late Spring and Autumn periods until by the early Warring States period the calcium oxide content had increased to between 17.09 and 17.7 wt% and the phosphorus pentoxide content had increased to 1.08–1.36 wt%. By the early Warring States period the iron (III) oxide content had also dropped further to around 2.58 to 3.51 wt%.

One of the most significant advances in the development of porcelain was the improvement of kiln design during the eastern Han period c.25–220 CE. Dragon kilns were probably introduced into china during the late Xia to early Shang period. Dragon Kilns were built on hillsides and consisted of a firebox that extended to an inclined kiln floor that ran uphill towards a flu and vent. They would have been pivotal in enabling the transition from earthenware ceramics to the proto-porcelains fired at 1100–1280 °C. This gave rise to the green glazed Yue wares of the late Eastern Han produced in Zhejiang Province (Wu & Gao, *Research on the kiln site protection and utilization of early Yue in Zhejiang based on protection first*, 2021; Wu, Li, Zhang, Cheng, Chun Chen, Zheng, 'PIXE Analysis of Early Yue Wares from Shangyu and Nanjing (2nd–6th centuries AD)', Nuclear Instruments and Methods in Physics Research Section B: Beam Interactions With Materials and Atoms, 2017). The body of this early celadon ware mostly consisted of aluminium (III) oxide and silicon oxide (92 wt%) at a ratio of 1:4.4, a calcium oxide content of around 2.0 wt% around 0.25 wt% a phosphorus pentoxide content 0.78wt%, an iron (III) oxide content of 2.05wt% and a titanium dioxide content of 0.81 wt%. These bodies were glazed in a glaze consisting of around 72 wt% aluminium (III) oxide and silicon oxide in a ratio of 1:4.7, an iron (III) oxide content of around 4% a high calcium oxide of around 16 wt% and a high potassium oxide content of 1.27 wt% indicating the addition of potash to the glaze. The high iron content of these high temperature calcium glazes was particularly important in that in the reducing conditions prevalent in the cooling dragon kilns it was the reduction of iron (III) oxide that gave rise to the typical greenish, greyish yellowish coloration of the celadon glaze. However, during the late Eastern Han period kilns became steadily longer allowing kiln temperatures to increase to over 1300 °C and this allowed the proto-porcelain to evolve into the first true Chinese porcelains. These greyish white porcelains were stronger than previous ceramics with very low porosity and associated low water absorption reflecting a more compact glassy micro-structure with fine rounded quartz with some porcelainite (Mullite) crystals (Li, Leung, Yu, Wu, 'Ceramics in China'. In: Selin, H. [Eds.], *Encyclopaedia of the History of Science, Technology, and Medicine in Non-Western Cultures*, 2008) which would indicate high a temperature during firing.

The porcelains of the late Eastern Han period are very different from the western view of what constitutes porcelain. The Western perception of porcelain is that of a high temperature white porcelain, white ceramics didn't start to appear until the late Sui to Tang Dynasties of Northern China. Up until that point manufacture of porcelains was believed to have been mostly restricted to the green to grey glazed celadons of Southern regions in China, with ceramic production being limited to lead glazed low temperature fired earthenwares in the North (Li, Leung, Yu, Wu, 'Ceramics in China'. In: Selin, H. [Eds.], *Encyclopaedia of the History of Science, Technology, and*

Medicine in Non-Western Cultures, 2008; Huang, Freestone, Shi, Qian, 'The Path to Porcelain: Innovation in Experimental White Stoneware from Luoyang', *Journal of Archaeological Science*, 2023). It seems that the methods developed at the celadon producing kilns in Zhejiang during the late Eastern Han were adopted by potters in the North of China in Henan province but using locally available raw materials. These included kaolin rich secondary sedimentary clays and the potassium, calcium rich alumina-silicate feldspar. This gave rise to the Northern kaolin/quartz/feldspar celadon bodies consisted of up to 27 wt% aluminium (III) oxide to 68 wt% silicon dioxide content (Aluminium (III) oxide to silicon dioxide ration of 1:2.5) and 1.2 wt% titanium oxide. As evidenced by recent work looking at sherds derived from the Yiyongjie workshop waste pit near the Sui to Tang period Jiangzuojian imperial court (Huang, Freestone, Shi, Qian, The Path to Porcelain: Innovation in Experimental White Stoneware from Luoyang', *Journal of Archaeological Science*, 2023), it appears that by the late Northern Sui and into the Tang dynasties the potters of Northern Henan region had an experimental approach towards ceramic production. This allowed them to develop the first white porcelain with glazes that in some cases had a high potassium oxide content almost exceeding that of calcium oxide giving rise to the Northern white alkali calcium glazes. These were famously used in the first white Chinese porcelains produced at the Xing, Gong and Ding kilns of Henan province c.608–907CE.

Developments similar to those in the Northern potteries also occurred in the southern kilns of Jingdezhen during the late Song and Early Yuan (1127–1279) periods as kaolin clays started to be introduced into the manufacturing process. However, the body of these southern white porcelains differed from those of the Northern porcelains in that they consisted of kaolin, quartz and the lamina potassium and aluminium rich phyllosilicate muscovite rather than the kaolin, quartz and feldspar of the Northern porcelains. The manufacture of these porcelains was, according to Giovani Battista Ramusio 1550–1559, described by Marco Polo in his travels (1271–1295) as involving the mining of earth into large heaps which were then refined by weathering for 30–40 years before being used to make porcelain (Yule & Cordier, *Travels of Marco Polo*, 1903). These Southern white porcelains, the so called Qingbai wares, were glazed using glazes with a bluish green colouration probably arising as in celadon because of reduction firing at high temperatures (Hao, Wang, Qiu, Liu, Feng, 'Study on Additives Assisted Reduction of Celadons', 2022).

3.2 The Early Chinese Trade in Porcelain

It was the celadon and Qingbai wares of the late Song dynasty and early to mid-Yuan Dynasty that were widely exported from China via the port of Zayton (modern day Quanzhou) to Asia, North Africa and then from there in the late thirteenth century to Europe via European city state merchants trading in the Levant. It is evident, assuming one can rely on the veracity of contemporary sources such as the writings of Marco Polo, that porcelain could be bought very cheaply in the port of Zayton (Yule & Cordier, *Travels of Marco Polo*, 1903). There has been much debate around the accuracy of Marco Polo's travel journals, however, this comment regarding the cheapness of Chinese ceramics in China is very human and equally very akin to a modern traveller touring Asia commenting on the low price of modern-day electronic goods. The comment is indicative that Marco Polo was familiar with Chinese porcelain but as expensive luxury products. This would also imply that he may have encountered Chinese ceramics as a Venetian merchant prior to his travels, which in turn could indicate that there may have been some limited Venetian trade in Chinese porcelain prior to his voyages.

There is evidence supporting this early European trade in Chinese ceramic. Sherds of porcelain bowls dating to around the late tenth century have been found in the Alcazaba (Islamic fortress) of Almería in Spain, the palace of which was built by the Caliph of Cordova Abd al-Rahim III in 955CE. Up until it sacking 1147 Almería was the main Spanish Islamic port in which luxury goods such as oils, silks and other fine cloths were exchanged for silver, dyes and brocades and presumably diplomatic gifts such as pieces of Chinese porcelain imported from North Africa, Egypt and Syria (Gutiérrez et al., 'The earliest Chinese ceramics in Europe? *Antiquity*, 2021).

Chinese porcelain shards were also found at Lucera castle. Built by the Holy Roman Emperor Frederick II in 1233 following the 6th Crusade, Lucera represented the last Islamic colony in Italy (Abulafia, 'The Last Muslims in Italy', *Dante Studies, with the Annual Report of the Dante Society*, 2007). The colony was established by Frederick II from 1223 in response to Muslim rebellions in Sicily and Malta. The formation of the colony involved the relocation of between 15,000 and 30,000 people from Sicily from 1223 until around 1246, and the later transport of more families from Malta 1249–1270. Lucera therefore represented a society lead by Christian rulers whose Muslim inhabitants were employed as artisans, shopkeepers, farmers and soldiers. The latter served in the armies of Frederick II until 1250, Manfred King of Sicily (1258–1266) and with the Angevin King Charles I when he conquered the Kingdom of Sicily in (1266–1285). Thus, it is probable that the Chinese sherds found in Lucera found their way there via this population movement.

3.3 The Mongol Empire and the Silk Routes

This influx of Chinese goods into Europe via Islamic states was evidently mediated first by the silk road and later superseded by maritime routes. The silk routes or later Silk Roads initially described a network of routes connecting regions within China with Mongolia, Afghanistan, India, Pakistan, Turkmenistan, Iran (Persia), Iraq, Syria, Turkey and Istanbul (Constantinople), that had its origins in the Han Dynasty in around 200 BCE (Peters, 'The Ancient Silk Road and The Birth of Merchant Capitalism,' *Educational Philosophy And Theory*, 2021). Its early development involved the establishment of a series of interlinked trading posts via negotiation and the development of alliances between neighbouring tribes. This allowed a steady trickle of goods across Asia into India, Persia and North Africa. However, its later development into a network of true roads occurred during the period of the Mongol empire (1206–1368).

Ghengis Kahn (1162–1227) was the founder and first Khan of the Mongol Empire. Following his defeat of the Northern Jin Dynasty in 1210, he employed a policy of extending his conquests westward by following the silk routes from city to city and then transformed those routes into true roads enabling the rapid transport of military resources, goods and people. Ghengis Khan also established a system whereby each member of his family was provided with a Khubi, which to all intents and purposes represented a corporate share of the Empire. This share entitled each member of the family to a portion of the wealth generated in the form of tribute from its distant components and taxes arising from the huge increase in trade between China, its dominions and surrounding neighbours. This was all mediated by a significant increase in traffic along the upgraded Silk Road. That tribute and trade would have included furs, silver, gemstone, soldiers, steel, furs, spices and horse, all of which would have been exchanged for Chinese goods including silks and Porcelain (Weatherford, 'The Silk Route From Land and Sea', *Humanities*, 2018). By 1279 when Marco Polo was traveling in China, the Mongol emperor Khubilai Khan's power extended across the entirety of China, Korea, Mongolia Kyrgystan, Kazakhstan, Tibet, Afghanistan, Iran (Persia) and Turkmenistan along with parts of Russia, Ukraine, Romania Pakistan, Iraq, Syria and Turkey. This allowed the Silk Roads to extent to ports in India, Persia Syria and Turkey. The Mongol emperors not only upgraded, extended and used the silk routes for trade, but also used them to spy on neighbouring regions.

Historically Venice had a long history of trading with the Byzantine Empire, particularly Constantinople and the Levant (Israel, Jordan, Palestine, Lebanon, and Turkey). However, Byzantine power had failed by the time that the Mongol Empire had started to develop. Venice by this stage had developed into a marine power in the Eastern Mediterranean with a substantial navy. Venice played an active part in the 4th Crusade, and it was Venetians who were first to enter Constantinople by force at its fall in 1204 (Puga & Treffler, 'International Trade and Institutional Change: Medieval Venice's Response to Globalization', *The Quarterly Journal of Economics*, 2014). As a result of this Venice acquired a large number of colonies in the Aegean and Eastern Mediterranean, which within 50 or so years had become home to over 70,000 Venetians and generated a large highly effective trading network within what became the Venetian Empire (Puga & Trefler, 'International Trade and Institutional Change: Medieval Venice's Response to Globalization', *The Quarterly Journal of Economics*, 2014). This meant that Venetian merchants were very well placed to take advantage of the developing Mongol empire and the huge increase in trade along the upgraded Silk Road. By 1221, Venice had established a trade agreement with Genghis Khan, which allowed them to import Chinese goods transported via the Silk Roads directly to Venetian ports where Chinese goods could be exchanged for goods derived from Europe, North Africa and the Levant (Weatherford, 'The Silk Route From Land and Sea', *Humanities*, 2018). Direct trade with China didn't occur, for example, Marco Polo in his Journey to China, didn't sail Directly to China from

Venice in 1271, he instead travelled to Persia to seek a ship from Hormuz to China. There was no sea route, so Marco Polo's party of travellers had to take the Silk Road (Yule & Cordier, *Travels of Marco Polo*, 1903).

As a result of the silk road goods from China including porcelain flowed from the ports in the Levant to Venice. This would have allowed Chinese silks, spices and some porcelain to reach other parts Europe. There is physical evidence to support this, the so-called Marco Polo Jar held in the Treasury of San Marco in Venice. The jar is approximately 12 cm high and 8.1 cm in diameter and is described by Meicun and Zhang as having a hard, a white, thin body which is glazed with the typical Qingbai blueish green glaze (Meicun & Zhang, 'A Chinese Porcelain Jar Associated with Marco Polo: A Discussion from an Archaeological Perspective', *European Journal of Archaeology*, 2018). The jar consists of a low foot and a short neck with four loop shaped lugs. The vase is decorated in four zones: two bands of moulded floral scrolls in the middle between two bands (superior and inferior) decorated with a moulded petal-like design. The jar was initially believed to have been acquired during the fall of Constantinople in 1204 by the then Dodge of Venice Enrico Dandolo. Re-evaluation by Raphael indicated that the original dating was probably incorrect, and based on the shape, the jar was more akin to those manufactured in the Dehau kilns of Fujan Province in the late thirteenth century (Raphael, 'Chinese Porcelain Jar in the Treasury of San Marco, Venice', *Transactions of the Oriental Ceramic Society*, 1932). This view is further supported by similarly shaped stoneware pieces having been found at kiln sites in Fujian Province. However, Raphael made a further leap of faith, suggesting that based on the probable date of manufacture it was altogether possible that the jar was brought back to Venice by Marco Polo himself as one of the treasures acquired during his travels. This has led to sherds of Qingbai similar in design to the Marco Polo Jar found at multiple archaeological sites across the Indian Ocean to becoming known by archaeologists as the Marco Polo wares. Circumstantial evidence supporting the attribution of the Marco Polo vase to Marco Polo's collection of treasures could be acquired by a more precise dating of the jar.

Two major Dehau Kiln sites (Wanpinglun and Qudougong) dating to around the 12th to thirteenth centuries and the late thirteenth to fourteenth centuries respectively have been excavated. Of the two the Qudougong site yielded moulded designs very similar to the Marco Polo Jar, along with kiln furniture inscribed with the Phags-pa script introduced by to China in around 1269 by Kuhblai Khan. Other artefacts were also discovered at Qudougong, including clay boxes marked with a date mark indicating the date of either 1307 or 1367. There is evidence derived from looking at the relative distributions of the Dehau Qingbai and blue and white wares that the earlier date of 1307 is more likely. The famed Sinan shipwreck discovered in 1977 provides some of this evidence. This ship carried a cargo including 16,800 pieces of Chinese ceramics, 10,000 of which were fine porcelains and none of them were decorated in blue and white. One of the most important finds at the wreck were packaging slips indicating that the ship sailed from Fujian in 1323 and was heading for the Japanese port of what is now Fukuoka. Thus, the Sinan wreck has been used as evidence that the Chinese were not making blue and white at that point in history (Gerritsen, *The City of Blue and White*, 2020). Further evidence supporting the earlier date of 1307 comes from looking at the presence or lack of Chinese blue and white porcelain in the Minab region of Southern Iran and Hormuz Island. Chinese Qingbai wares have been found in the Minab region of Southern Iran, but no blue and white Chinese porcelain from Jingdezhen has been found in this region. However, the reverse is the case in the Island of Hormuz, which was re-settled by the Kingdon of Ormus in 1325 (Meicun & Zhang 'A Chinese Porcelain Jar Associated with Marco Polo: A Discussion from an Archaeological Perspective', *European Journal of Archaeology*, 2018) where blue and white Chinese porcelain has been found rather than Qingbai ware. Both these pieces of evidence suggest that the shipment of Dehua ware declined, to be replaced with shipments of Blue and white porcelain before 1367 which in turn could indicate that the Marco Polo Vase did potentially date to the time of his Travels (Meicun & Zhang 'A Chinese Porcelain Jar Associated with Marco Polo: A Discussion from an Archaeological Perspective', *European Journal of Archaeology*, 2018). Although useful in dating the jar, it doesn't specifically confirm that the Jar was part of Marco Polo's collection of treasures. It is probably that there may actually have been a trickle of similar wares into Venice during the mid to later part of the thirteenth century, after all, by the late 1290s Venetian merchants had been trading Chinese goods transported to the Levant and Persia via the Silk Roads since 1221. Other pieces of a similar period to that of the Marco Polo jar were also arriving in Europe by the fourteenth century. One of the most well-known examples is a white porcelain vase with a blue/green Qingbai glaze from Jingdezhen region, known now as the Fonthill Vase (see Chapter 4). This vase is currently in the care of the National Museum of Ireland and is believed to date to around 1300. The vase is said to have been brought back from China by the Franciscan monk Giovanni Marignolli who was sent as one of four legates to the court of Mongol Chinese Emperor Togon-Temure between 1339 and 1353. On his return to Europe, he served as Chaplain to the Holy Roman Emperor, Charles IV in Prague (1354–1355). The vase is documented to have passed to Louis the Great of Hungary, who had it mounted with a silver handle and base before passing on to the King of Naples (Charles III of Durazzo), the vase then passed via the Kings of Aragon to the son of Louis XIV of France (the Dauphin) and was eventually acquired by William Beckford of Fonthill Abbey. In 1882, the vase was sold at auction and acquired by the Museum for the sum of £28 and 7 shillings. By this time, it had lost its silver handle and base.

3.4 The Genesis of Blue and White Chinese Porcelain

During the Yuan Dynasty porcelain manufacturers in centres such as Jingdezhen continued to experiment with pastes and glazes. More heavily potted porcelains started to appear with a much thicker less translucent luanbai or eggshell white glaze carrying impressed moulded decorations in the form flowers and Dragons and incised characters such as shu and fu, giving rise to these porcelains being referred to as shufu type wares believed to have been made for the Yuan central government. Also, during the early to mid-period of the thirteenth century porcelain manufacturers in Jingdezhen developed the technique of decorating their products with the underglaze colours, blue (cobalt oxide) and red (coper oxide). The technique was most probably derived from Mongolian controlled Persia whose potters had been decorating their tin glaze earthenwares with underglaze pigments since the ninth century. In the case of cobalt oxide this led to the genesis of the stunning Chinese blue and white wares exemplified by the famed Chinese David vases dating from the Yuan period, c1351, which are beautifully decorated in the typically Persian blue and white style. This form of decoration very soon dominated the Chinese home and export markets (Gerritsen, *The City of Blue and White*, 2020). The source of cobalt for these early Chinese blue and white wares has been a bone of contention, however, (Chen, Guo, Chen, 'Discussion for the Origin of the Cobalt-Pigment on Chinese Yuan Blue-and-White Porcelain', *Chinese Ceramic*, 1993) seems to indicate that the wares of the Yuan period to the Yongle period used a source of cobalt that lacked copper (Cu) and nickel (Ni) but had a high iron (Fe) content along with traces of arsenic (As) and sulfur (S). It appears from comparison of the pigments from the Yuan to Yongle periods with blue decorated earthenware from Persia and Syria, that both shared a common source of cobalt (Cowell & Zhang, *Analyses and Source of the Cobalt-Blue Pigments Employed on Chinese Ceramic, Ming Ceramic*, 2001; Feng & Baoru, 'Further Study of Sources of the Imported Cobalt-Blue Pigment Used on Jingdezhen Porcelain from Late 13 to Early 15 Centuries', *Sci China Ser E-Tech Sci*, 2008), most probably Al Amarah and Al Hayy both of which supplied cobalt to the ceramics manufacturers of Samarrain, North of Baghdad in what is now Iraq. During the Yuan period, this region would have been a dependency of the Mongol Empire, and it would only be natural for Mongol Emperors to make use of this valuable commodity and transport cobalt via the terrestrial and maritime silk routes to South Eastern China where it could be used in the manufacture of blue and white porcelain in Jingdezhen. Post Yongle (Cowell & Zhang, *Analyses and Source of the Cobalt-Blue Pigments Employed on Chinese Ceramic, Ming Ceramic*, 2001; Feng & Baoru, 'Further Study of Sources of the Imported Cobalt-Blue Pigment Used on Jingdezhen Porcelain from Late 13 to Early 15 Centuries', *Sci China Ser E-Tech Sci*, 2008), the source of cobalt used in the decoration of blue and white porcelain seemed to change to one that was high in manganese (Mn) rather than iron.

Trade with China via the Silk road and growing shipping links continued to increase during the thirteenth century into the fourteenth century. According to Weatherford, by the 1295 the Chinese Mongol Navy had developed into the first Naval superpower which protected the ocean routes to ports as far South as Hormuz and the Arabian coast and as far North as Siberia (Weatherford, 'The Silk Route From Land and Sea', *Humanities*, 2018). This meant that Marco Polo, who had to make his way from Hormuz to China via the land-based Silk Roads was able to sail back to Hormuz in 1295 from Zaiton in the company of a fleet of fourteen ships. To further improve the efficiency of maritime transport the Mongols built larger four-masted junks that were said by Marco Polo to have had up to 60 cabins and a crew of up to 300 men. Needless to say, the opening up of the maritime Silk Road by the Chinese Navy greatly increased the speed and efficiency of maritime transport. This is exemplified by Marco Polo's journeys. His outward journey towards china took around 4 years and involved him having to take the landward silk routes due to an unavailability of safe maritime routes from Persia. His return journey back via the newly established Mongol shipping routes only took two years. The development of the maritime silk routes resulted in a huge increase in trade, expanding the range of shipped goods across East indies, India, North Africa and the Levant to the extent of enabling the import of exotic animals such as live elephants and rhinoceri.

3.5 The Emergence of Ming and the Ming Gap

The Mongol empire was very alien to China. The Mongol Emperors had a flexible approach to religion and established their own courts which mostly excluded the native Chinese nobles. The Mongols were also very focused on trade, whereas the Chinese population was mostly Buddhist and heavily influenced by almost a thousand years or so of Confucian ideology. This Confucian ideology was closely associated with a belief that China, the Middle Kingdom, was self-sufficient and as such external trade was unnecessary and un-wanted. During the early thirteenth century famine and drought began to generate unrest in the populations of Northern and Central China, this led to a series of rebellions. One of the leading figures in these rebellions was Zhu Yuanzhang, formerly Zhu Chongba, who joined the ranks of rebels lead by Guo Zixing. Zhu Yuanzhang became a very successful military leader and as such rose quickly through the ranks to become second in command to Guo Zixing's. Zhu Yuanzhang later married the adopted daughter of Guo Zixing, which meant that on Guo Zixing's death in 1355 he became the leader of the rebel armies. In 1353 Zhu Yuanzhang captured Chuzhou, just Northwest of Nanjing and some 328 miles North of Jingdezhen (Encyclopaedia Britannica). This unrest was thought to have played a part in the Porcelain Bureau of Jingdezhen suspending the manufacture of porcelains in 1352 (Tai, Daly, Mckinnon, Parnell, Feener, Majewski, Ismail, Sieh, 'The Impact of Ming and Qing Dynasty Maritime Bans on Trade Ceramics Recovered from Coastal Settlements in Northern Sumatra, Indonesia', 2020). Manufacture started up again at Jingdezhen in 1361 once Zhu Yuanzhang gained control of the city. Eventually in 1368 Zhu Yuanzhang overthrew the last Chinese Mongol emperor Togon-Temir to become Hongwo, the first Emperor of the Ming Dynasty.

The rebellions and war leading to the formation of the Ming Empire had a devastating effect on trade and the manufacture of porcelain. The Chinese Mongol navy collapsed, maritime routes became unsafe again due to the activities of pirates and the landward Silk Roads remained under Mongol control and so less available to Chinese merchants. Policies imposed by the new Ming emperor also had a deleterious impact on the availability and export of porcelain. Although Hongwo enabled the reopening of the manufacture of porcelain in Jingdezhen, he designated a proportion of kilns of Jingdezhen (Jiangxi Province) and Longquan (Zhejiang Province) as imperial kilns with the remit of producing ceramics exclusively for Imperial use. All the porcelain made in these kilns was dispatched to the Imperial Palace for selection and those not perceived as being of a high enough quality were destroyed. All these kilns produced blue and white porcelain. As a consequence, the commercial output of porcelain manufactories dropped which also reduced the availability of porcelain for trade.

Hongwo also initiated an embargo on maritime trade in 1371. There were several reasons for this embargo, including Confucian ideology and the fact that post the collapse of the Mongol Chinese Navy, Japanese pirates had started raiding the Chinese coast. The edict banning maritime trade prohibited setting a sail of any kind included fishing boats and was formally put into law in 1391. However, some exchange of porcelain was continued via elicit smugglers, and legitimately in diplomatic spheres where the Ming Emperors used porcelain along with other luxury goods as Imperial gifts to diplomatic missions from vassal states. This all led to a sudden and very dramatic decrease in the trade of Chinese Porcelain. Tom Harrison (Harrison, 'The Ming Gap and Kota Batu, Brunei', *Sarawak Museum Journal*, 1958), curator of Sarawak museum was one of the first to notice this. He commented that it was impossible to find any Ming Chinese ceramics along a stretch of 100 miles of the Borneo coast, despite there being examples of ceramics from earlier and later Chinese periods. He referred to this unavailability of Ming porcelain as the Ming Gap. This perception was later supported by Roxanna Maude Brown (Brown, 'The Ming Gap and Shipwreck Ceramics in Southeast Asia: Towards a Chronology of Thai Trade Ware', *Siam Society*, 2009). She, by examining ceramics found at shipwreck sites across South East Asia, determined that the Ming gap extended from around 1350 right the way through to 1567 and was associated with an increased trade in ceramics from, Thailand and Vietnam (Tai, 'Ming Gap and The Revival of Commercial Production of Blue and White Porcelain in China', *Bulletin Of The Indo-Pacific Prehistory Association*, 2011; Weatherford, 'The Silk Route From Land and Sea', *Humanities*, 2018; Tai, Daly, Mckinnon, Parnell, Feener, Majewski, Ismail, Sieh, 'The Impact of Ming and Qing Dynasty Maritime Bans on Trade Ceramics Recovered from Coastal Settlements in Northern Sumatra, Indonesia', *Archaeological Research in Asia*, 2020: Gerritsen, *The City of Blue and White*, 2020). However, the Ming Gap didn't represent a continuous period of over 200 years of limited trade in porcelain and other Chinese goods. There were periods of time when Ming Emperors once more promoted maritime trade. For example, between 1405 and 1433 successive Ming Emperors Yongle, Hongxi and Xuande tried to re conquer the maritime routes mediated by Zheng He the Eunuch Admiral and diplomat of the great Ming Treasure Fleets.

Zheng He, initially called Ma Saboa, was a Muslim born in Yunnan Province in Southwestern Chin. Yunnan was the last Mongol province to be taken by the Ming forces in 1381. The young boy Ma Saboa, later to become Ma He and then Zheng He, was captured and like other boys of his age, was castrated and conscripted into the Ming army. Army life suited him, and he was able to become influential in the newly established Yongle court (1402–1424) the emperor of which gave him the name of Zheng and the task of establishing a maritime trading fleet. That fleet was huge, said to have reached around 300 ships in which the largest, the nine-masted treasure ships were up to 120 m long with a capacity of up to 1500 to 3000 tonnes and a crew of up to 600 men. Zhenge undertook seven voyages, the first in 1405–1407 with a fleet consisting around 28,000 men in 62 ships. This fleet travelled to Southern Vietnam, Thailand, Melaka, Java, Sri Lanka and Kozhikode on the Southwestern Indian coast. In subsequent voyages Zheng He extended the range of his voyages, so that by the fifth and sixth of his voyages (1417–1419 and 1421–1422 respectively) his fleet sailed to ports as far away as Hormuz, Aden, Mogadishu, Baraawe and Malini off the African coast. By the last voyage (1431–1433) Zhenge He had further extended the journey by also taking in Jiddah off the coast of Arabia, near Mecca. The ships were loaded with huge quantities of goods. Some of the goods were used as diplomatic gifts, but the vast majority was exchanged for silver, spices, gems and other important raw materials such as cobalt from Persia for the manufacture of imperial blue and white porcelain. Unsurprisingly, Chines porcelain made up one of the major components of the cargo, so much so that the Kilns of Imperial Jingdezhen are documented as having had to provide approximately 443,500 pieces of porcelain (Gerritsen, *The City of Blue and White*, 2020) for the seventh and last of these extraordinary voyages. Some of the goods reaching Persia (Iran) and Arabia would no doubt have found their way to merchants in the Mediterranean and from there into Europe. It appears that there were some within the Ming Imperial court who continued to oppose the voyages and associated costs, despite the income generated. After the final voyage in 1433 the treasure ships were abandoned completely and their wood was sold for fuel, as the Ming Empire turned away from maritime trade and put restrictions on all forms of external travel and communications (Finlay, 'The Treasure-Ships of Zhenge He: Chinese Maritime Imperialism in the Age of Discovery'. *Terrae Incognitae*, 1991).

3.6 Cobalt and the Implications of the Ming Gap

There is evidence that these restrictions on trade had a significant influence on the manufacture blue and white porcelain in Jingdezhen. During the late Hongwu to early Yongle period the production of underglaze blue decorated porcelain decreased at Jingdezhen, whilst the production of underglaze copper red porcelain increased. This may have been driven by a shortage of cobalt from the by then traditional Persian sources resulting from the maritime shipping ban. It may have been one of the factors that influenced Yongle in his wish to re-open trade routes with potential suppliers. What is strange is that post the Yongle period the source of cobalt used in the decoration of blue and white porcelain seemed to change to one that was high in manganese (Mn) rather than iron indicating a switch from imported Su Ma Li Qing cobalt from Persia to Su Bo Ni Quing cobalt imported from Sumatra (Cowell & Zhang, *Analyses and Source of the Cobalt-Blue Pigments Employed on Chinese Ceramic, Ming Ceramic*, 2001; Feng, & Baoru, 'Further Study of Sources of the Imported Cobalt-Blue Pigment Used on Jingdezhen Porcelain from Late 13 to Early 15 Centuries', *Sci China Ser E-Tech Sci*, 2008). According to Chan (Chan, 'The "Chinese Barbarian Officials" in the Foreign Tributary Missions to China During the Ming Dynasty', *Journal of the American Oriental Society*, 1968) and Feng et al.

(Feng & Baoru, 'Further Study of Sources of the Imported Cobalt-Blue pigment Used on Jingdezhen Porcelain from Late 13 to Early 15 Centuries', *Sci China Ser E-Tech Sci*, 2008) a number of southeast Asian states, including the Majapahit Kingdom of Java, Siam, Malacca and Sumatra sent Tributary missions to the Chinese court during the fifteenth and sixteenth centuries. In the case of Sumatra, missions were sent to the Youngle court in 1407, and the Xuande court in 1431 and 1435. The tribute arising from some of these missions is thought to have included cobalt with a high manganese content rather than a high iron content. No doubt sourcing cobalt from Sumatra was perceived as less expensive than sourcing from far away Persia. This new source of cobalt may have been one of the factors that contributed to the cessation of Zhenge He's maritime trade activities. By the Emperor Chenghua's reign (1465–1487) the decorators at the imperial kilns of Jingdezhen had switched once more to using Peng Deng Qing, a locally sourced cobalt from Jiangxi Provence which was also high in manganese and low in iron. The problem with this form of cobalt was that the cobalt tended to diffuse into the glazing during firing, which reduced the fineness of the decoration. The use of cobalt with a high Manganese content continued right up to the reign of Kangxi in 1662 until a new source of cobalt was adopted as a result of European influences.

3.7 The Portuguese, the East Indies and the Genesis of European Armorial Porcelain

By the establishment of the Ming dynasty in 1368, Southern Europe was well aware of the terrestrial and maritime spice routes. Merchants in the Italian City States such as Venice were trading goods transported to the Levant via these routes in exchange for spices, silks and other luxury items such as the occasional piece of Chinese porcelain. There was no known means of sailing directly from Western Europe to the source of these far Eastern goods, but after the Portuguese capture of Ceuta in 1415, Prince Henrique of Portugal (1394–1460) (Henry the Navigator) developed a strong interest in maritime exploration and dispatched ships on voyages of exploration. By 1419–1420 the Portuguese had colonized Madeira 350 miles off the coast Morocco, they also colonized the Azores in 1439 and had discovered the Senegal river on the West African coast of Guinea in 1446. They later colonised the Cape Verde Islands in 1460 and then after Henry the Navigator's death in 1460 continued their voyages; a squire of King John II of Portugal (1455–1495), Diogo Cão, discovering the mouth of the Rive Congo in 1485 and, Bartholomeu Dias in 1488 unknowingly sailing round the Cape of Good Hope in a storm to make landfall at Algoa bay on the Western end of the South African South coast before headed on to what is now Mossel Bay. It was only on his return journey, that he actually set eyes on the Cape of Good Hope. This later journey set the scene for King Manuel of Portugal (1469–1521), who soon after becoming King in 1495 appointed Estevado da Gama to lead an expedition around the Cape of Good Hope in search of a route to India. Estevado died but his third son Vasco da Gama took on the appointment. He set sail with a small fleet of four ships, the *Santo Gabriel* (120 tons) as his flagship, the Santo Raphael (120 tons) captained by his eldest brother Pancho along with the Berrio (50 tons) and a store ship of 200 tons on the 8 July 1497. Comparison of these ships with those of Zhenge He is quite interesting. It has been pointed out that da Gama's ships were tiny in comparison and could have theoretically been carried in the hold of one of Zhenge He's 9 masted treasure ships (Finlay, 'The Treasure-Ships of Zhenge He: Chinese Maritime Imperialism in the Age of Discovery', *Terrae Incognitae*, 1991). Despite the smallness of his fleet and ships, after an eventful journey da Gama reached the Southwestern Indian port of Calicut on 20th May 1498 (Tate, 'Mediaeval Navigator: Vasco da Gama' *Journal of the Royal African Society*, 1944; O'Rourke & Williamson, 'Did Vasco da Gama Matter for European Markets?' *The Economic History Review*, 2009).

Although an able seaman Vasco da Gama was ill prepared for a diplomatic mission. It seems that on arrival at Calicut he put together a woeful selection of gifts for the local ruler that, according to an official in Calicut, could have been bettered by even the poorest of merchants from Mecca or other parts of India. As a result of this Vasco da Gama met the local King with no gifts at all claiming that his mission was one of discovery and not trade (Wolff, 'Da Gama's Blundering: Trade Encounters in Africa and Asia during the European 'Age of Discovery,' 1450–1520', *The History Teacher*, 1998). Despite this rather shaky start, da Gama was given permission to build a fortified trading post at place called Chinacota.

Whilst Vasco da Gamma was on his first journey, bad news reached the Portuguese court from a Florentine merchant about rumours regarding ships of white Christians visiting Calicut some years prior to da Gama's voyage. This suggested that somebody else representing another European maritime power had beaten the Portuguese to finding a navigable ship route to India. On Vasco Gamma's return, it was discovered that these white Christians were in actual fact Chinese sailors, and the sailors of the Malabar coast claimed descent from them. As a consequence of this the Malabar cost sailors were known in Persian slang as the Chini-bachaqan or China boys. The Chinese white Christians were sailors manning the fleet of Zeng He and they initially established Chinacota as a trading post or Fortress of the Chins during the earlier voyages of the Chinese Yongle period some 80 to 90 years before da Gama established a trading post there (Finlay, 'The Treasure-Ships of Zhenge He: Chinese Maritime Imperialism in the Age of Discovery'. *Terrae Incognitae*, 1991). That the Chinese had sailed to Calicut would have had a profound impact on da Gama, King Manuel I of Portugal and his courtiers. After all, if the Chinese had historically established trading relationships with the states in the Indies, surely the Portuguese could do also and not only cut out the Venetian merchants trading in pepper and other spices such as ginger via the Levant, but also the Venetian's Muslim intermediaries (Wolff, 'Da Gama's Blundering: Trade Encounters in Africa and Asia during the European 'Age of Discovery,' 1450–1520', *The History Teacher*, 1998), potentially even forming direct trading links with the Chinese.

Da Gama's first voyage wasn't a commercial success, however, he returned to Portugal in triumph having demonstrated a maritime route to Indies. He also gained vital information regarding the peoples and trading habits along with some information about what the peoples of the Indies wished to trade for their goods. The Zamorin of Calicut made this explicitly clear in a message he had da Gama deliver to the Portuguese court, he wanted gold, silver, corals and crimson cloth in return for the spices and precious stones of Calicut. The potential impact of da Gama's voyage to Calicut wasn't lost on the trading nations of the Mediterranean. The Venetian diarist Girolamo Priuli feared that as a result of da Gama's discovery of a maritime route to the Indies, the Venetian monopoly of the spice trade from Cochin and Calicut to Alexandre and Beirut in the Levant would be superseded by the Portuguese with the Portuguese King and Lisbon becoming the new centre of the European spice trade (Chadhuri, *The Portuguese Maritime Empire, Trade, and Society in the Indian Ocean During the Sixteenth Century*, 1992). For the Portuguese, da Gama's success led King Manuel I of Portugal in 1500 to send a second larger fleet consisting of 13 ships loaded with silver, merchandise, and 1200 soldiers and sailors to Calicut under the command of Pedro Alvares Cabral. By this time, the Portuguese had learnt to dress to impress, so officers were equipped with expensive uniforms of the finest materials (Nowell, 'The Discovery of Brazil—Accidental or Intentional', *The Hispanic American Historical Review*, 1936). The Fleet sailed on the 8 March in 1500. Cabral initially followed a route similar to that of da Gama to the Cape Verde Islands, but instead of a South-easterly route like da Gama, he headed Southwest until discovering the coast of South America. Historically, this discovery was believed to have been accidental, and possibly caused by storms. However, that his Captains included the likes of Dais and others who had considerable experience of sailing along the West African coast, this seems unlikely. They would have been aware of the need to head Southeast from the Cape Verde Islands, so the southwestern detour was obviously deliberate. It was also very unlikely that storms drove them further West than they intended, because those storms would have had to have been North Easterlies, or Easterlies, or South Easterlies, to do this. It is an unfortunate truth that storms in the South Atlantic in March and April come from the West, so they would have blown Cabral's fleet away from the south American coast (Nowell, 'The Discovery of Brazil—Accidental or Intentional', *The Hispanic American Historical Review*, 1936). This leads to the conclusion that a component of Cabral's mission was to act on some prior knowledge of a large expanse of land to the West and claim it for the Portuguese Crown. This he did. The first evidence of land was noted in a very matter of fact way in their logs on the 21 April 1500, and they anchored off the coast of what is now Brazil on the 22 April. After moving North along the coast to what became Baia Cabralia, Cabral decided to send one ship back to Portugal to report his discovery, and then having erected a single wooden cross in the care of two convicts (degredados) to keep watch, set sail on the 2 of May 1500 for the Cape of Good Hope. What is interesting is the dispatch sent with the ship back to Portugal in which there is reference to an old map belonging to a Pero Vaz Basagudo that the King could consult to see the location of the newly discovered land. This indicated that Cabral's South-westerly diversion and discovery of Brazil was planned and functioned to confirm a sighting of land by a previous Portuguese expedition.

On leaving Brazil, Cabral's fleet made good time, but he lost four ships with all hands whilst rounding Cape of Good Hope on the 29 May. The remains of his fleet made Calicut by the 13 September where the Zamorin let him build a fortified trading post. However, Cabral had to move on from Calicut because of fighting between his crew and the local Muslim population. He sailed on to buy cargos of spices at the ports of Carangolas and the ports of what are now Kochi (Cochin) and Kanmur (Cannanore). At Cochin, Cabral formed an alliance with the Hindu Prince of Cochin (the Raja Trimumpara) against the Zamorin of Calicut and established a factory before he set sail for Portugal on 16th January 1501 (Shokoohy, 'The Town of Cochin and Its Muslim Heritage on the Malabar Coast, South India', *Journal of the Royal Asiatic Society*, 1998). Cabral lost a further two ships on the way. On the 23 June 1501 his four remaining ships loaded with spices made it back to Portugal.

Manuel I of Portugal was apparently pleased with the outcome of Cabal's voyage. The spices he returned with proved profitable and yet another fleet was commissioned to revisit the Indian Coast under the command once more of Vasco da Gama in 1502. Da Gamma, after a failed attack on Calicut sailed on to Cochin to be welcomed by Raja Trimumpara. The Zamorin of Calicut was left no choice but take control of Cochin on da Gama's departure to Portugal inducing Trimumpara to flee to the Island of Vypin. However, the Zamorin of Calicut's forces had to retire from Cochin on the arrival of a large Portuguese fleet under the command of Afonso de Abuquerque in 1503, which allowed Trimumpara to return. He gave Abuquerque permission to build a fort at Cochin in 1505 but insisted as did succeeding Rajas of Cochin that the local Jewish and Muslim population came under his protection as his subjects. Cochin thus became the centre of Portuguese operations in India until being superseded by Goa in around 1530. Despite this, Cochin remained a major trading post (Shokoohy, 'The Town of Cochin and Its Muslim Heritage on the Malabar Coast, South India', *Journal of the Royal Asiatic Society*, 1998). Afonso de Abuquerque went on in 1511 to conquer the Malaysian costal State of Malacca (Chin-Keong, 'Trade, the Sea Prohibition and the "Folgji" 1513–50', *Boundaries and Beyond: China's Maritime Southeast in Late Imperial Times*, 2017) which served as a key trading post for the Portuguese right up until the Dutch liberated it from the Portuguese in 1641.

The conquest of Malacca was very significant. Da Gama was known to have brought back a dozen or so pieces of Chinese porcelain with him from his voyages to India, which he presented to Manuel I of Portugal. Dom Manuel I of Portugal has been described as one of the first western Princes to have been infected by what Samuel Johnson described as the 'contagion of China-fancy' (Finlay, 'The Pilgrim Art: The Culture of Porcelain in World History.' *Journal of World History*, 1998). Chinese porcelain had for centuries been admired in the Middle East, as having a jewel like form and according to the Muslim merchant Sulayman in 851 CE as having a transparency akin to glass so that a porcelain vessels contents were visible through their body. This was perceived as miraculous bearing in mind the that such porcelain vessels were made from Clay. It is apparent that porcelain had the same impact on Dom Manuel I. When Malacca was first discovered by the Portuguese in 1509 by Diego Lopes de Sequeira, it seems that he carried instructions from Manuel I to find out as much as he could about the Chinese merchants, their ships, their merchandise and their country. On reaching Malacca, de Sequeira found a number of Chinese Junks manned by friendly Chinese sailors, which suggested that the Portuguese had some prior knowledge that Chinese ships were active in the region of Malacca before their initial voyage of discovery. When Afonso de Abuquerque arrived in 1511 with the aim of conquering Malacca, he specifically spared all the Chinese ships he found in the harbour. He also went out of his way to establish good relations with the Chinese sailors in order to learn more a about China and the Chinese. In return, the Chinese ferried Portuguese envoys from Malacca to Siam and also carried back with them favourable reports about the Portuguese to China. Portugal's first envoy to China, Tomé Pires arrived at Malacca in 1512 and he quickly amassed information for his text the *Suma Oriental* about what the Chinese sought from Malacca and the goods the Chines sold at Malacca. Pires also learnt that the coastal city of Guangzhou (Canton) functioned as the trading gateway to China. Soon after Pires arrival in Malacca, a series Portuguese trading missions were sent via Chinese Junks returning to China, and although Portuguese were unable to land, they were able to trade their goods at considerable profit. These trading trips to China on Chinese Junks, were vital in that it allowed the Portuguese to gather first-hand experience of trading with the Chinese whilst also allowed them to establish themselves as new and potentially useful trading partners. The trips also allowed the Portuguese to estimate the potential profits that could arise from trading directly with China, which according to one estimate indicated that pepper could be sent from Malacca to China at profits of around 300% (Chin-Keong, 'Trade, the Sea Prohibition and the "Folgji" 1513–50', *Boundaries and Beyond: China's Maritime Southeast in Late Imperial Times*, 2017). These early trade experiences paved the way for a Portuguese fleet led by Fernão Peres de Andrade to set sail with Tomé Pires for Guangzhou in 1517. The Portuguese were initially well received by the Chinese authorities at Guangzhou, and in May of 1520 the Portuguese Envoy Tomé Pires arrived in Nanjing and arrangements were made for the for then Chinese Emperor Zhengde to meet the Portuguese in Beijing the following February. It was during this period of apparent amicable communication between the Chinese authorities and the Portuguese c1517-19, that Dom Manuel I of Portugal commissioned the very first European Armorial piece of Chinese porcelain, what can be considered as the true European genesis of the concept, a blue and white ewer decorated with an armillary sphere, the Royal Standard of Manuel I (Fig. 2.28), which eventually became the Portuguese symbol indicating the global expanse of their empire. Relations between the Portuguese and the Chinese rapidly soured as a result of the arrival of Simão de Andrade, brother to Fernão Peres de Andrade. Simão de Andrade apparently ignored the Chinese officials, was accused of piracy, stealing Chinese girls and oppressing the Chinese. He even built an illicit fort at Tunmen, an Island in the Pearl River Delta of the coast of Guangzhou. At about this time the Chinese Emperor Zhengde died (20 April 1521) to be replaced with the 14-year-old Emperor Jaijing on the 27 May 1521, who ordered all emissaries back to Guangzhou including Tomé Pires. Just before the arrival of Pires at Guangzhou, the Chinese evicted the Portuguese from Tunmen in a long bloody sea battle and Tomé Pires was arrested and later died in a Chinese jail. The Portuguese attempted to re-establish trade in 1522, but their initially peaceful overtures were forcibly rejected and in the ensuing battle, the Portuguese lost 35 men killed and 43 captured. These captives were later executed by Imperial decree. Despite this, Portuguese traders continued an illicit trade with merchants along the Chinese coast with the collusion of local officials incentivised by bribes until around 1554 when Capitão-Mor Leonel de Sousa established a verbal agreement with the Coastal Surveillance Vice-Commissioner of Guangdong, Wang Dong, allowing the Portuguese to trade once more under the guise of representatives from the tributary state of Saim. One of the driving forces for this was that by this time the Portuguese had established an efficient triangular trading network exchanging Japanese copper and silver for Chinese Porcelain, gold and silks. This trade was first enabled by error rather than design. In 1542 three Portuguese traders, Antonio da Mota, Francisco Zeimoto et Antonio Peixoto sailing in a Chinese Junk were blown off course in a storm and made landfall on the small Japanese Island of Tanegashima to the South of Kyushu. Their ship carried a cargo of porcelain and silks (Kammerer, 'La Découverte de La Chine Par Les Portugais Au XVIème Siècle et La Cartographie Des Portulans', *T'oung Pao*, 1944). Like the Portuguese, Japanese merchants were officially excluded from trade with China in around 1523 (Chin-Keong, 'Trade, the Sea Prohibition and the "Folgji" 1513–50', *Boundaries and Beyond: China's Maritime Southeast in Late Imperial Times*, 2017) and although Japanese traders, like the Portuguese, continued an unofficial trade with China as smugglers and pirates (the Wokou pirates) their supply of Chinese goods to Japan was outstripped by demand. As a consequence of this, the first Portuguese merchants to arrive in Japan returned to Malacca having exchanged their Chinese goods for a valuable cargo of Japanese goods. They also apparently introduced firearms to Japan. They were soon to be followed by others including Captain Jorge Álvares who wrote an account in 1547 of his visit to Japan (Willis, Captain Jorge Alvares and Father Luis Frois S.J. Two Early Portuguese Descriptions of Japan and the Japanese, 2012). This account was passed on to Saint Francis Xavier, co-founder of the Jesuits just before he led the first Christian mission to Japan in 1549. This mission eventually led to the establishment of thriving Japanese Christian communities in Kagoshima and Hirado. By 1557 the Portuguese had managed to establish a land base in Macao near the mouth of the Pearl River and had also established the port of Hirado as the point of entry of the goods shipped to Japan (Chin-Keong, 'Trade, the Sea Prohibition and the "Folgji" 1513–50', *Boundaries and Beyond: China's Maritime Southeast in Late Imperial Times*, 2017; Cooper, 'The Mechanics of the Macao-Nagasaki Silk Trade', *Monumenta Nipponica*, 1972). The majority of trade was associated with shipping Chinese goods to Japan and vice versa via Malacca in a triangular trade called the nanban trade. A small proportion of these goods including porcelain were shipped back to Lisbon.

By 1580, Phillip II of Spain, son of Charles V King of Spain and Holly Roman Emperor and Isabella of Portugal, ascended to the Portuguese Throne. He had a strong interest in porcelain and amassed considerable collection of around 2000 pieces, some of which were decorated with armorials. Examples of Chinese Ming Wanli period underglaze blue porcelain bottles carrying Philip II's coat-of-arms can be found in at least two museums. One example, known as the Bottle of King Philip II, is located in the Museu-Biblioteca da Casa de Braganca—Paco Ducal e Castelo de Vila Vicosa (Museum and Library of the House of Braganza in the Vila Vicosa) in Portugal and consists of a flattened circular body supported on a trapezoid base with a truncated neck finished with a copper lid. A similar example is also held in the British Museum. The later bottle like the Bottle of King Philip in Portugal is of a flattened circular form supported on a trapezoid base. These two bottles are almost identical, they are both decorated in underglaze cobalt blue with Philip's coat-of-arms to the centre which is thought to have been copied from Spanish coinage of the period. However, the example in the British museum has a longer neck and lacks the copper lid, thus it is possible that the example held at Museum and Library of the House of Braganza in the Vila Vicosa may have been damaged and the copper lid may represent a repair which converted the bottle into a drinking vessel. This indicates that even in Spanish Royal Circles, these bottles were prized possessions, so much so, if damaged they we repurposed. There is also evidence that Philip II of Spain was in constant fear of being poisoned. However, there was a belief that Chinese porcelain was not only very beautiful, but it was also believed to have a very important magical property, that of cracking if its contents contained poison. Thus, Philip II only drank from Chinese porcelain vessels, and it is altogether possible that the Chinese porcelain bottles carrying his coat-of-arms were manufactured specifically for his use.

The Portuguese of Macau only discovered that Philip II had ascended to the throne. In 1581 on the arrival of Philip's first magistrate of Macao. This caused a certain degree of consternation because the Chinese had granted Macau to the Portuguese throne not the Spanish throne. If the Chinese authorities discovered that the Portuguese had become subjects of the Spanish King, they may have insisted on evicted from them from Macau. To prevent this, they continued flying the Portuguese flag. Trade between Japan and China via the Portuguese thrived. By around 1587 Philip II changed the status of Macau to that of becoming a city and by 1595 Macau had become one of the major commercial ports in the far East. Things were about to change, however, as a result of the Dutch expanding their fleet's operations into the East Indies.

3.8 The Arrival of the Dutch in the East Indies

The Spanish Netherlands or what the Spanish called Flandes included Western Germany, modern day Belgium and Luxembourg, parts of France and the Southern Netherlands. These states were all part of the Holy Roman Empire held by the Holy Roman Emperor Charles V. When he abdicated in 1556 these states passed to Philip. Non-conformism was being adopted in the the Netherlands at this time but catholicism had been imposed on the population by Philips Father Charles V, the Holy Roman Emperor by the Introduction of the Inquisition. Philip on inheriting the Netherlands gave the Inquisitors further powers in 1559 with the aim of eradicating the heretic non-conformists. To aid this Philip II introduced new Bishoprics. At about this time Philip II was at war with the Ottoman Empire in the Mediterranean and had only just concluded a conflict with the French. Philip II still had troops stationed along the Southern Netherlands border with France and his leading minister in the Netherlands Cardinal Granvelle directed affairs in the provinces of the Netherlands via a Council of State consisting of what were called Stadtholders. However, there was growing opposition to Philips policies and Cardinal Granvelle amongst the Stadtholders of the Council of State, one of the most powerful of which was William the Silent of Orange. They wanted the troops and the Cardinal to Leave and a return to the times of Charles V when they had almost complete independence. This would enable them to abolish the meddlesome Bishoprics and provincial inquisitors. The Spanish troops were withdrawn in 1560 following a catastrophic defeat of the Spanish fleet at Djerba by the Ottoman forces, and Cardinal Granvelle was recalled to Spain in 1564. Both events were thought to have been precipitated by a proportion of the Stadtholders threatening to boycott all public duties should their demands not be met. Philip II even indicated that perhaps a Netherlands committee of Churchmen could be formed to consider alternative and less severe ways of handling the non-conformist heretics. Philip II reneged on his word after the defeat of the Ottomans and the lifting of the Siege of Malta in 1565 and instead reaffirmed his support for the inquisition and the New Bishoprics. The response in the Netherlands was dramatic. The Stadtholders had convinced themselves that their proposals had met with Philip's approval. Their disappointment and subsequent demands led to Philip sending a force of 60,000 troops under the Command of the Duke of Alva to the Netherland to put down the revolt of 1566–1568. This force was opposed by a force put together by William the Silent Prince of Orange and Count Louis of Nassau. The Prince of Orange was defeated and exiled, but the 1568 revolt was followed by the larger Revolt of 1572 which occurred just as the Spanish had to divert resources once more to counter the Ottomans in the Mediterranean. Thus, this rebellion soon secured a strong base and Spain with its military resources spread over two different and distant fronts eventually ran out of money and had to admit defeat submitting to the demands of the Dutch rebels in the February of 1577. Philip also made peace with the Ottomans and once more reneged on his word and reinstated hostilities in the Netherlands. However, he was again distracted, this time by the death of Dom Sebastin King of Portugal in 1578. Philip was the heir to the Portuguese throne and had to divert his troops from the Netherlands to Portugal (Parker, 'Spain, Her Enemies and the Revolt of the Netherlands 1559–1648', *Past & Present*, 1970). Although the Spanish made headway in recapturing Southern provinces of the Lowlands in 1580, the Spanish armies were not sufficiently financially supported to reconquer the entirety of the Netherland. In 1585 the possibility of Spain doing so became more remote when Queen Elizabeth of England in the Treaty of Nonsuch committed herself to keeping seven thousand three hundred and fifty foot and horse in Holland in support of the Dutch in their war of independence from Philip II. She also committed to pay a quarter of the costs of the war. This led Philip to the mobilization of funds to build what the Spanish considered as the Invincible Armada, which was fended off by Drake in 1587 and eventually pretty much completely destroyed in 1588 representing a huge loss of financial resources. The following year, 1589, Henry III of France died. His obvious successor was a Protestant, Henry of Navarre, who allied himself with the Dutch. As a result, Philip II weighed in to support the French Catholics in a war that continued until 1598 and cost the Spanish treasury heavily and once again diverted resources from the Netherlands.

Despite the war between Spain and the emerging United Netherlands, trade continued between the Netherlands and Spanish controlled ports. One of the major commodities required in huge quantities by the Dutch was salt. It was used for the preservation of herring, in the dairy industry and also domestically. Historically the Dutch had acquired salt from the Bay of Bourgneuf and Brouage in France along with Setubol and Cadiz on the Iberian Peninsula. That trade seemed to continue through the war with the Spanish post 1568 but was badly interrupted in 1585 when Philip II of Spain ordered the confiscation of Dutch, German and English ships moored in the Iberian Ports in order to bulk out the Armada being assembled against England. The Dutch lost a large number of ships, there is evidence that the Dutch City of Hoon alone lost around 30 ships at Lisbon and Setubal. The immediate impact of this was a significant decrease in the availability of salt in the Netherlands and a significant increase in its price (Sluiter, 'Dutch Maritime Power and the Colonial Status Quo' *Pacific Historical Review*, 1942). This led Dutch traders to look elsewhere for supplies opening up new trade routes. These new trade routes included South America and in 1595 the East Indies when the first Dutch fleet of four ships commanded by Cornelis de Houtman, followed the example of Edward Lancaster 1601 (Keay, *The Honourable Company*, 1991) and made its way around the Cape of Good Hope on route to the East Indies. Between 1598 and 1601 a further 12 fleets sailed to the east indies. During this period 6 competing Dutch East India companies began to emerge, but because of their disunity they could have potentially had a negative impact on the Dutch war with Spain. So, on 20th March 1602, the Dutch followed the example of the English and the Establishment of the English East India in Company in the December of 1600, and united the 6 competing Dutch East India companies into the Verenigde Oostindische Campagnie (VOC) otherwise known as The United Dutch East India Company (Gaastra, 'The Dutch East India Company A Reluctant Discoverer', *The Great Circle*, 1997; Ricklefs, *A History of Modern Indonesia Since 1300*, 2001). The VOC was capitalised with around seven million Florins and granted a monopoly charter. The VOC served two functions, the generation of a powerful and commercially viable merchant fleet, and the generation of a military force that could compete with and harry the Spanish/Portuguese fleets, allowing the Dutch to break into the highly profitable pepper and spice trade whilst simultaneously further depleting Spanish resources. Interestingly, the motivating force behind the English East India Company wasn't that dissimilar from those of the VOC but was considerably less well funded and in relation to the pepper/spice trade, destined to being outcompeted by the Dutch (Keay, *The Honourable Company*, 1991).

The Dutch first encountered Macau in 1601, a Dutch fleet of six ships led by Admiral Jakob van Neck, who was surprised at seeing a Spanish looking city at the Mouth of the Pearl River (Hao, *Macau History and Society*, 2011). He sent two boats to investigate, which along with around twenty sailors were captured by the Portuguese. Of those sailors captured the Portuguese are said to have treacherously murdered seventeen and later sent the remaining three to Goa. In reprisal in 1603

Jakob van Heemskerck who assumed command of the Dutch fleet on Van Neck's return to the Netherlands, captured the huge 1400–1500 tonne Portuguese Carrack the Santa Catarina as it sailed from Macau to Malacca. This ship was loaded with Chinese silk and Porcelain (Hao, *Macau History and Society*, 2011; Blussé, 'Brief Encounter at Macao', *Modern Asian Studies*, 1988) and caused quite a stir when it arrived back in Amsterdam in 1603 where its highly valuable cargo of 60 tonnes of porcelain, 70 tonnes of gold and 1200 bales of raw silk was sold (Brook, 'Trading Places: The Santa Catarina Brought the First Large Shipment of Porcelain from the Far East to Europe, creating a demand for new luxury goods, which Feature in Paintings of the Dutch Golden Age', *Apollo*, 2015). The proceeds of the sale were considerable. There is evidence that the sale raised the equivalent of around half the funds the VOC was originally capitalised with on its foundation. Despite this a proportion of VOC shareholders voiced concerns about the apparent act of piracy, and potential diplomatic repercussions from their allies, the English and the French. Thus, the Heeren XVII (the Seventeen Gentlemen or directors) of the VOC commissioned a treatise by the young 20-year-old Lawyer Hugo de Groot entitled *De Jure Praedae Commentariu* (or A Commentary of the Right of Booty) in which the many wicked crimes of the Portuguese were recounted, particularly those associated with the execution of the seventeen members of de Neck's crew at Macau. However, it seems that this treatise wasn't required. The English and the French were far from unhappy about the liberation of goods from the Portuguese. Their envoys instead indicated that the James I of England and Henry IV of France would be more than happy to receive gifts of porcelain from the ill-gotten gains. There is even evidence that the stepmother of Prince Maurice of Orange, Louise de Coligny, helped the French envoy, Paul Choart de Buzanval, to select one of the finest porcelain dinner services along with a table and two chairs from the cargo of the Santa Catarina. This meant that de Groot's treatise was pretty much forgotten about and that both the English and the French were perfectly comfortable with the concept of piracy in the East Indies or elsewhere if aimed at the Portuguese or the Spanish. This is presumably because both the French and the English were equally interested in profiting from similar activities. In fact, English sailors had already been active in waylaying Portuguese ships in the East Indies, for example Edward Lancaster in his first ill-fated voyage of 1591–1594 to the East indies is known to have foregone trade for the far easier option of preying on the Portuguese shipping to fill his hold. Sir Francis Drake in 1587 is believed to have commented on the ease with which the huge lumbering Portuguese carracks could captured in his raids on the Eastern Atlantic (Keay, *The Honourable Company*, 1991).

Up until this point the flow of Chinese porcelain into western Europe could barely be called a trickle, more of an infrequent drip. The arrival of the Santa Catarina and other Carracks cargoes liberated by the Dutch from the Portuguese in the East Indies represented the opening of the flood gates. The Dutch found a ready market, within a few years Dutch painters were depicting still life scenes containing underglaze blue and white Chinese kraak porcelain (carrack porcelain now known as kraak porcelain Fig. 3.1) for example, Floris van Dijck c1615 and his still life 'Fruit Bread and Cheese' exhibited at the Rijksmuseum Amsterdam had his fruit and olives displayed on Kraak Porcelain. For the Dutch, the Chinese blue and white porcelain became highly fashionable and collectible. In 1632 Louise de Coligny, mother of Fredrick Hendrik, Prince of Orange is documented as having a collection of 283 pieces of kraak porcelain kept in a closet in her apartment. A good proportion of this porcelain was no doubt derived from captured Portuguese carracks including the Santa Catrina. Louise de Coligny's daughter in law Princess Amalia van Solms (1602–1675), Fredrick's wife was also an ardent collector of fine Chinese porcelain and Japanese wares (Treanor, '"Une Abondance Extra Ordinaire": The Porcelain Collection of Amalia van Solms', *Early Modern Women*, 2014). By 1673 she had built a collection of over 1200 pieces of fine Chinese porcelain. However, Amalia viewed porcelain differently to her mother-in-law. Instead of sequestering it away in a closet with other valuable curios, she displayed it, using the porcelain to dress a room, with a preference for large, massed displays. She is known to have had an entire room added to her quarters at the Oude Hof (now Noordeinde Palace in The Hague) c.1648, in which she displayed around five hundred pieces of kraak porcelain. Such a display would have made a very clear statement regarding her status and good taste. It seems she displayed other components of her porcelain collection in a similar manner in her other residencies. This architectural style of dressing a room was also adopted by Mary daughter of James II of England. At the age of 15, she married William III Prince of Orange in St James's Palace in November 1677. William himself was the only grandson of Charles I of England and the match between the two of them functioned by uniting the two strongest claims to the throne of England excluding James II himself and his son also James (Price, 'An Incomparable Lady: Queen Mary II's Share in the Government of England, 1689–94', *Huntington Library Quarterly*). Following the wedding William and Mary returned to The Hague in December of that year. Mary would have been familiar with Chinese porcelain, but probably not in such an abundance as she encountered in the Court of the House of Orange. However, she like Amalia van Solms and Louise de Coligny, developed quite a taste for it. She also adopted the Dutch architectural style of displaying her porcelain collection. Mary's apartments in The Hague and in her nearby country house contained porcelain rooms, and her sumptuously decorated audience chamber was furnished with Chinese furniture and pictures along with a mirrored ceiling and a chimney piece laded with precious porcelain (Cocks, 'The Nonfunctional Use of Ceramics in the English Country House During the Eighteenth Century', *Studies in the History of Art*, 1989). The porcelain was arranged so that chimney piece and the porcelain fitted together as one architectural structure and all was beautifully reflected by the mirrored ceiling, which no doubted added to grandeur of the visual feast. Mary brought the garniture style to England with her in 1688, when she acceded the throne with William III.

Mary's father, James II, succeeded his older brother Charles II in 1685, despite him having converted to Catholicism in 1669. After his succession, he was perceived as introducing pro Catholic policies when he appointed Roman Catholics to high office. James then had a son, James, with his second wife, Mary Modena, who he had married by proxy in a Roman Catholic ceremony in 1673. Prior to the birth of the young James, all William III of Orange and his wife Mary had to do to

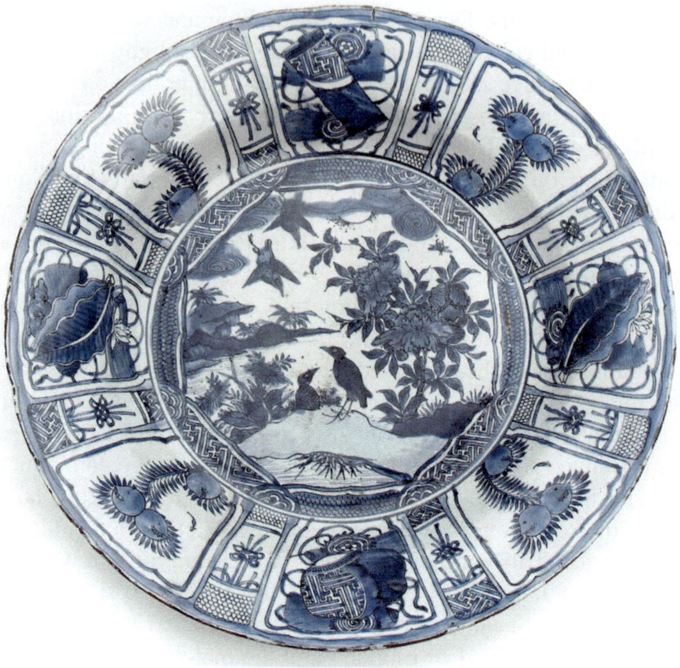

Fig. 3.1 A typical early seventeenth century Ming Dynasty Kraak porcelain underglaze cobalt blue dish with a flared scalloped rim decorated with alternate panels of peaches and auspicious motifs around a central octagonal panel depicting a river scene with crows and trees. *Reproduced with the Courtesy of Bonhams Auctioneers and Valuers, Bond Street, London*

inherit the throne of England was to wait. James II was perceived as prematurely elderly and wasn't believed capable of producing an heir. For William III of Orange, the birth of the young James represented the loss of the English Crown and also potentially the establishment of an unfriendly Roman Catholic state just across the North Sea and the Channel. Mutual dislike of the situation led Dutch and English conspirators to consider the Orange alternative (Speck, 'The Orangist Conspiracy Against James II', *The Historical Journal*, 1987). On November 5, 1668, William III of Orange landed with his army at Brixham in Devon and whilst marching unopposed towards London, James II fled with his wife and son to France. On January 22, 1689, James's escape to France was accepted as an abdication and William and Mary were offered the English Crown by Parliament with a Declaration of Rights giving them joint sovereignty, which in both cases was accepted resulting them becoming King William III and Queen Mary II of England. A Bill of Rights was also passed giving succession to Anne, Mary's sister, should William and Mary have no issue. The bill also barred Roman Catholics from the throne of England.

William and Mary first took up residence at Hampton Court whilst the Earl of Nottingham's House in Kensington was converted into a palace by Sir Christopher Wren. Mary brought her ceramics with her which, according to Daniel Defoe, consisted of a collection of delftware and fine Chinaware, the like of which had never been seen before. Apparently, she displayed her ceramics in the Long Gallery on every available surface. This was something very new in England and had a very significant impact. The concept of using fine porcelain garniture to decorate a room had landed. Defoe also notes that this habit was rapidly adopted by 'lesser mortals' who had started to acquire porcelain collections themselves which they piled upon specially constructed shelves, chimney pieces and any other surface available at huge cost (Cocks, 'The Nonfunctional Use of Ceramics in the English Country House During the Eighteenth Century', *Studies in the History of Art*, 1989). This 'contagion of China-fancy' as Samuel Johnson described it, spread like wildfire and the demand for porcelain in England increased almost as fast as the purses of those who bought it emptied (Finlay, 'The Pilgrim Art: The Culture of Porcelain in World History.' *Journal of World History*, 1998). This process was augmented by another import from China, namely tea.

3.9 The Power of Tea

Prior to the seventeenth century the choices available for refreshing drinks in Western Europe were somewhat limited, however as the seventeenth century was ushered in so were 3 new beverages: cacao (chocolate), coffee and tea. The first, cacao, was first encountered by Columbus in 1502 when he met 25 men paddling a Mayan trading canoe off the coast of what is now Honduras. Columbus believed that the cacao beans, which he called nuts were used as a form of currency. It was only later in 1523 the Spanish discovered these beans were used by the Aztecs to create an infused drink which was often further flavoured with vanilla, herbs, spices such as chili and flowers. By 1580, amongst the elite of the South American New Spain, women had become quite addicted to what was described as black chocolate. By the 1620s this addiction had spread to Spain and thus on to Spanish provinces. Demand was met by a shipment of around 5 tonnes of cacao beans a year from Plantations established by the Spanish in what are now Ecuador and Venezuela (Jamieson, 'The Essence of Commodification: Caffeine Dependencies in the Early Modern World', *Journal of Social History*, 2001). The Dutch became major importers of chocolate from the 1630 s and chocolate was eventually first advertised in England in 1640 via a translation of a Spanish document (Loveman, 'The Introduction of Chocolate into England: Retailers, Researchers, and Consumers, 1640–1730', *Journal of Social History*, 2013).

The concept of coffee houses was adopted in Western Europe from the coffee houses of Cairo and Istanbul. In these Eastern coffee houses, coffee was consumed socially, but women were excluded. Coffee was first commercially shipped into Europe by the Venetians from Makha (in the Yemen) in around 1640 as a medicinal drink, however, within a short space of time, coffee was consumed in coffee houses which like those in the near east excluded women. Coffee mostly came to England via independent immigrant Eastern merchants. For example, the first coffee house in England opened in around 1650 in Oxford, and it was established by what were described as Levantine merchants. At this time, coffee houses were also being opened across Europe by Eastern merchants. By the late seventeenth century there were lots of coffee houses and they had started to replace ale houses. The coffee houses steadily became meeting places of sober urban professional where business was conducted. For example. Lloyds of London was established in the Lloyds coffee house and the London Stock Exchange was launched in Jonathan's coffee house (Jamieson, 'The Essence of Commodification: Caffeine Dependencies in the Early Modern World', *Journal of Social History*, 2001). Despite chocolate and coffee both being available in coffeehouses, chocolate was not as popular, it has been postulated that this was partly due to the belief that chocolate had aphrodisiac properties, which suggested that it could lead to venery, however, it is probable that cost played a major factor. Chocolate was more difficult to prepare and was often made with milk or half milk with the addition of another expensive commodity, sugar. Thus, a dish of chocolate in a chocolate house or coffee house cost up to three times more than a dish of coffee and it is likely that this cost limited the demand.

Tea is first mentioned in Western texts by a Venetian writing in relation to a Persian merchant named Chaggi Memet (Hajji Mohammad in Persian), who arrived in Venice in the 1550s with a large cargo of rhubarb which he had bought in what is now Jiuquan, in Gansu province China. He apparently also described a plant he called Chia Catai, which was dried and widely used in China to make a drink by boiling it in water. The next reference to tea comes from Portuguese Dominican Friar Gaspar da Cruz (c1520–1570). He obtained permission in 1556 to visit Canton, where he stayed for over a month. In his book *Tractado em que se cotam muito porestenso as cousas da China* he recounted his experience of tea. It seems that visitors, when entering a house as guest were first offered a porcelain cup filled with a warmish red medicinal drink called cha, which was made from a bitter herb (Zanini, 'Before the Arrival of Tea in Europe. The Chinese Beverage in Western Sources Prior to the 7th Century', *The Tea Gardens of Dazhangshan*, 2019).

The habit of drinking tea was introduced to Europe by the Dutch, they made their first shipment of tea to the Netherlands in 1610, and by 1637, the VOC directors were requesting that some jars of China could be transported with the tea in every ship. In that year some 25,000 of these China jars (Chinese porcelain tea bowls) we ordered, clearly demand for tea was increasing as was the demand for Chinese porcelain to drink it from (Volker, Porcelain and the Dutch East India Company as Recorded in the Dagh-registers of Batavia Castle, Those of Hirado and Deshima, and Other Contemporary Papers, 1602–1682, 1954). By 1641, Dutch naval power had reached its peak in the far East having finally broken the Iberian monopoly of trade by taking Malacca from them, something they failed to do three times previously. Tea first arrived in England in 1650, and Samuel Pepys in his diary of that year wrote that he had tried '*tee, a china drink*' in a coffee house because it as a drink had come to England to stay as had the cups to drink it from along with other tea wares such as saucers, jugs and basins. By this point huge shipments of tea, along with porcelain of every form, were arriving in the Dutch ports.

In around 1621 as the 1607 truce between the United Netherlands and Spain came to an end, the Dutch were very aware that they needed a base close to Guangzhou similar to that of Macau. They initially came to the conclusion that the obvious thing to do was take Macau by force from the Portuguese, basically killing two birds with one stone, eliminating a competitor in the market they happened to be at war with, whilst gaining an already thriving trading centre. The Dutch attempt on Macau 1622 failed abysmally, despite the Portuguese defenders being heavily outnumbered (Porter, '"The Past Is Present": *The Construction of Macau's Historical Legacy', History and Memory*, 2009). As a consequence, the Dutch sought an alternative and instead established a colony on Formosa (Modern Day Taiwan) in 1624 from which they harried the Portuguese ships and establish a competing nanban trade in silk and silver between the Chinese and the Japanese. This was initially quite difficult seeing as some of the indigenous population, particularly those from the village of Mattau, resisted their arrival, and Japanese pirates had also established themselves in Formosa ('The Rise and Fall of Dutch Taiwan, 1624–1662: Cooperative Colonization and the Statist Model of European Expansion.' *Journal of World History*, 2006; Andrade, *How Taiwan Became Chinese Dutch, Spanish and Han Colonization in the Seventeenth Century: Dutch, Spanish, and Han Colonization in the Seventeenth Century*, 2008). The impact of the Japanese pirates was significantly reduced in c1635, as a result of the Tokugawa bakufu (Military government of the Japanese Shogunate) under the Shogun Tokugawa Lemitsu prohibiting on pain of death Japanese subjects from leaving or re-entering Japan. Under this ban the Japanese encouraged trade through non-Japanese partners via formal agreement with Korea, the Ryukyu Island chain which includes Okinawa, and the Dutch in Formosa. The Portuguese were not included as trading partners because of their activities in promoting Catholic Christianity in Japan mediated by the Jesuits (Hang, 'The Shogun's Chinese Partners: The Alliance between Tokugawa Japan and the Zheng Family in Seventeenth-Century Maritime East Asia', *The Journal of Asian Studies,* 2016). This allowed an unopposed Dutch takeover of the triangular silk for silver nanban trade established between Nagasaki and China by the Portuguese following Saint Francis Xavier's visit to Japan in 1549. The Dutch acquisition of this trade enabled them to become the sole suppliers of Japanese goods and Chinese goods to Western Europe. This in turn led to further increases in the supply of Kraak porcelain to the United Netherlands, which further fed the growing Western fashion for collecting porcelain (Hang, 'The Shogun's Chinese Partners: The Alliance between Tokugawa Japan and the Zheng Family in Seventeenth-Century Maritime East Asia', *The Journal of Asian Studies*, 2016). Unfortunately, internal trouble had been brewing in the Ming Empire since the early seventeenth century in the form of unrest in Manchuria.

Manchuria at this time was inhabited by various tribes who claimed decent from the Jurchens. In 1600 the assorted Jurchen tribes of Manchuria in North-eastern China were united by Nurhaci (1559–1626) (Moll-Murata, 'Tributary Labour Relations in China During the Ming-Qing Transition (Seventeenth to Eighteenth Centuries' *International Review of Social History*, 2016). Nurhaci declared war on the Ming Empire in around 1618. In many respects this was a fortuitous time to declare war on the Ming Empire. A series of drought, floods, famines and contagious epidemics struck China between 1627 and 1643 resulting in a peasant's uprising, which allowed the Manchu now under Hong Taiji to conquer Beijing in 1644. Hong Taiji died before reaching the city and his son Fulin became the first Qing emperor to rule China from Beijing in 1644 (Rowe, *China's Last Empire: The Great Qing*, 2012). The Manchu's continued to expand their Chinese empire South, as the Ming forces retreated. However, the native Ming were supported by the Zheng family headed by Zheng Zhilong. He was born in the Southern Fujisan prefecture of Quanzhou, he spent some time in Macau, before arriving in Hirado and marrying Tagawa Matsu the daughter of a minor samurai retainer of the powerful shuinsen merchant (Japanese licenced trader) Li Dan. Zheng Zhilong then entered the service of the VOC in Formosa as a translator but maintained close links with Japan eventually inheriting the business of Li Dan, which he then expanded to the extent that it rivalled that of the Dutch in Formosa. Zheng Zhilong used the funds this trade generated to support the Ming cause and it was his son Koxinga by his Japanese wife, who in 1662 led the final stages of the Ming overthrow of the Dutch in Formosa. Formosa remained under Ming control supported by the Zheng family and their immense trading activities until 1683 when the Qing forces final took it (Andrade, 'The Rise and Fall of Dutch Taiwan, 1624–1662: Cooperative Colonization and the Statist Model of European Expansion', *Journal of World History*, 2006; Hang, 'The Shogun's Chinese Partners: The Alliance between Tokugawa Japan and the Zheng Family in Seventeenth-Century Maritime East Asia', *The Journal of Asian Studies*, 2016). The Dutch enterprise in China never really recovered from the loss of Formosa in 1662. Following the Qing capture of Formosa in 1683. The Dutch returned to trade in Canton in 1685, when the Quing emperor Kangxi opened the port to trade from all nations, but by this time the triangular trade between China and Japan had ceased due to diminishing stocks of silver in Japan (Morse, 'British Trade in China', *Journal of the Royal Society of Arts*, 1919; Hang, 'The Shogun's Chinese Partners: The Alliance between Tokugawa Japan and the Zheng Family in Seventeenth-Century Maritime East Asia', *The Journal of Asian Studies*, 2016) making their original trade no longer viable. However, they did continue a trade in tea and porcelain.

3.10 The English East India Company (EIC) in China

Up until the opening of Canton for trade with foreign Nations in 1685, the English East India East India Company was conspicuous by its absence in China. The EIC sent only a small number of ships to Guangzhou (Canton) between 1600 and 1685, one in 1637, a second in 1664, a third in 1674 followed by 2 sent to Amoy, one in 1670 and the second in 1677. By 1700, the EIC had started to send ships every year to Canton. This was driven by tea becoming an exceedingly popular drink in the English market. Tea was perceived as a civilising drink more likely to be consumed by women, in elegant tea gardens such as the Golden Lion established by Twinings of London in 1706, and at home. There was even a fear in Britain that men who drank tea at home with the ladies could undergo feminisation reducing the country's military capacity.

By 1720 the EIC and the VOC between them shipped around 180 tonnes of tea, by 1740, that total had risen to 360 tonnes and by 1765 the two companies shipped a total of 720 tonnes of tea from Canton. Following the opening up of Canton in 1685, other East India companies started to trade with the Chinese (Jamieson, 'The Essence of Commodification: Caffeine Dependencies in the Early Modern World', *Journal of Social History*, 2001). By 1728 the French East India Company had a factory in Canton. They were followed by the East India companies of virtually every Western nation including the Danish, Swedish and by the North Americans following their successful war of independence from the British c1775-1783. Of the Nations sending ships to Canton, the British acquired a much larger share of the trade, a Swedish correspondent in 1751 recorded that of the ships anchored just off the coast of Canton at Whampoa Island, nine were British, four were Dutch two were Swedish along with one Danish ship (Morse, 'British Trade in China', *Journal of the Royal Society of Arts*, 1919). The English ships would have included a good proportion of the huge Indiamen capable of carrying up 1200 tonnes of cargo meaning that the English ships accounted for around three fifths of total tonnage due for export. Grabbing the larger share of tea from China became an official policy of the EIC in around 1730, with the companies Court of Directors issuing instructions that the EIC should outcompete their European competitors making them 'sick of their voyages for tea'. The EIC were very successful in doing this, by 1789 there are records indicating that there were 86 British ships, 15 American, 5 Dutch, 1 French, 1 Danish and 3 Portuguese ships anchored at Whampoa Island off the coast of Canton. In terms of tonnage, the British share of goods in those ships had risen to around 75% (Morse, 'British Trade in China', *Journal of the Royal Society of Arts*, 1919). The consumption of tea in England continued to increase, by around 1801–1810 80% of the tea transported by ship from Canton ended up in Britain via the English ships. The consumption of tea in Britain continued to rise but the British share of the market decreased a bit because of competition with the USA (Chung, 'The Britain-China-India Trade Triangle (1771–1840)', *Proceedings of the Indian History Congress*, 1974). By 1830, and 1850, the Dutch and the British had established their own tea plantation in Java and India respectively ending the Chinese monopoly of Tea (Jamieson, 'The Essence of Commodification: Caffeine Dependencies in the Early Modern World', *Journal of Social History*, 2001). A number of factors drove the expansion of the tea market. Firstly, there seemed to be an insatiable thirst for tea in the British market. Tea had initially been introduced as a rare and expensive commodity consumed by the elite, however, it became a drink of the masses and eventually provided inexpensive and non-alcoholic refreshment for the growing population of poor factory workers during the Industrial revolution. Secondly, the margins on Tea were high, up to four-fold. For example, between 1771 and 1780 £2,649,589 worth of tea bought in Canton could be sold in London for £10,277,592. Thirdly, tea could be imported with all of the Chinese porcelain paraphernalia of tea drinking, which could be sourced cheaply in Canton and sold at a considerable profit back home in Britain. So, the English, like the Dutch, not only imported tea from China, but they also imported Chinese luxuries such as silks and porcelain in support of the tea trade and to meet the requirements of a rapidly expanding home market. Vast quantities of porcelain were shipped into

Britain by the EIC and privately by their officers. The profit margins were very large, with porcelain being sold in England at up to four times the price paid for it in Canton. The import of tea also helped to enhance the trade of commodities other than just tea and porcelain. For example, tea was commonly sweetened with West Indian sugar.

To fund the purchase of Chinese goods, the EIC functioned within a new and highly profitable triangular trade system whereby good acquired by the company in India such as cotton and opium were shipped to Canton to be sold in exchange for tea, porcelain and other Chinese goods. These Chinese goods were then shipped back to England (Morse, 'British Trade in China', *Journal of the Royal Society of Arts*, 1919); Chung, 'The Britain-China-India Trade Triangle (1771–1840)', *Proceedings of the Indian History Congress*, 1974). Private employees of the EIC could also take advantage of this triangular trade system and invest in Indian products such as cotton and opium, which was then sold in Canton in exchange for tea, silks and porcelain, this in turn was shipped back to England. Vast fortunes were generated in this manner and some directors of the EIC became incredibly wealthy. It was these individuals that provided the impetus for the genesis of English armorials on ceramics by commission Chinese decorators in Canton to produce English armorials on Chinese porcelain. Sir Henry Johnson was the first to do this, he commissioned 6 large underglaze blue and white jardinières decorated with eight armorials, 4 each for himself and his second wife Martha Lovelace, on or around the date of their wedding in 1692. Sir Henry was the son of Sir Henry Johnson, an EIC shareholder and Royal Navy/EIC shipbuilder who was buried in the EIC Chapel following his death in 1683. The younger Henry like his father retained shares in 33 company ships and no doubt used his influence with the company to acquire and fund his armorial porcelain. What is interesting about the decoration of these jardinières is that although the arms are recognisable, they are not really very finely painted. This probably reflects one of the problems associated with underglaze Chinese cobalt blue and white porcelain at this time. By the late seventeenth century, the Chinese were still using locally produced cobalt high in manganese to decorate their ceramics. This cobalt had a tendency to run a little bit during firing which reduced edge definition of the decoration. The underglaze blue also had other drawbacks, armorials normally consist of a range of colours and those colours are very important. Underglaze blue armorials are monochromatic and not really the ideal medium to use in the generation of polychromic armorial ceramics. This didn't seem to matter to those who wished to demonstrate their wealth and power in the late seventeenth century. Henry Johnson was making a statement with his impressive Chinese armorial jardinières.

The shipment of vast quantity of goods was achieved by the EIC building huge ships, the East Indiamen. By nature, these ships were a bit top heavy and a bit leaky, and the removal of sea water via bilge pumps would have been an unending task. Bearing in mind that one of the major commodities transported from Canto to England was tea and the large volumes of tea are not overly heavy, what was required for the return Journey was a heavy, water-resistant saleable commodity to accompany the tea. Porcelain was the perfect product, although reasonably fragile, it is heavy, and very resilient to the damaging effects of saltwater. Thus, porcelain was used as additional ballast. It was carefully packed into cases that fitted tightly together into a stable dry floor over the ballast already contained within the ships, so although bilge water would accumulate in the hold, the floor formed from the porcelain packing cases would remain dry and thus provide the perfect surface onto which the more perishable goods starting with the lowest grade course black tea (Bohea tea), followed the Singlo tea (late season green tea), other fine teas and finally silks, along with other less resilient commodities.

3.11 Qing and the Adoption of European Enamels

A special landmark in the history of Chinese porcelains was the preparation and adoption of a compatible cobalt blue enamel which would not diffuse and thereby spoil the finely executed applied decoration when the porcelain was fired.

Ming blue and white porcelain, although beautiful had some drawbacks, and most of these revolved around the cobalt their potters used to decorate their ceramics. During the Yuan period when blue and white decoration first appeared on Chinese porcelain, Su Ma Li Qing cobalt was sourced in Persia. As previously discussed, this cobalt had a high iron (Fe) content along with traces of arsenic (As) and sulfur (S). However, due to isolationist policies during the later Ming Emperor Xuande's reign (1424–1435), the supply of cobalt was first changed to Su Bo Ni Qing cobalt from the vassal state of Sumatra and then changed again during the Ming Emperor Chenghua's reign (1465–1487) to Peng Deng Qing, a locally sourced cobalt from Jiangxi Provence. Both these sources yielded cobalt high in manganese and low in iron, but cobalt rich in manganese was of a paler blue and was prone to some diffusion into the glaze during firing.

The kilns at Jingdezhen were destroyed in the civil war that raged in China after the collapse of the Ming Dynasty in the mid-seventeenth century and the influx of Japanese porcelains imported into Europe by the Dutch VOC then started to make their impact, firstly through their blue and white decorated Arita porcelain but even more so with their polychrome, red, blue and gold, Imari and Kakiemon wares (Medley, *The Chinese Potter: A Practical History of Chinese Ceramics*, 1989). In the late 1680s, following the civil war hostilities, the re-establishment of the Jingdezhen manufacturing kilns and the creation of new Imperial enamelling workshops in the Forbidden City in Beijing were undertaken by the Emperor Kangxi, the second Emperor of the Qing Dynasty in China. This was further enhanced around 1700 by the adoption of a new superior deep underglaze blue cobalt pigment consisting of an iron rich cobalt with traces of arsenic probably arising from mines in Saxony, along with the use of other over glaze coloured enamel pigments first revealed in China by the French Jesuit mission to Beijing. The Chinese first encountered these enamels on metal and glass diplomatic gifts from King Louis XIV presented to the Emperor Kangxi (reigned 1661–1722), as recorded by Father Xavier d'Entrecolles SJ. The Jesuits shared the secrets of these enamels with the Chinese, which directly facilitated the introduction of their use in the production of the new Chinese polychrome enamelled porcelain. This allowed the Chinese to compete directly with the Japanese polychrome Arita and Kakiemon wares that were then becoming so desirable in Europe (Curtis, *Glass Exchange Between China and Europe, 1550–1800*, 2000; Xu, *Europe-China–Europe: The Transmission of the Craft of Painted Enamels in the 17th and 18th Centuries*, 2015; Le Corbellier & Frelinghuysen, "Chinese Export Porcelain" 2003). In the West these polychrome porcelains acquired the names of *famille jaune* (yellow), *famille noire* (black), *famille rose* (pink) and *famille verte* (green) (Kerr, Needham Wood, *Science and Civilisation in China: Volume 5, Chemistry and Chemical Technology, Part 12, Ceramic Technology*, 2004). It is believed that the transition to polychrome enamelling from the standard blue and white decoration began in earnest in 1729 when the Emperor Yongzheng instructed the enamellers at Jingdezhen to utilise some 18 new pigments and enamels, he had acquired from the French Jesuits for the decoration of their porcelains. Clearly, the advent of polychrome decoration led to the genesis of tinctured heraldic armorial bearings on Chinese porcelains in the early eighteenth century which hitherto had been restricted to blue and white designs only.

3.12 The Demise of English Trade in Chinese Porcelain

The EIC involvement in the Chinese porcelain trade continued up until 1798, by which time a number of factors had contributed to Chinese porcelain and Chinese armorials becoming less popular. Firstly, by around 1850 English ceramics manufacturers had started to manufacture porcelain and other ceramics that could be considered as viable alternatives to Chinese Porcelain. Some of the English manufacturers by the early nineteenth century employed specialist decorators for the manufacture of armorial Porcelain. For example, according to Twitchett (Twitchett, *Derby Porcelain 1748–1858: An Illustrated Guide*, p. 104); (Denyer et.al., *The Pendock Barry Porcelain Service: A Forensic Evaluation*, 2023) Thomas Soar undertook the majority of gold decoration on heraldic works at the Derby factory until 1810, when he then set up an independent business in which continue decorating porcelain services with armorials for the local gentry. Secondly, the delay between ordering a Chinese porcelain service decorated with a specific armorial and receiving it could be up to three to 4 years or more if based according to Copeland's (Copeland, *Spode's Willow Pattern & Other Designs After the Chinese*, 1980) estimate of 22 months for the round trip plus a year or so for the communication of the commission to Jingdezhen and the delivery of the finished product back to Canton. Thirdly Chinese decorators tended to make errors, which for a generic pieces of tea ware would not be a problem but could be for an armorial in which an error could impact on the meaning of the armorial. Finally, tax, the duty on Chinese porcelain in 1787 was 47.5%, by the time of the Napoleonic Wars in 1799, the tax had risen to over 109% of the actual value of the porcelain. This made Chinese porcelain even more expensive than it had been. Needless to say, demand dropped, as evidenced by the EIC's own records which documented a drop in their own sales of Chinese porcelain from £23,630 in 1763 down to £7,332 in 1796. Figures for private trade in imported Chinese porcelain for the same period were even worse with a drop from £49,351in 1763 to £1,613 in 1796. However, tea trade remained unaffected, and although very heavily taxed at a rate of 119% in 1784, continued to arrive in Britain via tea smugglers, which resulted in the EIC being heavily undercut by a black market in tea. William Pitt the Younger dealt with this problem by introducing the Commutation act in 1784 which reduced the tax on tea to 12.5% but compensated for the loss in revenue by introducing the window tax. Immediately following this change in taxation, the consumption of tea in Britain rose from around 12 million lbs (5 million via the EIC and 7 million via the smugglers) to around 15 million lbs per annum. By this time tea had become an alternative beverage to alcohol and is credited with lowering alcohol consumption in all social classes (Copeland, *Spode's Willow Pattern & Other Designs After the Chinese* 1980).

References

Abulafia, D. (2007). The Last Muslims in Italy. *Dante Studies, with the Annual Report of the Dante Society, 125*, 271–287.

Andrade, T. (2006). The Rise and Fall of Dutch Taiwan, 1624–1662: Cooperative Colonization and the Statist Model of European Expansion. *Journal of World History, 17*(4), 429–450.

Andrade, T. (2008). *How Taiwan Became Chinese Dutch, Spanish and Han Colonization in the Seventeenth Century: Dutch, Spanish, and Han Colonization in the Seventeenth Century.* Columbia University Press.

Blussé, L. (1988). Brief Encounter at Macao. *Modern Asian Studies* 22(3), 647–664.

Brook, T. (2015, November). Trading Places: The Santa Catarina Brought the First Large Shipment of Porcelain from the Far East to Europe, Creating a Demand for New Luxury Goods, Which Feature in Paintings of the Dutch Golden Age. *Apollo, 182*(636), 70.

Brown, R. M. (2009). *The Ming Gap and Shipwreck Ceramics in Southeast Asia: Towards a Chronology of Thai Trade Ware.* Siam Society.

Chan, H. L. (1968). The "Chinese Barbarian Officials" in the Foreign Tributary Missions to China during the Ming Dynasty. *Journal of the American Oriental Society, 88*(3), 411–418.

Chadhuri, K. (1992). The Portuguese Maritime Empire, Trade, and Society in the Indian Ocean During the Sixteenth Century. *Portuguese Studies, 8*, 57–70.

Chen, X. C., Guo, Y. Y., & Chen, H. (1993). Discussion for the Origin of the Cobalt-Pigment on Chinese Yuan Blue-and-White Porcelain. *Chinese Ceramic*, 5, 57 (in Chinese).

Chung, T. (1974). The Britain-China-India Trade Triangle (1771–1840). *Proceedings of the Indian History Congress, 34*(1974), 77–91.

Cocks, A. S. (1989). The Nonfunctional Use of Ceramics in the English Country House During the Eighteenth Century. *Studies in the History of Art, 25*, 195–215.

Cooper, M. (1972). The Mechanics of the Macao-Nagasaki Silk Trade. *Monumenta Nipponica, 27*(4), 423–433.

Copeland, R. (1980). *Spode's Willow Pattern & Other Designs After the Chinese.* Cassell Ltd.

Cowell, M, & Zhang F.K. (2001). *Analyses and Source of the Cobalt-Blue Pigments Employed on Chinese Ceramic, Ming Ceramic* (p. 604). The British Museum Press.

Demandt, M. H. S. (2019). Tracking Proto-Porcelain Production and Consumption in the Dongjiang Valley of Bronze Age Lingnan. *Cambridge Archaeological Journal., 29*(4), 671–689.

Denyer, R. L., Denyer, M. C. T., & Edwards, H. G. M. (2023). *The Pendock Barry Porcelain Service: A Forensic Evaluation.* Springer Nature.

Li, Y., Zhang, B., Cheng, H., & Zheng, J. (2015). The Earliest Chinese Proto-Porcelain Excavated from Kiln Sites: An Elemental Analysis. *PLoS ONE, 10*, 11.

Feng, D., & Baoru, S. (2008). Further Study of Sources of the Imported Cobalt-Blue Pigment Used on Jingdezhen Porcelain from Late 13 to Early 15 Centuries. *Science in China Series E: Technological Sciences, 51*(3), 249–259.

Finlay, R. (1991). The Treasure-Ships of Zhenge He: Chinese Maritime Imperialism in the Age of Discovery. *Terrae Incognitae, 23*, 225–241.

Finlay, R. (1998). The Pilgrim Art: The Culture of Porcelain in World History. *Journal of World History, 9*(2), 141–187.

Gaastra, F. S. (1997). The Dutch East India Company A Reluctant Discoverer. *The Great Circle, 19*(2), 109–123.

Gerritsen, A. (2020). *The City of Blue and White*. Cambridge University Press.

Gutiérrez, A., Gerrard, C., Zhang, R., & Guangyao, W. (2021). The Earliest Chinese Ceramics in Europe? *Antiquity, 95*(383), 1213–1230.

Hang, X. (2016). The Shogun's Chinese Partners: The Alliance Between Tokugawa Japan and the Zheng Family in Seventeenth-Century Maritime East Asia. *The Journal of Asian Studies, 75*(1), 111–136.

Hao, Z. (2011). *Macau History and Society*. Hong Kong University Press.

Harrison, T. (1958). The Ming gap and Kota Batu, Brunei. *Sarawak Museum Journal, 8*, 273–277.

Huang, S., Freestone, I. C., Shi, Z., & Qian, G. (2023). The Path to Porcelain: Innovation in Experimental White Stoneware from Luoyang. *Journal of Archaeological Science, 158*, 105839.

Jamieson, R. W. (2001). The Essence of Commodification: Caffeine Dependencies in the Early Modern World. *Journal of Social History, 35*(2), 269–294.

Kammerer, A. (1944). La Découverte de La Chine Par Les Portugais Au XVIème Siècle et La Cartographie Des Portulans. *T'oung Pao, 39*, 38–39.

Keay, J. (1991). *The Honourable Company*. Harper Collins.

Kerr, R, Needham, J., & Wood, N. (2004). *Science and Civilisation in China: Volume 5, Chemistry and Chemical Technology, Part 12, Ceramic Technology*. Cambridge University Press.

Le Corbellier, C., & Frelinghuysen, A. C. (2003). Chinese Export Porcelain. *The Metropolitan Museum of Art Bulletin,* 60(Winter), 1–60.

Li, J., Leung, P. L., Yu, K. N., & Wu, R. (2008). Ceramics in China. In H. Selin (Eds.), *Encyclopaedia of the History of Science, Technology, and Medicine in Non-Western Cultures*. Springer.

Liu, L., & Chen, X. (2012). The Archaeology of China: From the Late Palaeolithic to the Early Bronze Age. *Cambridge University Press, 2012*, 253–296.

Loveman, K. (2013). The Introduction of Chocolate into England: Retailers, Researchers, and Consumers, 1640–1730. *Journal of Social History, 47*(1), 27–46.

Meicun, L., & Zhang, R. (2018). A Chinese Porcelain Jar Associated with Marco Polo: A Discussion from an Archaeological Perspective. *European Journal of Archaeology., 21*(1), 39–56.

Moll-Murata, C. (2016). Tributary Labour Relations in China During the Ming-Qing Transition (Seventeenth to Eighteenth Centuries). *International Review of Social History, 61*, 27–48.

Morse, H. B. (1919). British Trade in China. *Journal of the Royal Society of Arts, 3497*(68), 17–28.

Chin-Keong, N. (2017). *Trade, the Sea Prohibition and the "Folgji" 1513–50*. China's Maritime Southeast in Late Imperial Times. NUS Press.

Nowell, C. E. (1936). The Discovery of Brazil—Accidental or Intentional. *The Hispanic American Historical Review, 16*, 311–338.

O'Rourke, K. H., & Williamson, J. G. (2009, August). Did Vasco da Gama Matter for European Markets? *The Economic History Review, 62*(3), 655–684 (New Series).

Parker, G. (1970). Spain, Her Enemies and the Revolt of the Netherlands 1559–1648. *Past & Present, 49*, 72–95.

Peters, M. A. (2021). The Ancient Silk Road and The Birth of Merchant Capitalism. *Educational Philosophy and Theory, 53*(10), 955–961.

Porter, J. (2009). "The Past Is Present": The Construction of Macau's Historical Legacy. *History and Memory, 21*(1), 63–100.

Puga, D., & Treffler, D. (2014, March). International Trade and Institutional Change: Medieval Venice's Response to Globalization. *The Quarterly Journal of Economics, 129*(2), 753–821.

Raphael, O. (1932). Chinese Porcelain Jar in the Treasury of San Marco, Venice. *Transactions of the Oriental Ceramic Society, 10*, 13–15.

Price, R. (2012). An Incomparable Lady: Queen Mary II's Share in the Government of England, 1689–94. *Huntington Library Quarterly, 75*(3), 307–326.

Ricklefs, M. C. (2001). *A History of Modern Indonesia Since 1300* (3rd ed.). Palgrave Macmillan.

Rowe, W. T. (2012). *China's Last Empire: The Great Qing*. Harvard University Press.

Shokoohy, M. (1998). The Town of Cochin and Its Muslim Heritage on the Malabar Coast, South India. *Journal of the Royal Asiatic Society, 8*(3), 351–394.

Wu, S., & Gao, S. L. (2021). Research on the Kiln Site Protection and Utilization of Early Yue in Zhejiang Based on Protection First. *E3S Web of Conferences, 267*, 01033.

Wu, S., Li, Y., Zhang, B., Cheng, H., Chun Chen, C., & Zheng, J. (2017). PIXE Analysis of Early Yue Wares from Shangyu and Nanjing (2nd–6th centuries AD). *Nuclear Instruments and Methods in Physics Research Section b: Beam Interactions with Materials and Atoms, 412*, 180–184.

Sluiter, E. (1942). Dutch Maritime Power and the Colonial Status Quo. *Pacific Historical Review, 11*, 29–41.

Speck, W. A. (1987). The Orangist Conspiracy Against James II. *The Historical Journal, 30*(2), 453–462.

Tai Y. S. (2011). Ming Gap and The Revival of Commercial Production of Blue and White Porcelain in China. *Bulletin of The Indo-Pacific Prehistory Association, 31*, 85.

Tai, Y. S., Daly, P., Mckinnon. E. E., Parnell, A., Feener, R. M., Majewski, J., Ismail, N., & Sieh, K. (2020). The Impact of Ming and Qing Dynasty Maritime Bans on Trade Ceramics Recovered from Coastal Settlements in Northern Sumatra, Indonesia. *Archaeological Research in Asia, 21*.

Tate, H. R. (1944, April). A mediaeval navigator: Vasco da Gama. *Journal of the Royal African Society, 43*(171), 61–65.

Treanor, V. (2014). "Une Abondance Extra Ordinaire": The Porcelain Collection of Amalia van Solms. *Early Modern Women, 9*(1), 141–154.

Twitchett, J. (2002). *Derby Porcelain 1748–1858: An Illustrated Guide* (p. 104). Antiques Collectors Club, Ltd.

Volker. T. (1954). *Porcelain and the Dutch East India Company as Recorded in the Dagh-Registers of Batavia Castle, Those of Hirado and Deshima, and Other Contemporary Papers, 1602–1682* (p. 49). Mededalingen Van Het Rijksmuseum Voor Valkenkunde.

Weatherford, J. (2018). The Silk Route from Land and Sea. *Humanities, 7*, 32.

Willis, C. (2012). Captain Jorge Alvares and Father Luis Frois S.J. Two Early Portuguese Descriptions of Japan and the Japanese. *JRAS,22*(2, Series 3), 391–438.

Wolff, R. S. (1998). Da Gama's Blundering: Trade Encounters in Africa and Asia During the European 'Age of Discovery', 1450–1520. *The History Teacher, 31*(3), 297–318.

Hao, Y., Wang, F., Qiu, B., Liu, P., & Feng, B. (2022). Study on Additives Assisted Reduction of Celadons. *Open Ceramics, 12*, 100312.

Yule, Sir H., & Cordier, H. (1903). *Travels of Marco Polo*. 3rd ed. John Murray.

Chapter 4
European Armorials in Porcelain and Earthenware from the Sixteenth Century

Abtract The earliest examples of Chinese armorial porcelains imported into Europe and their European armorial analogues in porcelain, faience and delftwares that were manufactured in simulation from the sixteenth century are reviewed. The use of the earliest Chinese and European porcelains and ceramics as diplomatic gifts and the reasons underlying their adoption for this purpose are proposed. A survey of the earliest European manufactories of porcelain and ceramics at Florence (Medici), Fulham, Rouen, St Cloud and Meissen is made, and some noted early armorial artefacts and services that have survived are highlighted. Special named armorial services in Meissen porcelain, and other armorials such as the *Stepney* service, *Gripsholm* service and the *Horatia* service are considered in some detail.

Keywords Chinese porcelain imports · Early European porcelain manufactories · Porcelain diplomatic gifts · Medici porcelain · Fulham Pottery · Rouen porcelain · St Cloud porcelain · Meissen porcelain · *Gripsholm* service · *Stepney* service · *Horatia* service

> There is nothing more deceptive than an obvious fact.
>
> Sherlock Holmes, The Boscombe Valley Mystery.
>
> Sir Arthur Conan Doyle, 1891

> In solving a problem of this sort, the grand thing is to be able to reason backwards. There are few people who if you told them the result would be able to evolve from their own inner consciousness what the steps were which led up to that result. This is what I mean when I talk of reasoning backwards, or analytically.
>
> Sherlock Holmes, A Study in Scarlet.
>
> Sir Arthur Conan Doyle, 1887.

R. L. Denyer et al., *Armorial Porcelain*, https://doi.org/10.1007/978-3-031-63745-2_4

4.1 Early Porcelain Manufacture in China and Its Importation into Europe

Although the manufacture of earthenwares had been known in several cultures for several thousand years hitherto, the first production of porcelain in China is generally accredited to the Tang Dynasty (618–906 CE) although there is an ongoing debate that several attempts must have been made at its manufacture before this time but with varying degrees of success. Certainly, during the following Song Dynasty (960–1279 CE) quality porcelain ceramics were starting to appear outside China and these new materials were also being talked about by explorers and adventurers to the Far East. Marco Polo (1254–1324), the Venetian adventurer and traveller (Marco Polo, *Il Milione*, 1928), who arrived in China in 1271 CE and was received at the Yuan Dynasty Court of the Mongol Emperors Kublai Khan-Shizu (1260–1294) and Temur Khan—Chengzong (1294–1307) is believed to be the first Westerner to describe Chinese porcelain and to have brought a specimen of *qingbai* porcelain back to Europe. However, the first literary record of the existence of Chinese porcelain outside China was reported much earlier by Suleiman, an Arabic merchant trading with China, who pre-dated Marco Polo by four centuries and who in 851 CE during the Chinese Tang Dynasty, said:

> They have in China a very fine clay with which they make vases which are as transparent as glass; water is seen through them. These vases are made of clay.

Clearly, Suleiman must be referring here to porcelain and not to glazed celadons and earthenwares which are not translucent (Edwards, *Porcelain Analysis and the Forensic Attribution of Ceramic Specimens*, 2022a). Although many ceramic historians believe that the early ceramics of the Tang Dynasty were more like faience in character and really, therefore, comprised a low temperature fired earthenware substrate with an applied vitreous metal oxide glaze, some production of a translucent ceramic was also being undertaken at higher firing temperatures successfully during this period but perhaps this was achieved only in very limited quantities (Medley, The Chinese Potter: A Practical History of Chinese Ceramics, 1989). C A recently excavated wreck of an Arab dhow which sank in 830 CE in the Singapore Straits, off Belitung Island, with a cargo of ceramics *en route* for the Abbasid Arabic Dynasty in north Africa yielded a large ceramic cargo of 60,000 pieces, comprising mostly Tang Dynasty glazed stonewares with some Yue celadons but contained about 300 pieces (representing 0.5% of the cargo) of a very fine Xing white porcelain from Heibei, which has been called the "*Tang Treasure from the Sea*". This physical evidence indicates that even at this early period in history the Arabs were importing novel Chinese ceramics, including porcelain, long before it was even known in Europe.

Persian inspired blue and white porcelain was first manufactured in China in the Yuan dynasty during the early to mid fourteenth century, as evidenced by the Chinese David vases dating from the Yuan period, c.1351 (Gerritsen, *The City of Blue and White*, 2020). This was a style the later Ming Dynasty adopted (1368–1644 CE), when some very fine quality Chinese porcelains decorated with applied cobalt blue on a white substrate of exceptionally good translucency were made. These porcelains were imported to Europe by the Honourable East India Company (HEIC; founded in 1600), the Dutch East Indies Company (VOC; Vereenijde Oost-Indische Compagnie, founded in 1602) and independent or chartered Portuguese navigators (following the demise of their own Companhia da India Oriental, CIO—also known as the Casa da India—in 1633). The porcelain proved very popular throughout the sixteenth century and beyond into the seventeenth and eighteenth centuries (Carswell, *Blue and White Chinese Porcelain*, 2000). There was a similar but smaller operation being carried out by the Swedish East India Company (SOIC; Svenska Ostindiska Companiet) but this only started much later in 1731.

One of the problems with Ming blue and white porcelain from the Ming Emperor Xuande's reign (1424–1435 CE) onwards was a change in the source of cobalt driven by isolationist policies. This led to a change from a Persian source of iron rich cobalt with a trace of arsenic to a more local source of cobalt low in iron and high in manganese (Cowell & Zhang, *Analyses and Source of the Cobalt-blue Pigments Employed on Chinese Ceramic, Ming Ceramic*, 2001). The later cobalt was paler and could run in the glaze and thereby spoil the finely executed applied decoration when fired onto the porcelain substrates. This problem was circumvented in the reigns of Kangxi (1661–1722 CE), Yongzheng (1723–1735 CE) and Qianlong (1735–1796 CE) of the Qing Dynasty by their use of a European source of high iron Cobalt with a trace of arsenic from Saxony, comparable with the cobalt used in the Yuan period from Persia, along with the use of 18 new polychrome enamel decoration. During the later Kangxi period these polychromes became known in the West as *famille jaune* (yellow), *famille noire* (black), *famille rose* (pink) and *famille verte* (green) (Kerr & Wood, *Science and Civilisation in China*, 2004). These advances, particularly those associated with the coloured enamels were of great significance in that they allowed the decoration of Chinese porcelains with tinctured heraldic armorial bearings.

4.2 History of Armorials on Porcelains and Earthenwares

The depiction of armorial devices on faience and earthenwares can be traced back to the fifteenth century, and porcelains to the sixteenth century, although the largest number of these on ceramics generally seems to have been initiated from the early eighteenth century with the phenomenal rise in importation of the armorial Chinese porcelains into Europe. The major point that needs emphasising here is that the earliest porcelain decoration would have been almost exclusively monochrome, using cobalt blue on an opacified white background. This would mean that any very early armorial decoration would also have necessarily been undertaken in monochrome blue.

4.2.1 *Porcelain as Diplomatic Gifts: The Gaignieres-Fonthill Vase*

In 1338, the last Mongol Emperor of the Yuan Dynasty (1206–1368), Togon Temur (also known as Huizong), who reigned from 1333 until he was overthrown by the rebellion that established the Ming Dynasty in 1368, sent a Chinese delegation of Christian Nestorian monks from Peking to Pope Benedict XII (1334–1342), who was then based in Avignon. On the way, it is alleged historically, that they visited King Louis the Great of Hungary and presented him with a piece of *qingbai* porcelain from Jingdezhen bearing his coat-of-arms and those of the Kingdom of Poland. This would certainly have been the earliest recorded piece of Chinese porcelain specifically made for the European market and, remarkably, it still survives as the *Gaignieres-Fonthill Vase*, which now resides in the National Museum of Ireland, having arrived there via a circuitous route from Charles III Durazzo, King of Naples, then via the Kingdom of Aragon and thence to Jean, Duc de Berry and Dauphin of France, the son of King Louis XIV. *Qingbai* porcelain was very highly prized and possessed a pale bluish-green glaze (which is thus reminiscent of Chinese jade)—which is known now as *yingqing*—and dates from the Song Dynasty kilns in Jingdezhen which first commenced its manufacture around 1127. This event seemed to start the trend for using porcelain as "diplomatic gifts", a desirable practice which was carried on in Europe thereafter and throughout the seventeenth and eighteenth centuries.

This intriguing story has been adopted into the annals of early European porcelain history but actually suffers some rather questionable links in chronology which cannot be sustained upon closer forensic examination. King Louis the Great of Hungary (Lajos Nagy, reigned 1342–1382) did not accede to the throne upon the death of his father King Charles I until 21st July 1342 at the age of sixteen and Pope Benedict XII also died the same year, in April 1342! This means chronologically that the delegation of Chinese Nestorian diplomats could not have appeared physically at the Court of King Louis in 1338 as alleged as he was not even King of Hungary at that time. Also, King Louis of Hungary did not assume the title of King of Poland until the death of his uncle in 1370, so the alleged depiction of his coat-of-arms as King of Hungary and King of Poland (where he adopted the name of King Ludwig) could not have appeared on a piece of porcelain as early as 1338! The earliest description of this porcelain vase is that of Barthelemy Remy, valet of Francois Roger de Gaignieres in 1713, when the vase was in the possession of Francois Lefebvre de Caumartin of Chalons-en-Champagne, who was Bishop of Vannes and Blois. The vase by that time had been fitted with silver mounts, namely a foot and a silver handle, which bore the coat-of-arms of Louis of Hungary, the Angevin court at Naples and the Kingdom of Poland (Fig. 4.1). It is now believed that these silver mounts had been fitted to the vase later, either by Queen Joanna of Naples or by her successor King Charles III Durazzo, in the 1380s to cement their alliance with King Louis of Hungary and Poland after some years of war: an inscription on the silver handle of the vase gives the name "JEHENNA", which has been translated as "Joanna". The vase then found its way into the Angevin line of French kings as described above and eventually into the collection of William Beckford at Fonthill Abbey, whence it was acquired by John Farquhar, the gunpowder millionaire, in a Phillips auction sale in 1823. At the sale of the effects of Alexander, the 10th Duke of Hamilton, at Hamilton Palace in 1882, which realised the sum of £400,000 (equivalent to £20 Million today) the National Museum of Ireland purchased the vase for the sum of £28 with an unrecognised provenance, as by that time the silver mounts had been stripped away to leave the vase in its original, unmodified condition which does not in fact bear any armorial motifs at all, whether relating to King Louis of Hungary or anyone else! Initial research has established that the *Gaignieres-Fonthill Vase*, as it is now known, is certainly of Yuan *qingbai* porcelain with floral mouldings, with the characteristic blue-green glaze, and dates from the period 1300–1330. It seems reasonable to conclude therefore that the story of its presentation by the Chinese diplomats to the Hungarian court of King Louis is fabricated (Arnold, 'Princely Gifts and Papal Treasures: The Franciscan Mission to China and Its Influence on the Art of the West, 1250–1350', in *Princely Gifts and Papal Treasures; the Artistic Legacy of China in Europe*, 1999; Lane, *Italian Porcelain*, 1954; Lane, 'The Gaignieres-Fonthill Vase—A Chinese Porcelain of About 1300', *The Burlington Magazine*, 1961; Horvath, 'The Pedigree of Louis the Great's Ewer', in *Louis the Great, King of Hungary and Poland*, 1986) and it is more reasonable to propose that it had probably been presented to the Angevin kingdom of Naples by the Nestorian envoys in 1338 on their way through to Avignon to meet with Pope Benedict XII. It is not recorded if the Nestorian mission emissaries also presented the Pope with a similar diplomatic gift of porcelain, but they could well have borne several exemplars for presentation to the noble dignitaries that they encountered *en route* to Avignon. This account is a classic example of how an earlier supposition about an historical piece of porcelain which was made inappropriately and incorrectly on the basis of a flawed chronology and in the absence of accompanying documentation has still persisted into modern history, despite compelling evidence being uncovered to the contrary. Nevertheless, the *Gaignieres-Fonthill Vase* is of a very high importance and is unique historically as the earliest recorded piece of Chinese porcelain that was expressly made for the European market as a diplomatic gift from a Chinese delegation from a Yuan Dynasty Chinese Emperor and it can now be seen on display in its original form, without its later silver mount embellishment. The *Gaignieres-Fonthill Vase* as it now appears in the National Museum of Ireland, Collins Barracks, Dublin is shown in Figs. 4.2 and 4.3, giving the front and side views, respectively.

It is known that the fifth Mongol Emperor of the Yuan Dynasty, Kublai Khan (reigned 1260–1294, also known as Shizu), who was interested primarily in fostering foreign trade with China, had also sent a Nestorian mission much earlier to the Byzantine court of Andronikos II Palaiologos in Constantinople but it is not recorded whether a diplomatic gift of porcelain had also featured in this visit: if so, it has not survived to the present day. Other later diplomatic gifts of Chinese porcelain from Ming Emperors were then made to the Ottoman sultans at the Topkapi Palace in Constantinople, which fell to a Muslim siege in 1453, and to the Mongol Emperors in Samarkand in the mid-fifteenth century.

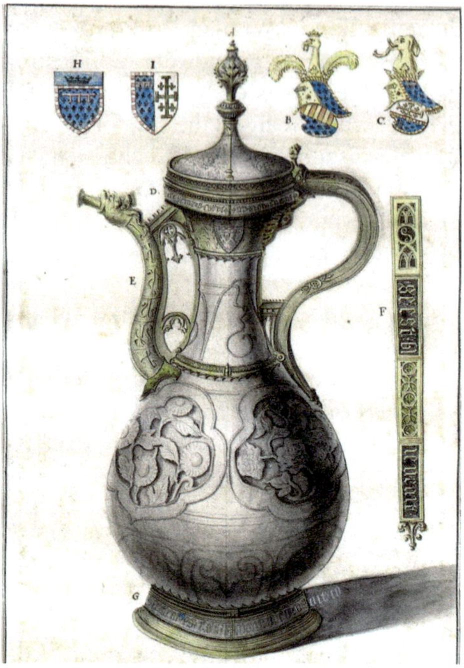

Fig. 4.1 The *Gaignieres-Fonthill Vase*: Yuan Dynasty, *qingbai* porcelain, *ca.* 1300–1330, the oldest piece of Chinese porcelain recorded definitively as a diplomatic gift from a Chinese mission sent from the Yuan Emperor to a European monarch. This picture shows a watercolour of the vase with additional silver mounts bearing the coats-of-arms of King Louis the Great of Hungary, the King of Poland and the Angevin Monarch of Naples. The watercolour painting was made in 1713 by Barthelemy Remy, valet of Francois Roger de Gaignieres from the vase which was owned at that time by Francois Lefebvre Caumartin. This vase is now in the National Museum of Ireland, Dublin, minus its associated silver mounts, where it was acquired from a sale of effects in Hamilton Palace in 1882. *Copyright, reproduced with permission from the National Museum of Ireland, Collins Barracks, Dublin*

This wonderful story, however factually questionable it now appears as recounted in the earlier ceramics literature, undoubtedly established the idea of presenting porcelain of the highest quality as diplomatic gifts and, as will transpire, this was readily adopted as a principle by the earliest European porcelain manufacturers and their aristocratic supporters. Of course, despite the fortunate survival of the beautiful and early *Gaignieres-Fonthill* vase, it cannot be recognised now as the first genuine piece of armorial Chinese porcelain as it appears that the armorials were added later in the form of silver mounts which have since been removed and lost.

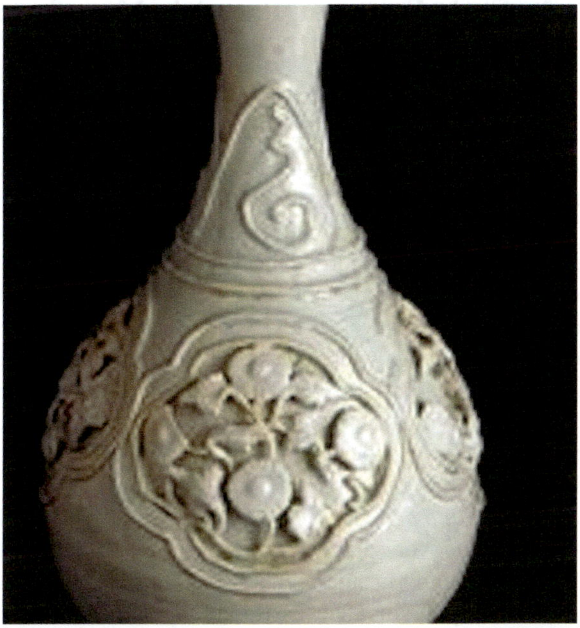

Fig. 4.2 The *Gaignieres-Fonthill Vase* as it is today, front view, minus its later silver mounts, demonstrating the superb quality of its porcelain; note that the vase itself does not possess any moulding which associates it with King Louis of Hungary. *Copyright, The National Museum of Ireland, Collins Barracks, Dublin, and reproduced with permission*

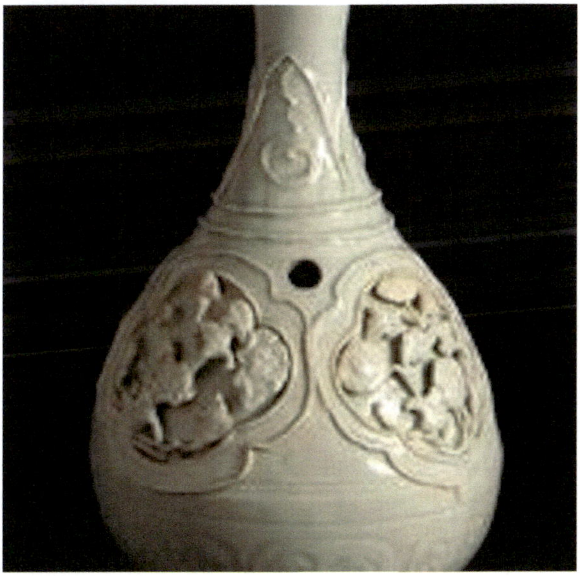

Fig. 4.3 The *Gaignieres-Fonthill Vase*, shown in Fig. 4.2, side view. *Copyright, The National Museum of Ireland, Collins Barracks, Dublin, and reproduced with permission*

4.3 Early European Porcelain Manufactories

4.3.1 The Medici Manufactory in Florence

Historically, the first European attempt at the manufacture of porcelain is deemed to have resulted from the establishment of the Medici manufactory in Florence. The artistic chronicler, Giorgio Vasari (*Le vite de piu eccelenti pittori, scultori e architettori*, 1568) recorded that Bernardo Buontalenti was experimenting with the art of manufacturing porcelain in Florence in the 1560s, but seemingly without achieving any success. However, a few years later in 1575, the Venetian ambassador to the Court of Tuscany, Andrea Gussoni, mentioned that Francesco Medici I, Grand Duke of Tuscany, had established a porcelain manufactory in the Casino of San Marco in Florence, having discovered the secret of the manufacture of the "*porcelain of India*" after 10 years of experimentation, but Buontalenti was not even mentioned as being involved in this enterprise. Gussoni said that the Medici porcelain was "*transparent, hard, light and delicate*" (Lane, *Italian Porcelain*, 1954). It is interesting at this point to try to discover what catalysed this interest in the Medici family that gave rise to their historically groundbreaking venture in the synthesis of porcelain, which was undoubtedly the first successful attempt at manufacturing this ceramic in Europe.

Lorenzo de Medici was given several pieces of Islamic "Levantine" porcelain from diplomatic envoys of the Mameluk Sultan of Egypt in 1478, described as being "*large vessels of porcelain, the like of which had never been seen before*" (Spallanzani, *Ceramiche alla Corte dei Medici nel Cinquecento*, 1994), which he held in the highest esteem, and which greatly impressed the visitors to his Tuscan Court. This seemed to catalyse the desire amongst the successive Medici Grand Dukes to manufacture porcelain themselves after his death in 1492 (Lane, *Italian Porcelain*, 1954; Kerr & Wood, *Science and Civilisation in China*, 2004). Cosimo Medici I (reigned, 1537–1574) was an avid collector of Chinese porcelain and a catalogue, dated 1553, listed over 400 pieces in his personal collection. Cosimo first initiated the effort to manufacture porcelain in his palace workshops, especially when he learned that the alchemists Camillo and Battista Gatti working in Ferrara for his rival, Duke Alfonso II d'Este, in 1561/1562 had apparently succeeded in the task of making porcelain there. If indeed any porcelain was made successfully there none has survived (Pollard, *Letter from China: A History of the Origins of the Chemical Analysis of Ceramics*, 2015). Cosimo started work on porcelain manufacture in his palace with Bernardo Buontalenti as director of operations and Giorgio Vasari, as we have already noted earlier, claimed he came very near to producing porcelain in 1568, but it was also claimed that he did not actually produce any finished items. In 1574, Cosimo's son, Francesco Medici, became Grand Duke of Tuscany and he constructed new workshops in the Boboli Gardens and in the Casino di San Marco in Florence. Francesco is recorded as working personally in these workshops alongside alchemists, potters and lapidaries to produce porcelain successfully. The Venetian ambassador, Andrea Gussoni, in 1575 reported that after 10 years of effort they had now succeeded in manufacturing good quality porcelain with the help of a

"Levantine" and several large pieces of porcelain were put on show there. Most of the porcelain in Francesco's lifetime was made at his Casino di San Marco workshop but there was a sporadic production carried on later at factories in Cafaggiolo and Pisa under his descendant, Ferdinand Cosimo II (reigned 1609–1620). Recent analytical studies of Medici porcelains (Colomban et al., 2004) support ambassador Gussoni's report of a Levantine influence in this porcelain production and there is a strong existence of a link with itinerant Islamic Iznik faience workers (Mack, *Bazaar to Piazza: Islamic Trade and Italian Art, 1300–1600*, 2002). Medici porcelain is a soft paste porcelain, which included a fine white clay in the raw materials from Vicenza, which fortuitously contained kaolin as a component. At that time the composition of Chinese porcelain and its kaolin content were unknown. Francesco Medici used his porcelain artefacts primarily as diplomatic gifts and his manufactory was never intended to be a commercial enterprise for the sale of porcelain: for example, he presented 20 pieces of porcelain to the d'Estes of Ferrara and some more to King Philip II of Spain (see below). Around this time there are also reports that Cipriano Piccolpasso working for Francesco Maria II, Duke of Urbino, had succeeded in making *la vera porcellana* in 1578/1579 (Mallett, 'The Earliest English Porcelains? Part 6: The Place of Buckingham Porcelain in Ceramic History', *Transactions of the English Ceramic Circle*, 2008). Clearly, there was much activity directed towards porcelain production in Italy, particularly in the Duchies of Savoy, Urbino, Ferrara and Tuscany.

It is believed that the Medici porcelain manufactory centred upon the manufacture of pieces which were intended as personal gifts and for "diplomatic" presents, which imitated the fashion of the trade in holy relics, which was also prevalent at that time and designed to secure favour and to cement alliances and support between nation states, rulers and the aristocracy (Freeman, *Holy Bones, Holy Dust: How Relics Shaped the History of Mediaeval Europe*, 2011). This suggestion is supported by the discovery of several pieces of Medici porcelain which bear the arms of the powerful King Philip II of Spain (*El Rey Felipe Segundo*, 1527–1598), a ruler who was also King of Naples and Sicily (1554–1598) and of all Spanish dominions in The Netherlands and America, King of Jerusalem (and *jure uxoris* King of England through his marriage to Mary Tudor, Queen Mary I, from 1554 until 1558) and also Duke of Milan (from 1540). The presentation of porcelain as "diplomatic gifts" was highly regarded in the sixteenth century and for this reason perhaps Francesco set out to make the highest quality porcelain product that he could. The substantial Medici bottles (height 28 cm, base 11 × 11 cm, volume 3.5 L) that were presented to King Philip II and which bear his coat-of-arms, include heraldic devices to signify his several titles, making them the earliest exemplars of armorial European porcelain. The King's escutcheon is bordered with a fettered chain bearing a sheep in deference to his title of Knight of the Golden Fleece and replacing the English Order of the Garter that was bestowed upon him by Queen Mary. His heraldic supporters in the form of the two beaked, sable eagles of St John are absent. The escutcheon is surmounted by the crown of a united Spain and Portugal. An interesting heraldic device can be seen beneath the escutcheon in the form of a crest that was occasionally used at this time with the Spanish coat-of-arms, depicting the Pillars of Hercules (*Fretum Herculeum*), which historically were believed to have been situated at Gibraltar and at Jebel el Musa on the North African mainland and, unusually, here these are placed beneath the escutcheon. The normal crest accompanying these arms would have been the embattlements of Castile and Leon with a rampant lion, which would have been located in its normal placement above the escutcheon on a helm and accompanied with a silk torse. The second porcelain bottle presents the arms but does not display this crest. The base of the first bottle depicts the cupola of the Duomo in Florence beneath which is an initial *F* for Francesco Medici—providing an example of a very early porcelain factory mark. Some estimate has been made of the total production of Duke Francesco's Medici porcelain at approximately 900 pieces and some 60–70 of these have survived to the present day in collections (Bowen Backus, *Asia Materialized: Perceptions of China in Renaissance Florence*, 2014), which is nevertheless a considerable number. These two Medici bottles currently reside in the Museum de Ceramique of the Sevres Porcelain Manufactory.

An example of a Medici porcelain gourd, although not bearing an armorial decoration, with blue and white floral decoration is shown in Fig. 4.4. Some early specimens exhibit the mark of Filippi Bruneschelli's dome of the Cattedrale di Santa Maria del Fiore, otherwise known as Il Duomo, a well-known landmark in Florence, along with a capital *F* for Francesco Medici. Others bear a Medici heraldic emblem known as the *alle*, comprising several balls contained within an heraldic escutcheon shield.

Francesco had itemised in his catalogue about 310 pieces of Medici porcelain in his personal collection at the time of his death in 1587: these were maintained initially by his descendants at their villa in Rome but after the due passage of time they became

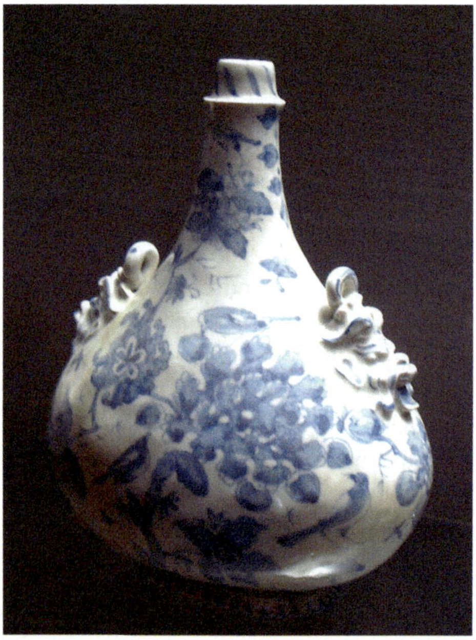

Fig. 4.4 Medici porcelain gourd decorated in underglaze blue with floral motifs, *ca.* 1575–1587. *Musee du Louvre, Paris. Creative Commons Attribution—Share Alike 3.0*

less well appreciated by his family in comparison with the finer porcelain pieces that were then starting to appear in Europe. Hence, despite its antiquity, the collection was gradually dispersed, but still left remaining the 60–70 pieces of Medici porcelain that are known today which are mainly now to be found in museum collections (Spallanzani, 1990; Cora & Fanfani, *La Porcellana dei Medici*, 1986). The Casino di San Marco manufactory ceased porcelain production immediately after Francesco's death and it is not known whether his recipe for porcelain manufacture died with him or if there was a disinterest in the continuation of his non-commercial factory operation by his descendants; it is known that Ferdinand Cosimo II did carry on with the manufactory for a while, it is thought possibly using the existing stocks of Medici porcelains and perhaps utilising the preserved seconds for decoration.

Medici porcelain has been universally categorised as a *soft paste* porcelain comprising feldspar, calcium phosphate (in the glaze), calcium silicate (occurring as *alpha*—and *beta*—wollastonite, $CaSiO_3$) with quartz sand and a white earth clay from Vicenza. However, as Colomban et al. (2004) point out from their detailed chemical analyses, this mixture certainly does not qualify its classification as a standard soft paste body and it appears that Medici porcelains should now be more strictly classified as a *hybrid* porcelain, coming between a *hard paste* and a *soft paste* body in their characteristics. The use of a tricalcium orthophosphate component ($Ca_3(PO_4)_2$) is of interest especially since this must be the earliest recorded usage of calcined bone ash in use for a ceramic in Europe, although from the analytical evidence this appears to be localised in the enamels and glaze rather than in the porcelain body as such (Colomban et al., 2004). The previous use of bone ash by Islamic faience producers to opacify their glaze would then explain the appearance of calcium phosphate localized in the Medici porcelain glazes, rather than in the body components, as has now been confirmed analytically (Colomban et al., 2004), and points to the Levantine (Iznik) faience influence that was recorded in earlier documentation.

The technical difficulty of achieving the high temperatures necessary and adequate kiln temperature control maintenance for the fusion of the paste components and their conversion into a translucent porcelain has been highlighted as a potential problem for Francesco Medici but several of his surviving pieces nevertheless seem to be well formed, delicately painted and glazed, although most are rather heavily potted and have suffered a surface pitting of the paste acquired during the glazing process. The style is rather eclectic and exotic, blending the Chinese and Islamic (Iznik) influences with the Italian Renaissance and employing a Westernised oriental decoration incorporating floral sprigs and scrolls. Medici porcelain is, hence, firmly established as an early European porcelain and surely can lay a fair claim to be the first successful European armorial porcelain production, even if it was never intended to be a commercial enterprise.

In most ceramic historical accounts, Italian porcelain production ceases with the Medicis in the late sixteenth century and does not then recommence until the foundation of the Vezzi manufactory in Venice in the early eighteenth century, but there have been several connections made to an enterprise that started in Padua and continued for just over ten years from 1627 (Mallett, 'Porcellana tenera del Seicento in Liguria? Due ciotole at Victoria & Albert', In *Atti XLV Courgno Internazionale della Ceramica*

Albisola, 1995; Lane, *Italian Porcelain*, 1954; Mallett, 'The earliest English porcelains? Part 6: The place of Buckingham porcelain in ceramic history', *Transactions of the English Ceramic Circle*, 2008). It appears that two pieces in the collection (accession numbers C.377-1919 and B.41-1905) of the Victoria & Albert Museum in London have been tentatively attributed to this Padua manufactory by John Mallett, both being marked in underglaze blue with *G/I.G.P.F* and a cross with dates of 1627 and 1638 and having a floral decoration. It is alleged that perhaps the *P.F* could mean either *Pisanus fecit* or *Padovano fecit*? The basis for the attribution of these pieces to Padua is not stated but, if confirmed, it undoubtedly represents another example of an early European porcelain manufactory site. In passing, it is perhaps noteworthy that the mention of the unsuccessful attempt by Duc Alfonso II de l'Este to create porcelain earlier in the sixteenth century has a geographical link in that the Duc's ancestral seat was located near Ferrara, which is only a few kilometres distance from Padua and it is not inconceivable that after the Duc died in 1597 some lingering expertise in faience and porcelain synthesis remained in the region which perhaps generated a prospective start-up in Padua some years later? A summary of these early seventeenth century Italian attempts to create porcelain is given by Wilson (*Western Decorative Arts*, 1993).

A possible early claim to being the first European-commissioned Chinese armorial porcelain service was that made some sixty years hitherto by King Manuel I of Portugal (Graca, 'The Portuguese Porcelain Trade with China', *Arts in Asia*, 1977; Rui, 'Portugal Embusca de China: Imagens e Miragens (1498–1514)', *Ler Historia*, 1990), who sent emissaries in 1517 to Canton and Peking; however, it is rather doubtful whether this early porcelain Portuguese armorial commission was actually completed as the emissaries of King Manuel charged with delivery of his armorial specification fell out of favour with the Ming Dynasty Chinese Emperor (Zhengde, 1505–1521) and were incarcerated for several years during which time King Manuel I died. We have already discussed the early example of Portuguese armorial Chinese porcelain in Chapter 1, which has been illustrated in Fig. 1.8 and this could well have been based upon the original commission that eventually was consigned to King Manuel's successor, King John. We have noted above that Francesco presented several "square" cross-section Medici porcelain bottles in 1581 to King Philip II of Spain, two of which are now in the ceramics collection of the Musee de Ceramique national de Sevres, France (acquisition numbers 5345 and 5778). These diplomatic gifts were much admired by Philip, who apparently was very attached to them and used them frequently: in sixteenth century Europe, porcelain was believed to possess a rather magical property which could act in response to the presence of toxins and could also heal minor skin afflictions—an idea which apparently carried over from its mystical use as a "tempered earth" in alchemy (Bowen Backus, *Asia Materialized: Perceptions of China in Renaissance Florence*, 2014)! The Italian antiquarian and historian Guido Pancirolli was concerned at the loss of knowledge from the ancient world in the sixteenth century and he wished to draw attention to lost knowledge (*deperdita*) during the explosion in the advancement of learning that appeared in the Renaissance in a book that was published in Latin (*Rerum Memorabilium Iam, Olim Deperditarium, Libro Duo*, 1599/1602) by his student posthumously (and translated into English in 1715 by John Nicholson as *The History of Many Memorable Things Lost*, 1715) (Keller, *Accounting for Invention: Guido Pancirolli's Lost and Found Things and the Development of Desiderata*, 2012; Keller, *Knowledge and the Public Interest, 1575–1725*, 2015). One of these *deperdita* selected by Pancirolli was porcelain and his description reveals how Chinese porcelain was considered as a "magical" creation in the mid- to late- sixteenth century:

> Never has porcelain been seen before, it consists of a paste of plaster, eggs and shells of marine locusts and of similar species, which after being well mixed is secretly hidden in the ground by the father of a family who then acquaints his children with its hiding place. It remains there for 80 years without seeing the light of day, after which the heirs recover it and, finding it in a fit state for manipulation, make of it those precious transparent vases so beautiful in form and colour architects can find no fault in them; amongst their inestimable virtues is that of breaking should poison be put into them. He who knows the substance never recovers it himself but leaves it to his children, nephews and heirs as a rich legacy from which they derive much profit; it is far more precious than gold.

Fanciful though some of these allegations and statements may now seem, this passage truly reflects the awe in which Chinese porcelain was held in Europe in the mid-sixteenth century with its associated magical powers. The idea that the raw material of a kaolin paste was needed to be buried for a considerable time echoes the recent discovery that in parts of China the kaolin/petuntse mixture is buried in large dung heaps, where after some years it can be excavated and is found analytically to have intercalated up to 70% w/w of absorbed urea, $CO(NH_2)_2$, from decomposed waste proteins into the layer silicates, with bonds formed between the hydroxyl groups of the kaolinite layers and the –CONH– amide groups of the urea, which assists in the malleability of the kaolin/petuntse mixture and in the achievement of its very fine potting capability (Seifi et al., 'Kaolin Intercalated by Urea. Ceramic Applications', *Construction and Building Materials*, 2016; Mako et al., 'Kaolinitye-Urea Complexes Obtained by Mechanochemical and Aqueous Suspension Methods', *Applied Clay Sciences* 2013). The historical analogy being that Chinese "eggshell porcelain" thereby emulates the behaviour of snakes, which observers noted used the warmth in dung heaps for the incubation of their very thin-shelled eggs.

4.3.2 The Fulham Pottery, London

There has been much debate amongst ceramic historians relating to the activities pursued by John Dwight at his Fulham Pottery between 1670 and 1703: what is perhaps surprising at first is that, aside from the rather sparse documentation and reference to sporadic attempts to manufacture porcelain in Italy in the early seventeenth century after Francesco de Medici's successful venture highlighted above, there was no follow-up made in Europe to combat the increasing market and demand for Chinese and Japanese porcelains. Chronologically, therefore, the Fulham Pottery of John Dwight was the next porcelain manufacturing venture to be well documented in Europe in the 1670s.

In contrast to the relatively large number of surviving pieces of Medici porcelain now known, which number approximately 60–70 in total as described above, there is not one complete finished piece extant that has been positively and definitively identified as porcelain originating from the Fulham manufactory, which was founded in London some one hundred years later in 1670. It is worth noting here that the Museum of London have on display a small damaged, glazed garniture porcelain vase which has been reconstructed from pieces found in their archaeological excavation of the site of the Fulham Pottery, and also have another similarly constructed damaged, unglazed vase, and these two items comprise the only known survivors of the Fulham manufactory found on site (Mallett, 'The Earliest English Porcelains? Part 6: The Place of Buckingham Porcelain in Ceramic History', *Transactions of the English Ceramic Circle*, 2008). The Fulham shards were clearly affirmed to be porcelain and dated to 1675–1680 (Green, The Fulham Pottery, 1999).

However, despite the absence of surviving pieces there is considerable documentary evidence for the manufacture of porcelain in England at the Fulham Pottery in 1670 where John Dwight with the technical scientific support of the Royal Society of London succeeded in making a *hard paste* china body that was both highly siliceous and refractory, yet which possessed a satisfactory translucency which enabled it to be termed "*porcellaneous*" by several independent contemporary scientific observers. Like the Medici porcelain, Fulham porcelain was essentially manufactured as a non-commercial operation but, unlike its Medici precursor which was made for diplomatic gifts, the Fulham product was primarily made for the scientific realisation of robust refractory materials that were capable of withstanding chemical attack by strong acids and fused alkalis at high temperatures and which perhaps could then be adapted for synthetic crucible manufacture for the extraction of metals from their ores.

The central character in the story of porcelain manufacture at Fulham was John Dwight, whose precise date of birth is unknown but is stated to be somewhere between 1633 and 1636, in Gloucestershire, United Kingdom. He studied natural philosophy at Oxford University and was appointed as a laboratory assistant there to Robert Boyle, the "Father of Chemistry" (*Authors*: this title has also been bestowed upon the French chemist, Antoine Lavoisier, who lived some 100 years later). He left natural philosophy/science upon his graduation in 1661, to work as a diocesan registrar in Chester; he is recorded as having served four successive Bishops of Chester in this role, the last of whom was John Wilkins, who served from 1668 to 1672. Apparently, John Dwight had a divisive falling out with Bishop Wilkins and then left to establish the Fulham Pottery, at the junction of the New King's Road and Burlington Road, not far from Putney Bridge on the River Thames. Recent archaeological excavations undertaken at the site of the Fulham Pottery by the Museum of London have revealed the presence of a ceramic kiln which has now been preserved on this urban site. Just prior to the foundation of his pottery at Fulham, John Dwight filed a patent in April 1671 describing his experiments in the use of clays and minerals to make a high quality "stoneware" and a phrase from this patent is particularly revealing, namely:

> … the discovery of the mistery of **transparent earthenware** commonly known by the name of **porcelaine** in China.

The wording of this patent is critical in that Dwight has been credited by several historians with the manufacture of stonewares only at Fulham and the granting of this patent for the manufacture of a "*transparent earthenware,*" i.e. "*porcelain*", rather than just the opaque so-called salt-glazed stoneware, is revealing although historians have explained this discrepancy incorrectly by assuming that his earthenwares must have themselves been transparent ! A full description of the activities undertaken by John Dwight at the Fulham Pottery under the close monitoring by interested members of the Royal Society has been given in a recent text (Edwards, *Porcelain Analysis and Its Role in the Attribution of Ceramic Specimens*, 2022a), which contained their many observations on Dwight's china products. The role of the Royal Society of London in English porcelain research and manufacture has been mentioned in an earlier ceramics work by Solon (*The Old English Potter*, 1883). Daniels (*The Origin and Development of Bow Porcelain, 1730–1747, Including the Participation of the Royal Society, Andrew Duche and the American Contribution*, 2007) has reopened the history of the Royal Society's scientific interest in the monitoring of Dwight's experiments with her interpretation of several key documents (Haselgrove & Murray, *John Dwight's Fulham Pottery, 1672–1978: A Collection of Documentary Sources*, 1979) which demonstrate clearly that in the mid- to late-seventeenth century, within a few years after its foundation during the Restoration of King Charles II in 1660, the Royal Society of London was actively encouraging and sponsoring research and experimentation in refractory materials with John Dwight at the Fulham Pottery and that this was occurring some 40 years prior to the successful Meissen porcelain manufacturing operation undertaken by King Augustus, Elector of Saxony. Because of the key scientific link between John Dwight and Robert Boyle, Fellows of the Royal Society were monitoring closely the high temperature synthetic ceramics work being carried out at the Fulham Pottery in the 1670s. The eminent scientist Robert Hooke FRS on February 1674 (in Haselgrove & Murray, *John Dwight's Fulham Pottery, 1672–1978: A Collection of Documentary Sources*, 1979) confirmed the successful production by Dwight of his "*English china*" including some figures and:

> Severall little Jars of severall colours all exceeding hard as flint, very light and very good shape.

Later, Dr. Robert Plot FRS in 1677 (Plot, *The Natural History of Oxfordshire Being an Essay Towards the Natural History of England*, 1677) noted that John Dwight had "*discovered the mystery of the Hessian wares and ways to make an Earth white and transparent as Porcellane*".

He also mentioned that Dwight had experienced problems with the glazing of this *white porcellane*, which was suspected by Ramsay & Ramsay (*The Evolution and Compositional Development of English Porcelains from the 16th Century to Lund's Bristol c. 1750 and Worcester c. 1752—the Golden Chain*, 2017) to arise from his adoption of a silica—alumina—calcareous glaze composition which required very high firing temperatures and which Dwight was unable to achieve reliably in his kiln at that time.

Dr. John Houghton FRS in 1693 (Haselgrove & Murray, *John Dwight's Fulham Pottery, 1672–1978: A Collection of Documentary Sources*, 1979) maintained that

"Mr Dwight of Fulham uses clay, the same earth China ware is made from and this is made by not lying long in the earth but in the fire …. we make as good China here as any in the world".

Tite et al. (1986) also inferred from optical vitrification measurements that a firing temperature of between 1200 and 1250 °C had been used for the white Fulham shards compared with a rather lower one of 1100 °C required for the stonewares (Tite et al., "A technological study of Chinese porcelain of the Yuan Dynasty", *Archaeometry*, 1984; Medley, *Yuan Porcelains and Stonewares*, 1974). It is also interesting that the firing temperature adopted for the Yuan Dynasty (1260–1368 CE) Chinese *yingqing* porcelains has been determined to be in the range 1200–1300 °C, which is in the same range as that now believed to have been used by John Dwight in his kilns at Fulham. What Tite et al. (Tite et al., "A Technological Study of Fulham Stonewares", Proceedings of the 24th International Archaeology Symposium, 1986) also found is that the Dwight porcelain, which perhaps can now be re-defined as really *proto-yinqing* porcelain, does not match the later Chinese Kangxi porcelains of the early eighteenth century in composition, as has been verified later by Spataro et al. (2008, 2009).

A most telling comment which has emerged from Daniels' research (Daniels, *The Origin and Development of Bow Porcelain, 1730–1747, Including the Participation of the Royal Society, Andrew Duche and the American Contribution*, 2007) is a note that refers to a visit made to the Fulham Manufactory made by Ehrenfriede von Tschirnhaus, who later "*transferred to the Meissen Factory*" in Saxony, where history now tells us incorrectly that European porcelain was invented! Von Tschirnhaus (1651–1708) arrived in London from Paris in May 1675, with an introductory letter from the philosopher Spinoza and met with influential member scientists of the Royal Society, possibly including Dwight, Hooke and Boyle, where they could well have discussed their novel experiments on refractory porcelains which then stimulated von Tschirnhaus' recorded visit to the Fulham Pottery to view the experimental products for himself. On his return to Paris, Ehrenfriede von Tschirnhaus immediately commenced his own experiments on porcelain synthesis and eventually he set up with Johann Bottger at Meissen under the patronage of King Augustus, the Elector of Saxony. It could be argued, therefore, that Dwight's manufacturing successes at Fulham catalysed and stimulated the early experiments of von Tschirnhaus which were made after his visit there in 1675 and which culminated in the historical labelling of the first European synthetic porcelain manufactured successfully at Meissen in 1709/1710.

However, it is a fact that no porcelain was made at the Fulham Pottery after John Dwight's death in 1703 and we should seek a reason for this. Perhaps John Haslem has summarised the situation nicely (Haslem, *The Old Derby China Factory*, 1876):

> In 1671 John Dwight had a patent for having by his own industry and at his own proper costs and charges invented and sett up at Fulham the mistery of Porcelain or China and Persian ware. Again in 1684 he protected the mistery of porcelain and other wares never before made in England or elsewhere. It would seem that the Doctor was in advance of the requirements of his day as his invention met with such poor encouragement that he burnt his receipts and implements in disgust and it is doubtful whether there is a specimen of his Porcelain

> extant. It is also said that he buried all his models, moulds and tools in some secret place on the premises at Fulham, observing that the production of such materials was expensive and unremunerating and he wished therefore to put it out of the power of his successors to perpetuate the business?

This may be considered a rather speculative sequel from John Haslem, but much of what he recounts in his work has been supported by later historical assessment: what is certainly true is that in many early porcelain manufactories there was a culture of the founder and driving force taking his secrets to the grave, Francisco Medici, Duke of Tuscany, at Florence and Louis Poterat at Rouen being just two of these. This landmark, cutting-edge work of John Dwight has been largely ignored until recently and perhaps the only sensible appraisal of this situation is that there seemed to be little physical evidence of his manufacture of his translucent *porcellaneous* material surviving today apart from the shards and reconstructed artefacts excavated archaeologically at the manufactory site and which have been discussed above. However, John Dwight must have made a significant number of finished and completed specimens for Hooke, Sloane and several other observers from the Royal Society to describe its ceramic body properties so accurately in terms of its potting, robustness to mechanical shock, translucency and heat resistance—the self-same characteristics for which the Chinese hard paste china then being imported into Europe was so highly esteemed.

4.3.3 Rouen Porcelain

Rouen had been a centre for the production of faience and tin-glazed earthenwares in France since 1540; in 1548 a Masserat Abaquesne was recorded as being an enameller on pottery tiles there and in 1647 King Louis XIV granted Nicholas Poirel, seigneur of Grandval, and Edme Poterat of the Rouen faience manufactory a 50 years monopoly to manufacture faience, which rendered any competition unfavourable, although other contemporary faience factories did exist in France and were already in production at that time. The earliest porcelain manufacture in France started in Rouen on a different site but clearly benefitted from the ceramics background and workforce skills that had been already acquired in faience manufacture there. The manufacture was of a *soft paste* porcelain (*pate tendre*), which again never seemed to aim ostensibly for a commercial output and only some 9–10 genuine Rouen porcelain pieces have now survived to the present day to attest to its successful manufacturing operation. Cyrille Froissart has recorded details of all existing surviving specimens of Rouen porcelain but one or two of these are even now believed possibly to be of a doubtful provenance and await analysis for confirmation (Froissart, *Collection d'un Amateur, Rouen 1680–1740: De la Premier Porcelaine de l'Age D'Or de la Faience*, 2008).

In 1674, Louis Poterat bought out Nicholas Poirel and started another factory in Rouen making porcelain exclusively (Dawson, *French Porcelain: A Catalogue of the British Museum Collection*, 1994; Solon, *The Rouen Porcelain*, 1905). Edme Poterat died in 1684 and his younger son Michel then continued with the faience arm of the Rouen business. Louis Poterat, with the authority and royal privilege granted by the King on the 1st October, 1673 for an initial period of 30 years, continued to manufacture porcelain at Rouen until his death from typhus in 1696. He took his secret porcelain recipe and details of his manufacturing processes with him to the grave and porcelain manufacture finished at Rouen very soon afterwards. The *privilege* granted by the King to the Rouen manufactory specified that the manufacture should be for

> …. any kind of dishes, pots and vases in porcelain similar to the Chinese (Author: i.e. in blue and white) and the production of polychromatic or blue and white faience in the Dutch style.

On the 16th June 1694, the Intendant of Rouen, Lefevre d'Ormessan, reported to Count Louis de Phelypeaux of Pontchartrain that Louis Poterat was now making quality porcelain and faience. In 1696, the Royal *privilege* was renewed for a further 20 years so the King must have been pleased with the quality of the porcelain production at Rouen which had been achieved thus far, but recognising the ongoing secrecy surrounding Louis Poterat's process and recipe, this renewal contract was made conditional upon his making the secret known to the King at the end of that time (Solon, *Rouen Porcelain*, 1905). However, Louis Poterat died of typhus the same year and the secret of the manufacture of Rouen porcelain died with him. Porcelain production effectively stopped in Rouen upon the death of Louis Poterat and the Royal patronage was removed therefrom, although there are reports that some wares were produced at Rouen later that year but that these were of a significantly inferior quality since the process details and presumably the recipe and formulation were unknown to the remaining workforce. Louis Poterat had always insisted upon carrying out the critical operations alone and he only ever made small quantities of porcelain at any time, so it can but be concluded that the Rouen wares that were still being issued after the death of Louis Poterat were perhaps factory "seconds" or stock that was defective in some way (Lacombe, *L'Apparition de la Porcelaine Tendre a Rouen chez les Poterat: L'Hypothese Protestante?* 2006; Grandjean, 'The Porcelain of Rouen', in *Discovering the Secrets of Soft Paste Porcelain at the St Cloud Manufactory, ca. 1690–1696*, 1999). The King was apparently pleased with the Rouen product which was considered to be a successful substitute for the Chinese porcelains, these becoming known as "*Porcelaine francaise – la veritable porcelain de Chine*" ("French porcelain – the true porcelain of China") (Munger & Sullivan, *European Porcelain in the Metropolitan Museum of Art*, 2018). However, the porcelain production was always limited, and an intriguing comment has been found in the records that in the kilns the relatively fragile porcelain items had to be surrounded by the more robust faience items during firing to protect them "*from the violence of the furnace*", which presumably refers to a critical inability to control the kiln temperature (de Plinval de Guillebon, *Faience et Porcelaine de Paris: XVIIIe-XIXe Siecles*, 1995).

Surviving examples of genuine Rouen porcelain are very rare and only some nine or ten Rouen soft paste porcelain artefacts are known in collections; it is indeed fortunate that one of these happens to be an armorial *moutardier* of 1695, decorated in underglaze blue and bearing the coat-of-arms of a French legal advocate and counsellor to the parliament of the Departement de Normandie, Comte Jacques Asselin de Villequiers (1669–1728), whose family was ennobled in 1643. This *moutardier* is shown in Fig. 4.5, from which the coat-of-arms is clearly visible in its blue and white decoration. These arms comprise an escutcheon with a chevron containing three tall ewers (known as *burettes* in French heraldry) supported by two gryphons rampant regardant and were first attributed heraldically to Jacques Asselin de Villequiers by Andre Pottier at the Musee de Sevres in 1847. Above the escutcheon sits a coronet which displays nine balls on spikes (also known as "pearls") which denotes the rank of Comte in French heraldry—equivalent to that of a Viscount in the heraldic lists of the British Isles. Undoubtedly, King Louis XIV embraced the emergence of porcelain as "the ambassadorial gift of preference in the eighteenth century" (Walton, *The Era of Louis XIV: A Turning—Point in the History of Diplomatic Gifts*, 1992; Walton, 'Diplomacy and Ambassadorial Gifts in the 16th and 17th Centuries', in *Gifts to the Tsars, 1500–1700; Treasures from the Kremlin*, 2001) and it may possibly be implied that other armorials were thereby created but seemingly no others have survived. It is not known whether this *moutardier* was presented to the Comte Jacques Asselin de Villequiers by the Rouen manufactory, or perhaps even by King Louis XIV, or whether it was purchased there personally by the noble family concerned, and there is no documentary record surviving which can corroborate any of these suggestions.

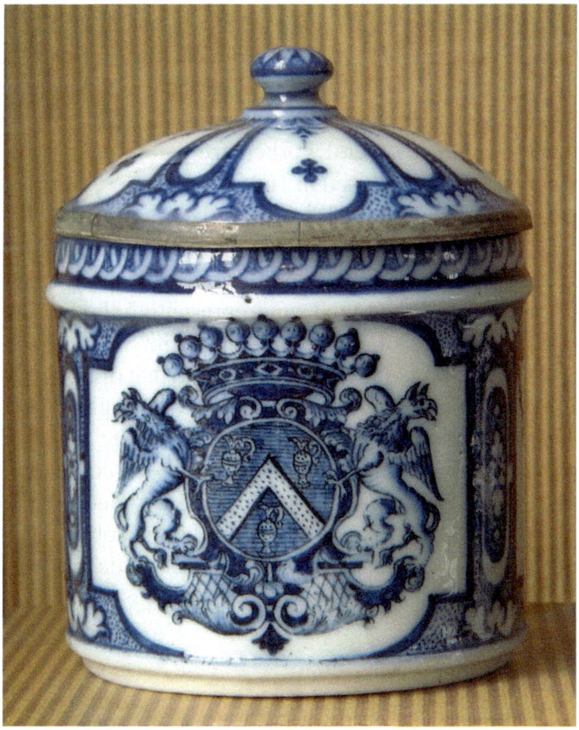

Fig. 4.5 Rouen porcelain mustard pot and lid, *ca.* 1695, bearing the coat-of-arms of Jacques Asselin de Villequiers, decorated in underglaze blue. *Musee national Ceramique de Sevres*. CC BY-SA 3.0

4.3.3.1 St Cloud Porcelain

The soft paste porcelain manufactory at St Cloud, in a suburb of Paris, commenced in 1693 under the protection of Philippe, Duc d'Orleans (1640–1701), younger brother of King Louis XIV, and so overlapped that at Rouen chronologically for some three years, and whilst Rouen was still operating under its royal privilege, St Cloud nevertheless became the first French porcelain manufactory to make porcelain commercially, obtaining an official patent for this in 1702. Already in 1700, a St Cloud factory inventory registered that 3146 pieces of finished and completed porcelain were held in stock at the site, so clearly the production of porcelain must have been undertaken there for some while before the patent was formulated. Dr. Martin Lister FRS (1639–1712), an English naturalist and natural scientist, reported on his visit to the St Cloud porcelain manufactory in 1698 (*Journey to Paris in the Year 1698*, 1699) that:

> They (Authors: St Cloud) were producing already a manufacture of porcelain as white and translucid as the one that came from the East.

The brothers Francois and Claude de Reverend, late of the faience factory in Rouen, had set up a competitive faience factory in St Cloud in the 1660s under the direction of Jean-Baptiste Colbert specifically to manufacture "*the Indian style porcelain*" and employed Pierre Chicaneau to decorate their faience: Chicaneau took over the faience factory in 1674 and also experimented with the production of porcelain (Solon, *St Cloud Porcelain, Part I and Part II*, 1906a, 1906b). After his death, his widow, Barbe Coudray, married Henri Trou from Nevers, squire to the Duc d'Orleans: they bought the factory in 1683 and along with the Chicaneau children in 1693 they decided to install a separate factory on the same site to manufacture porcelain. There is a report that they enticed several workers from the Rouen porcelain manufactory to join them when that factory closed in 1696. As Philippe, Duc d'Orleans, died in 1701, the royal patronage would most likely have been taken over by his elder brother, King Louis XIV, who, of course, had just lost the Rouen manufactory of which he was patron some years previously and this would explain the adoption from that time of the St Cloud motif of the rayed sun, representing the Sun King, as a mark on its porcelain (Louis XIV was of course known as The Sun King, *Le Roi Soleil*).

St Cloud porcelain is very similar to that of Rouen but is of an ivory colour (Honey, *European Ceramic Art*, 1952) with a satin glaze and the decoration is of a similar *camaieu* and *lambrequin* style in underglaze blue, so it is not surprising that in the past the output from both factories has been confused. By 1766, St Cloud was unable to compete with the more sophisticated products of Chantilly, and Vincennes so they filed for bankruptcy. However, unlike the solitary *moutardier* from Rouen, there is no specific evidence that St Cloud ever made and sold armorial porcelains.

4.3.4 Meissen Porcelain

German porcelain in the early eighteenth century is synonymous with the foundation of the Meissen manufactory, which is generally accepted historically to be the first in continental Europe to develop porcelain commercially and it has been alleged by several historians that the resounding success in this operation at Meissen spawned the creation of many other factories in Europe later in the century. The discovery of European porcelain is credited historically, and rather mistakenly, to Ehrenfriede von Tschirnhaus and Johann Bottger working together in the first decade of the eighteenth century under the patronage of Augustus the Strong, Elector of Saxony and King of Poland. The factory was established by King Augustus, which was entitled at that time the Royal Polish and Electoral-Saxon Porcelain Manufactory (Koningslich-Polnische und Kurfurstlich-Sachsisische Porzellan Manufaktur) at his castle in Albrechtsburg in Meissen, near Dresden. King Augustus called the porcelain material "*white gold*"—this nomenclature directly reflects the original purpose of Augustus bringing together these two experimental chemists to work for him, which was to assist in the alchemical transmutation of base metals into gold (Keating, *The Hunt for White Gold*, 2011). Several historians believe that von Tschirnhaus has not received sufficiently due credit for the discovery and eventual manufacture of porcelain at Meissen relative to the acclaim that has been later awarded to Bottger (Schonfeld, 'Was There a Western Inventor of Porcelain?' *Technology and Culture*, 1998), but von Tschirnhaus died before the production went commercial and it was left to Bottger to take it forward to completion. A detailed account of the interaction between von Tschirnhaus and Bottger is to be found in Simeon Shaw's book (Shaw, *A History of the Staffordshire Potteries and the Rise and Progress of the Manufacture of Pottery*, 1829). For quite some time the factory output was known as *Dresden* china in the English-speaking world (Honey, *Dresden China: An Introduction to Meissen Porcelain*, 1934). Von Tschirnhaus died of dysentery in 1708, before the advent of porcelain production commenced in Meissen but sufficient detail was left in his records so that the process could be developed to facilitate its entry into commercial production in 1710 by Bottger. The very first ceramic products at Meissen were actually a hard glazed red stoneware called "*Bottger wares*" (Bottgersteinzeugt), which appeared first at the Easter Fair in Leipzig in 1710, but the production of glazed white and painted porcelain soon followed, and the full commercial production of a *hard paste* porcelain appeared in 1713 (Schulle & Ulrich, 'Ergebnisse gefugeanalytischer untersuchungen an Bottgerporzellan', *Silikatechnik*, 1982; Schulle & Goder, 'Die erfinding des europanischen porzellans durch Bottger – eine systematische schopferisch entwicklung', *Keramische Zeitschrift*, 1982). To protect the goods, a crossed swords mark, the first on a commercial product, was introduced in 1720. Underglaze blue pigments, termed "*Meissen blue*", were introduced by Frederick Kottig and a wide range of other overglaze pigments in 1723 were adopted by Johann Herold. In the early 1730s, china wares were sold from the factory in the white to private workshops (Hausmalerei) in Augsburg and Bayreuth for independent retail and sale as finished, decorated Meissen items (Ducret, *German Porcelain and Faience*, 1962; Roentgen, *The Book of Meissen*, 1996).

King Augustus' passion for porcelain resulted in his construction of his Japanese Palace in time for the wedding of his son, Crown Prince Friedrich (who later became King Augustus II) to the Archduchess Maria Josephine in 1719. Augustus made some important acquisitions of early Japanese and Chinese porcelains to furnish his new Japanese Palace: in 1717, he exchanged a regiment of 600 elite soldiers for 151 choice pieces of Chinese porcelain from the Prussian Royal Collection, including the famous Dragoon vases. In 1722, he acquired a further 804 pieces of porcelain for the Japanese Palace, including several pieces of St Cloud porcelain which were catalogued as Meissen in his own Collection because no provision was made for porcelain categories other than Japanese, Chinese and Meissen! In the eighteenth and nineteenth centuries, Meissen was famous for its figures modelled in their eighteenth century rococo and classical poses and involving a menagerie of animals. The "*Meissen Temples*" were created as porcelain table decorations (Tafelaufsatz) around 1727–1728 for court dinner parties at King Augustus' Japanese Palace and there is a record of 6 "*Venus Tempel*" being made for this location alone in the 1720s. These porcelain temples were of complex construction and very few have survived. These temples were used at Meissen as table centrepieces for the royal weddings of the children of King Augustus III in the mid-eighteenth century from 1747 onwards.

Large Meissen dinner services were very much in vogue for local consumption and for export from the 1730s: for example, 17 Meissen dinner services were given as part of her dowry to Princess Maria Amalia of Saxony upon her wedding to Charles, King of the Three Sicilies, who later became King Carlos III of Spain. Other notable and very large services made at Meissen were the *Swan* service (*Schwanenservice*), 1737–1742, made for Count Heinrich von Bruhl, which numbered more than 2000 pieces, and the *Mollendorf* service of over 960 pieces made in 1762 and presented by King Frederick II the Great of Prussia to General Wichard von Mollendorf to celebrate his victories in the Seven Years' War (Cassidy-Geiger, *Princely Gifts in Meissen Porcelain*, 2007). Much blue and white Meissen porcelain copied the contemporary successful Chinese imports into Europe, such as the famous "*Blue Onion*" pattern (*Zwiebelmuster*), the Chinese motif used in the decoration actually representing a pomegranate—a fruit that was unknown in Europe at that time and was mistaken for an onion.

4.3.4.1 The Armorial *Schwanenservice* of Meissen Porcelain

The Meissen porcelain manufactory did make large services that were commissioned by Augustus III, Elector of Saxony, for his own use and also presented to the aristocracy: undoubtedly, the most famous of these was the *Schwanenservice*, which deserves special mention here as it was the largest armorial porcelain service in European porcelain created in the early eighteenth century. The service was commissioned by King Augustus III in 1737 as a wedding present for his Prime Minister of the Electorate of Saxony, Count Heinrich von Bruhl (1700–1763), Supervisor of the Royal Porzellan Manufactory at Meissen and later its Director, who married the Countess Franziska Kolowrat-Krakowsky in November 1737. Heinrich von Bruhl was a very influential statesman and a lavish entertainer who regularly served between 80 and 100 separate dishes at his large dinner parties in the *service a la francaise* mode, whereby much of the food was brought to the table all together and diners helped themselves from large tureens and platters at table in contrast with the alternative *service a la russe* style, where the food was served in individual plated portions by servants. The main dinner and dessert service comprised 100 place settings and was accompanied by matching tea and coffee services, which with a proliferation of centrepieces, candelabra, wall sconces, platters and tureens amounted in all to 2200 pieces of porcelain. The service was designed by Johann Jacob Kaendler, assisted by Johann Friedrich Eberlin, and took five years to complete, the last pieces being finished in 1742. It was most sumptuous in construction, in a high baroque style, the main feature of the porcelain being a moulded undecorated pair of swans swimming in reeds on a pond with scenes of still water life centred on an osier or wicker basket weave pattern at the edge: the lid handles of the tureens were modelled on neo-classical figures of Glaucus and Galatea riding on dolphins and the enamelled decoration was minimal to accentuate the fine quality moulded porcelain substrate. The only decoration comprised small delicate polychrome sprays of *indianische blumen* at the verge with minimal dentil edge gilding and a single armorial bearing: the latter consisted of the arms of Kolowrat-Krakowsky impaling those of Graf von Buhl. A dinner plate from the *Schwanenservice* is shown in Fig. 4.6, and a saucer from the matching tea service in Fig. 4.7, which both indicate the superb quality and fineness of the porcelain moulding. They bear the Meissen blue enamelled crossed swords mark and a dreher's mark (a dreher was a turner or lathe operator) of a quartered circle at the footrim. The service has been described variously as "*The most famous high baroque production in Meissen porcelain, a triumph of modelling and firing*" and "*... the most famous tableware ever conceived in porcelain*".

The armorial coat-of-arms which was enamelled at the verge, known heraldically as an *allianzwappen* is shown enlarged in Fig. 4.8: a baroque escutcheon holds the arms of Graf von Buhl at dexter, quartered 1st and 3rd with a double-headed armed eagle, per pile sable on argent and argent on gules, 2nd and 4th a chevron argent on an azure field, impaled by the arms of Kolowrat, comprising a double-headed eagle argent engorged with a three pointed coronet on a gules field. The escutcheon has two supporters, these being crowned lions rampant or regardant. The *allianzwappen* escutcheon is surmounted by a jewelled coronet with thirteen pearls and three small helms with closed visors and two emblems of a battlemented tower at sinister bearing three ostrich feathers and a stylised leaf frond at dexter. The crest is a double-headed armed eagle in argent and sable sitting atop a three-pearled coronet on the central helm. The escutcheon sits above a scallop shell emblem in gold. It is interesting that a Graf (equivalent to Count), the highest rank in Prussian nobility below a Duke, should display his *grafenkrone* (coronet) heraldically showing nine pearls visible

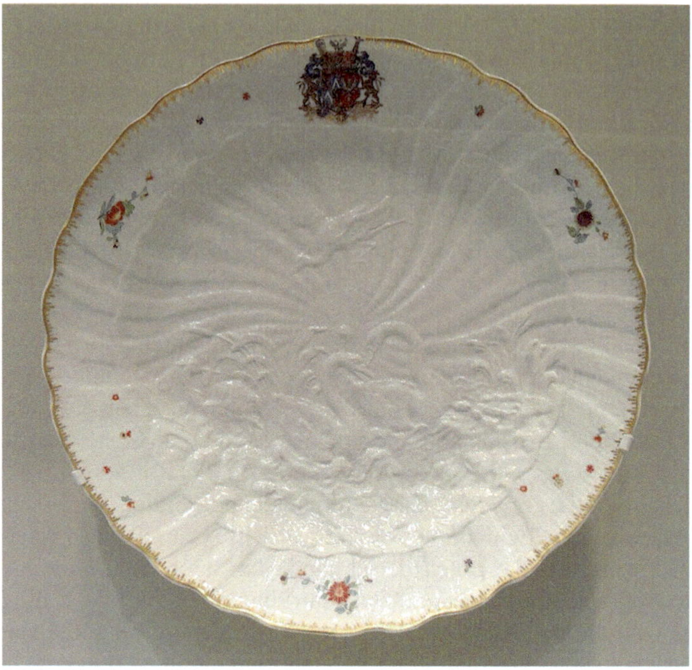

Fig. 4.6 Dinner plate from the *Schwanenservice*, Meissen porcelain, 1738, presented to Graf Heinrich von Bruhl, Prime Minister of the Electorate of Saxony and Supervisor of the Meissen Porzellan Manufaktory, Dresden, by King Augustus III of Saxony. The porcelain is beautifully moulded with a pair of swans swimming in reeds on a pond and only sparsely decorated with sprays of Indian flowers and dentil edge gilding. It bears the arms of von Bruhl impaled with those of Countess Kolowrat at the verge, in celebration of their marriage in 1737. *In the collection of the Art Institute of Chicago, Illinois, public domain reproduction*

from the front, but here Graf von Buhl's coronet shows all thirteen pearls visible from the front.

Graf von Bruhl had acquired the world's greatest collection of porcelain, along with some 62,000 books in his library and 600 paintings: after his death, the paintings were bought by the Empress Catherine II the Great of Russia for her Hermitage Palace in St Petersburg. Although much of the *Schwanenservice* survived into modern times at the family home in Schloss Pforten in Poland, it was mostly wantonly destroyed during the Russian invasion of the Second World War where the Russian troops used it as target practice and the only pieces that now appear on the market therefore command very high prices for this historically important early porcelain armorial service (Pietsch, *Schwanenservice – Meissner Porzellan fur Heinrich Graf von Bruhl*, 2000; Berling, *Das Bruhlsche Schwanenservice*, 1925).

The *Schwanenservice* was most certainly not the only large armorial service made by the Meissen Porzellan Manufaktory in the early eighteenth century and it is on record that armorial tea, coffee and chocolate services were made for Augustus the

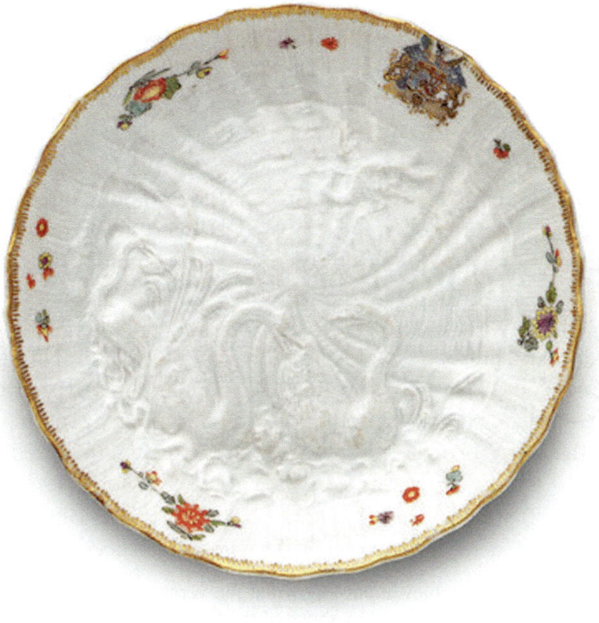

Fig. 4.7 Saucer from the *en suite Schwanenservice* tea set matching the plate in Fig. 4.6. *Reproduced courtesy of Fergus Gambon, Bonhams Antiques and Valuers, Bond Street, London*

Elector's personal use and as designated diplomatic gifts in the 1720s and 1730s: one of these will be described Sect. 4.3.4.2, namely the diplomatic gift made from King Augustus to the King of Sardinia. Other large Meissen porcelain services are known which are not armorial and an exemplar is the *Mollendorf* service, which was designed by King Frederick II the Great of Prussia (1712–1786) in 1762 for presentation to General/Field Marshal Wichard Joachim Heinrich von Mollendorf for his military achievements during the Seven Years' War (1756–1763) with Austria and the annexation of Silesia. Modelled on the *Schwanenservice* and created by Karl Jacob Christian Klipfil and Johann Joachim Kaendler it is also high baroque in style and profusely decorated with moulded nymphs and the goddess Minerva holding flowers, fruit, corn and pitchers of wine. It amounted to about 960 pieces, of which only 144 were dinner and dessert plates and 48 were soup plates.

4.3.4.2 The King of Sardinia Armorial Service

J.G. Horoldt joined Meissen in late 1724 and one of his first commissions in 1725 was the decoration of a prestigious armorial service which was destined as a diplomatic

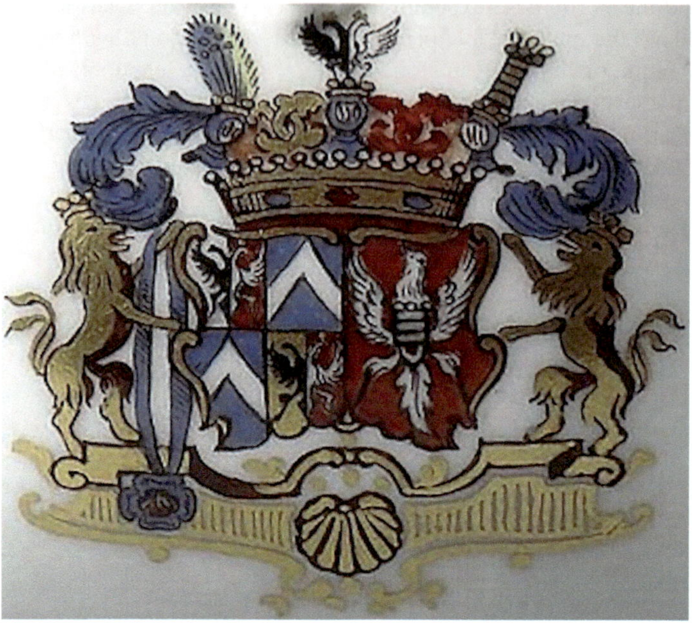

Fig. 4.8 Detail of the armorial achievements shown at the verge of the plate in Fig. 4.6, showing the arms of Graf von Bruhl being impaled by those of the Countess Kolowrat

gift for a personal friend of Augustus, namely Vittorio Amadeo II (1666–1732), Duke of Savoy and King of Sardinia. This comprised a large tea and chocolate service of 300 pieces, decorated with a fine polychrome chinoiserie of Japanese figures and the arms of the King of Sardinia. A part tea and chocolate group from this service was sold at Christie's in 2019 and realised £270 K, comprising six saucers and tea bowls, six chocolate beakers and saucers, a bowl and a teapot with a lid and a sugar bowl, in all 28 pieces. A chocolate beaker and saucer from this King of Sardinia armorial service and now resident in the Metropolitan Museum of Art, New York, is shown here in Figs. 4.9, 4.10 and 4.11. There are several features of interest in this item as it can be seen that the saucer displays the full escutcheon of the King of Savoy-Sardinia with his lion supporters and royal crown; this devolved from the original King of Sardinia coat-of-arms (Fig. 4.12, dating from 1720) in which the four moors sable on a red field with the white cross of Alcoraz (demonstrating the allegiance of Sardinia-Corsica with the Spanish house of Aragon-Castile) was replaced with the double eagle of the House of Savoy as shown here on the armorial porcelain saucer. Of even more interest is the armorial decoration on the chocolate beaker (Fig. 4.10), which does not match that of the saucer (Fig. 4.9) and instead displays the monogram of Vittorio Amadeo under his regal crown: unusually for Meissen too, the chocolate beaker does not have twin handles, nor does it possess a lid. Figure 4.11 shows the underside of the saucer bearing the blue enamelled mark *AR* of Augustus Rex.

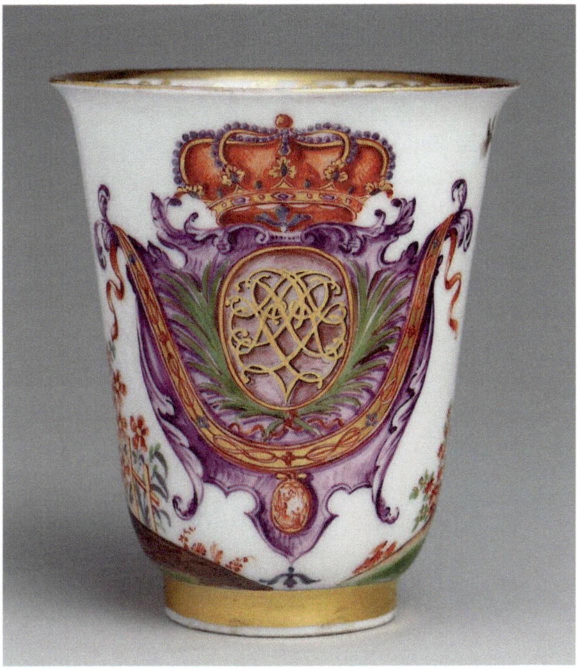

Fig. 4.9 Chocolate beaker from a Meissen porcelain armorial combined tea and chocolate service of 300 pieces manufactured as a diplomatic gift for Vittorio Amadeo II, King of Sardinia and Duke of Savoy, in 1725 by Augustus the Strong, Elector of Saxony and proprietor of the Meissen Porcelain Manufactory. The armorial shows the monogram of Vittorio Amadeo under his regal crown. Now in the Metropolitan Museum of Art, New York, the gift of R. Thornton Wilson in memory of Florence Ellsworth Wilson: accession number 54.147.75

4.3.5 Early and Mid-Eighteenth Century Armorials

Aside from diplomatic gifts, which were certainly still popular in the early eighteenth century as seen above for Meissen porcelain, numbers of services bearing arms are known which are predominantly composed of Chinese porcelain: typical examples are those of *Sir Thomas Stepney*, described by David Howard (*Chinese Armorial Porcelain*, 1974, p. 487) as "...*arguably one of the finest armorial services of this mid-Century date*", and the *Gripsholm Castle* service. These two service exemplars now deserve more detailed descriptions as follows.

4.3.5.1 The *Sir Thomas Stepney* Service

Sir Thomas Stepney Bt., 6th Baron Stepney of Prendergast, of Llanelly House in Llanelli, Carmarthenshire, who had multiple business and shipping interests, commissioned a large service of 366 pieces of Chinese porcelain from Canton in approximately 1760, it was certainly present in a detailed house inventory made by his wife Lady Elizabeth Stepney at Llanelly House in 1762. Although David Howard recorded that over 5000 armorial services in Chinese porcelain had been commissioned from British families in the 18th and early nineteenth centuries through the HEIC (The Honourable East India Company), this was only one of relatively few that

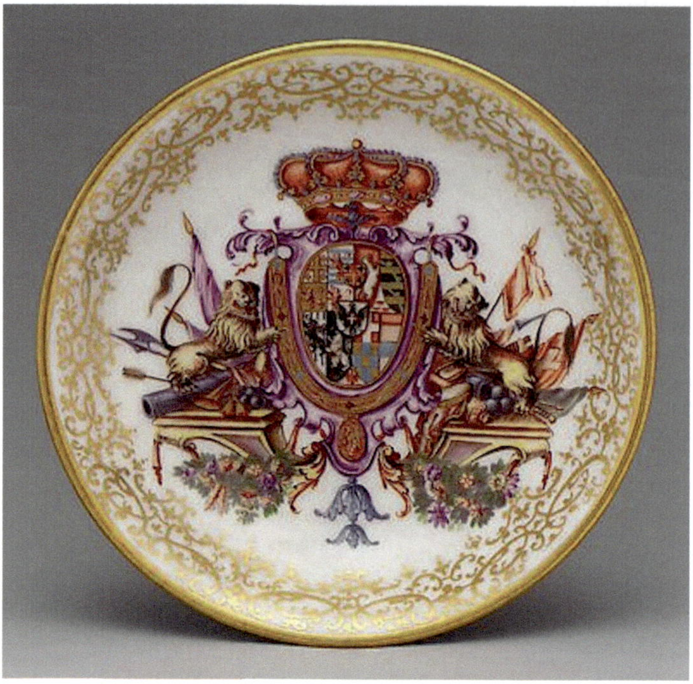

Fig. 4.10 Saucer from a Meissen porcelain armorial combined tea and chocolate service of 300 pieces manufactured as a diplomatic gift for Vittorio Amadeo II, King of Sardinia and Duke of Savoy, in 1725 by Augustus the Strong, Elector of Saxony and proprietor of the Meissen Porcelain Manufactory. The armorial shows the monogram of Vittorio Amadeo under his regal crown. Now in the Metropolitan Museum of Art, New York, the gift of R. Thornton Wilson in memory of Florence Ellsworth Wilson: accession number 54.147.76

were intended for Welsh estates. It was also one of the largest compound services commissioned and comprised two different types of similarly decorated dinner and dessert services with octagonal and circular shapes, as well as matching tea and coffee services. The items in Lady Stepney's inventory include the usual dinner and dessert plates and comports, salt trenchers, sauce boats, cruet bottles, mugs, a punchbowl, washbowls and water bottles (guglets) along with appropriate platters and stands. Lady Stepney was conscientious in her inventories of family expenditure and effects even though she rather grudgingly noted that "*I was steward to my husband, with this difference, that a steward has a salary and I had none*". An example of a plate from the octagonal *Sir Thomas Stepney* dinner service is shown in Fig. 4.13: the decoration shows the armorial escutcheon and supporters centrally located and the verge displays the crest and random sprays of garden flowers. Some 125 pieces of this service still survive and pieces occasionally come up for auction: the example shown here is reproduced with the kind permission of Bonham's Antiques of London, where it was described as being from the *Stepney* service which may have been commissioned by Sir Thomas Stepney, 4th Baronet, Sir John Stepney, 5th Baronet or Sir

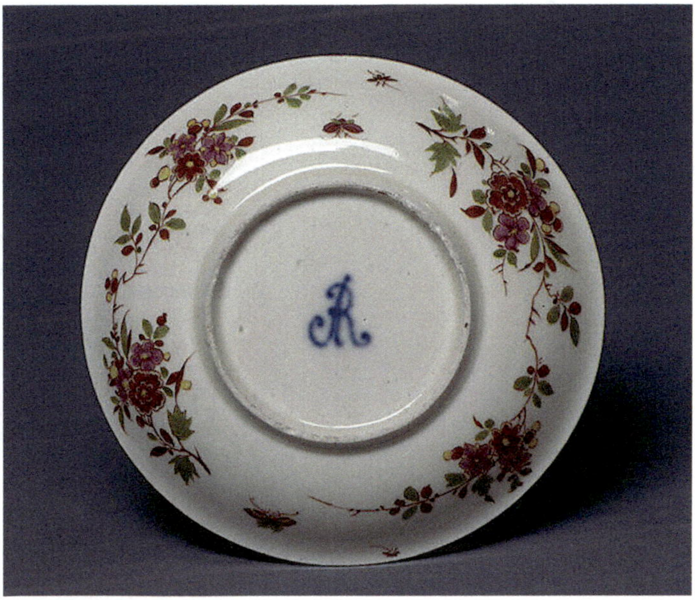

Fig. 4.11 Base of saucer from a Meissen porcelain armorial combined tea and chocolate service of 300 pieces manufactured as a diplomatic gift for Vittorio Amadeo II, King of Sardinia and Duke of Savoy, in 1725 by Augustus the Strong, Elector of Saxony and proprietor of the Meissen Porcelain Manufactory. The armorial shows the monogram of Vittorio Amadeo II under his regal crown. Now in the Metropolitan Museum of Art, New York, the gift of R. Thornton Wilson in memory of Florence Ellsworth Wilson: accession number 54.147.76

Thomas Stepney, 6th Baronet and that the armorial displays the arms of Stepney with Lloyd of Crogen in pretence. This rather general statement requires some additional genealogical and heraldic research to better define and properly attribute this important armorial service.

Elizabeth Eleanora Lloyd was born on the 28th March 1725 at Derwydd, Llandybie, Carmarthenshire, the only child of Thomas Lloyd of Trehir, Danyrallt, and Elizabeth Vaughan. She died on June 7th 1795 at Derwydd and is buried at Llandybie. She married Thomas Stepney in 1742 and he died in 1772, the baronetcy then passing to their son John. Consultation of the armories confirms that the armorial bearings represented on the plate shown in Fig. 4.13 are those of Sir Thomas Stepney with an inescutcheon of pretence which indicates that his wife was an armigerous heiress with no male siblings. The inescutcheon shows a lion rampant regardant sable, which is the armorial bearing of Lloyd of Trehir, Danyrallt, Carmarthenshire, which as expected would be borne by his heiress daughter, Elizabeth. However, The Bonham's description suggests that this belongs to Lloyd of Crogen: in the armories, Lloyd of Croghan (*sic.*), Roscommon, Ireland, bears a stag's head couped pp. the neck and charged with a laurel chaplet and this certainly does not match this assignment. The crest on the plate is the Stepney crest, namely, *a talbot's head erased gules, eared or, gorged with a collar chequy of the last and azure, holding in the mouth*

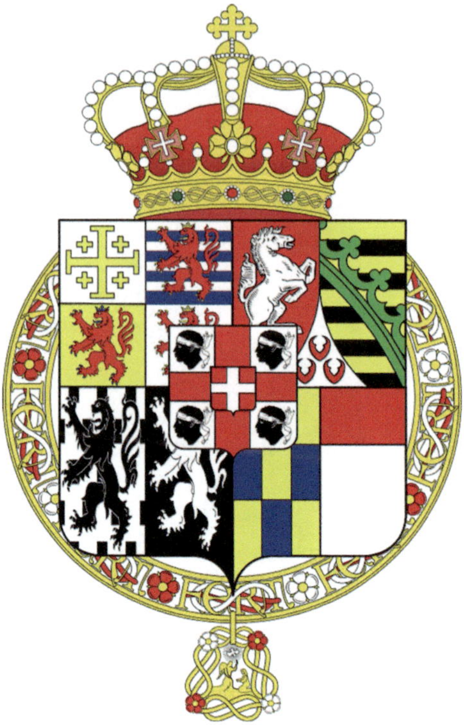

Fig. 4.12 Coat-of-arms of Vittorio Amadeo II, King of Sardinia and Duke of Savoy, ca. 1720 in which his allegiance to the Spanish House of Aragon and Castile is demonstrated with the emblazoning of the shield showing four moors head sable on a red field with the white cross of Alcozar; a substitution of this for the double-headed eagle of Savoy was made in 1724/5 to stress the new allegiance. *Public domain*

a buck's attire of the second. This accurately matches with the heraldic description provided in Fairbairn (plate 313/5) (Fairbairn, *Crests of the Families of Great Britain and Ireland and Their Kindred in Other Lands*, 1905): incidentally, there are 80 crests cited in Fairbairn for the surname Lloyd, so it is not a simple task to assign them meticulously. Also, the genealogy is quite clear that Elizabeth Lloyd was from the Derwydd, Danyrallt line and was not armorially related to the Lloyds of Croghan, Ireland. The motto "*Fide et Vigilantia*" is that of Stepney and translates as "*Trust and Vigilance*" and is credited in the armories to Lloyd of Danyrallt and not to Lloyd of Croghan (or Crogen as it is sometimes spelled). It transpires therefore that the Stepney service is truly that of Sir Thomas and Lady Stepney (nee Elizabeth Lloyd of Trehir and Derwydd), 6th Baron Prendergast.

4.3.5.2 The *Gripsholm Castle* Service

Although the Swedish SOIC operation was not as large as their analogous British HEIC or Dutch VOC companies, some detailed records exist which indicate the importance of the trade that they carried out importing Chinese porcelain into Europe in the eighteenth century (Koninckx, 'The maritime routes of the Swedish East India Company during its First and Second Charters, 1731–1766', *Scandinavian Economic*

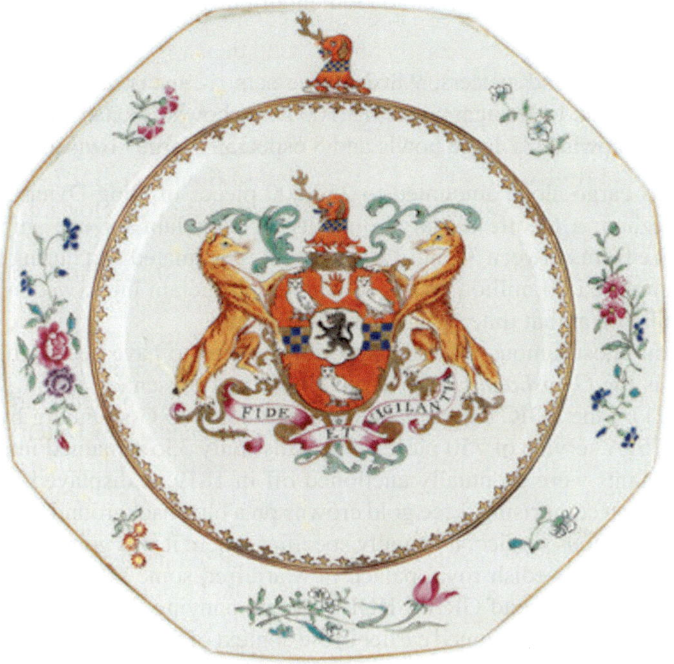

Fig. 4.13 Octagonal porcelain dinner plate from the *Sir Thomas Stepney* service in Chinese porcelain showing the arms of Stepney with those of Lloyd of Trehir and Derwydd in an escutcheon of pretence, *ca.* 1760. The crest is that of Stepney. *Reproduced with permission of Fergus Gambon, Bonhams Antiques and Valuers, Bond Street, London*

History Review, 1978). The first Swedish East Indiaman sailed in 1732 for Canton and returned in 1733 to Gothenburg. This first cargo of porcelain was auctioned off at a 100% profit after all expenses, which set the scene for future successful business ventures and expeditions of this sort. In all, the SOIC made 132 trips involving 37 different ships (with a loss of 8 ships and their cargoes in storms) in total up to their last voyage, which was accomplished in 1804. Colin Campbell, the supercargo on the first Swedish ship in 1732, ordered his own armorial service in Chinese porcelain, bearing his coat-of-arms and his motto, "*Memento Deus dabit Vela*" which translates as "*Remember it is God who gives us the Sails*", which is appropriate for a seafarer! Claes Grill, a director of a shipyard in Stockholm, ordered no fewer than eight porcelain Chinese armorial services with the SOIC between 1740 and 1780. In the 1750s, Count Frederik Gyllenberg commissioned an armorial service from the SOIC with his family coat-of-arms; unfortunately, the requisite watercolour drawing of the heraldic achievement provided to the supercargo for the guidance of the Chinese enamellers suffered some serious damage on the voyage out to China, but the Chinese enamellers in Canton did manage nevertheless to depict the coat-of-arms quite accurately, including the damaged areas. The cargo listing of the SOIC East

Indiaman "*Freden*" which returned to Gothenburg in 1746 comprised the following items:

> 149,668 teacups and saucers, 9,862 coffee sets, 129 sets tableware, 293 butter boxes and plates, 1840 sugar bowls, 13,160 white bowls, 23,036 sweetmeat plates, 41,140 rice bowls, 387 large bowls and 8 especially large bowls.

In all, this cargo alone amounted to 250,000 pieces of Qing Dynasty porcelain from Jingdezhen: a manifestly large quantity of porcelain for one ship. The sets ranged in size from between 37 and 120 pieces. It is estimated that during its lifetime the SOIC imported 50 million pieces of Chinese porcelain into Sweden, a country with a population at that time of just 2 million people.

One of the most famous Chinese armorial services to have been commissioned in Sweden is the *Gripsholm Castle* service: this was presented to King Gustav III (1746–1792) by the SOIC directors and brought home to Gothenburg by the *Terra Nova* in 1776. A service of 750 pieces originally, only 336 remained intact in 1792 and the remnants were eventually auctioned off in 1819. It displayed the Swedish royal escutcheon comprising three gold crowns on a blue background and, unusually for such a prestigious service, it actually specifies where it was going to be used—in Gripsholm Slot, a Swedish royal palace in Mariefred some 60 km west of Stockholm—with the gilt legend GRIPS HOLM accompanying the heraldic escutcheon (Fig. 4.14). It has been mentioned earlier in the context of the identification of heraldic crests and escutcheons in armorial services that this is indeed a major item of useful information for ceramic historians. The design of this service is still reproduced today in Sweden and is used for formal meals in Swedish embassies around the world.

4.3.5.3 The Richardson-Francis Armorial Punch Bowl

Independent research on armorial ceramic punchbowls of the eighteenth and nineteenth centuries (Edwards, *Welsh Armorial Porcelain: Nantgarw and Swansea Crested China*, 2022b) has revealed that several display their origin, the person who commissioned them, their place of use and the purpose of their creation in addition to their heraldic bearings, as shown in the *Gripsholm Castle* service described above, and of course this can be of an inestimable value for a ceramics historian and researcher. This is not confined solely to punch bowls, but being some of the largest pieces of ceramic ware made they afforded an opportunity to display legend and scenes which have relevance to their manufacture and commission: these often include a date, which is rarely encountered generally in armorial services, and the earliest piece of armorial porcelain of this type recorded is a Liverpool porcelain punchbowl dated 1758. An exemplar of the type of information that can be gleaned from such an armorial piece is the Richardson-Francis punchbowl from the Cambrian Pottery, Swansea.

This large Cambrian Pottery, Swansea, earthenware ceramic punch bowl manufactured by Lewis Llewellyn Dillwyn, of diameter 14.5″ and height 8″, displays the escutcheon of the arms of John Richardson, marking his position as Mayor of

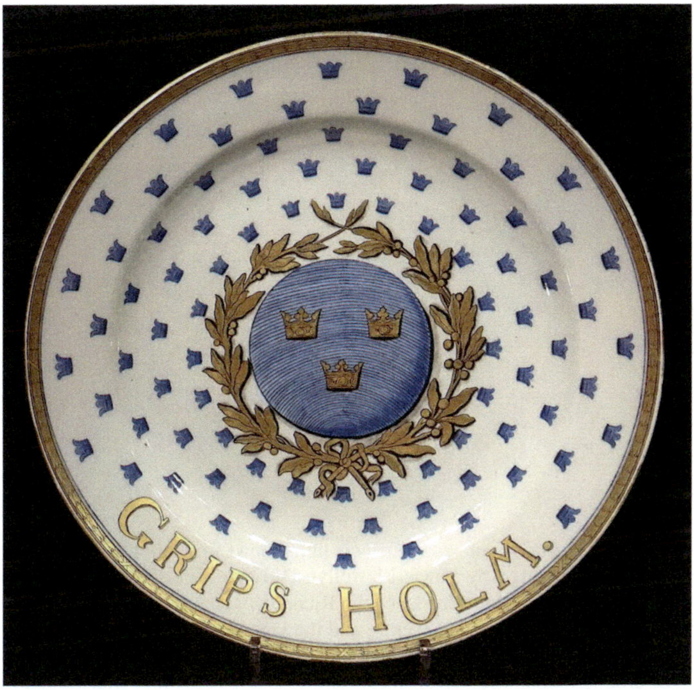

Fig. 4.14 Armorial plate from the *Gripsholm* service, *ca.* 1776, Qing Dynasty, Jingdezhen porcelain which was presented as a service of 750 pieces to King Gustav III of Sweden by the Directors of the SOIC, Gothenburg, the Swedish East India Company, for use in his royal palace at Gripsholm Slot, Mariefred, 60 km west of Stockholm. Now in the Ostasiatiska Museum, Stockholm

Swansea in 1845: his coat-of arms is *sable on a chief argent, three lions heads erased ermines langued gules* (Fig. 4.15). His crest as given in the armory is not shown on the bowl. Under the escutcheon is a statement "*Maiori vi(lle)ae Swansea*". An inscription on a continuous scroll inside the bowl (Fig. 4.16) reads

> This bowl is presented by Jno. Richardson Esqr. Mayor of the Antient Borough of Swansea to his grandson Jno. Crow Richardson-Francis, as a birthday gift this Vth day of February in the year of our Lord MDCCCXLV. Gaily still our moments roll whilst we quaff the flowing bowl.

The exterior is coloured with a deep cobalt blue ground underglaze with an overglaze enamelling of the arms of the County of Glamorgan, Swansea, and of John Richardson. Underneath the bowl is written in script a description of the arms and "Made at Dillwyn's Pottery, Swansea" and "*Painted by Stephen Dingley at the charge of Jno. Richardson Esqr. Designed by Geo. Grant Francis FSA*". Figure 4.16 shows the punch bowl from the other side depicting the coat-of-arms of the town of Swansea, comprising a fortified embattlemented gateway and portcullis surmounted by an osprey devouring a fish. John Richardson-Francis was the eldest son of (Colonel) George Grant Francis who was a respected early historian of the Swansea China

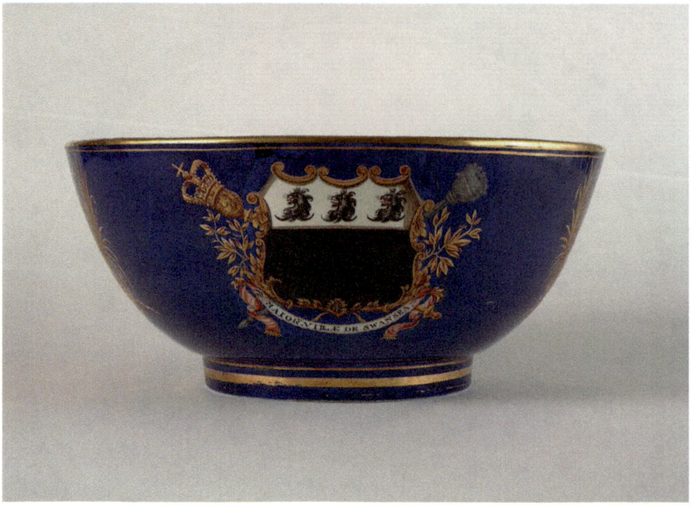

Fig. 4.15 The large armorial *Richardson-Francis* punchbowl in earthenware from the Cambrian Pottery, Swansea, manufactured by Lewis Llewellyn Dillwyn, son of Lewis Weston Dillwyn, in 1845, showing the escutcheon of the arms of John Richardson, comprising three lion's heads sable on a chief fess argent, with a mace and sceptre and the legend MAIOR.VILL.AE DE SWANSEA to celebrate his becoming Mayor of Swansea. The crest is absent. The bowl has a deep cobalt blue underglaze ground colour with overglazed enamels of the arms of Richardson, co. Glamorgan and Swansea. *Copyright and reproduced with permission from the National Museum of Wales, Amgueddfa Cymru, Cardiff*

Works and Pottery and whose personal interview of Henry Morris in the 1850s has provided anecdotal information about the Swansea China Works and its personnel, occasioning much debate and discussion from later ceramics historians (Edwards, *Porcelain Analysis and Its Role in the Forensic Attribution of Ceramic Specimens*, 2022a). His wife, Sarah, was the eldest daughter of John Richardson. It is believed by connoisseurs that Stephen Dingley was responsible for the enamelling of the actual armorial only as the painting is visibly superior to that executed for the accompanying border of grapes and vine leaves (Morton Nance, *The Pottery and Porcelain of Swansea and Nantgarw*, 1942). Cross-checking the armorial bearings in Burke (*Burke's General Armory—The General Armory of England, Scotland, Ireland and Wales: Comprising a Registry of Armorial Bearings from the Earliest to the Present Time*, 1884) and in Fairbairn (Fairbairn, *Crests of the Families of Great Britain and Ireland and Their Kindred in Other Lands*, 1905) reveals that John Richardson JP, of Pantygwydyr, co. Carmarthen, was descended from the Richardsons of Durham and had exactly the same escutcheon as that which is displayed on the bowl, with a crest (Fairbairn, *Crests of the Families of Great Britain and Ireland and Their Kindred in Other Lands*, 1905, Plate 48/12) of a mural crown or a lion's head erased and with a motto "*Praetio prudentia praestat*" which means "*Wisdom is beyond Price*". No fewer than 71 Richardsons appear in Fairbairn's lists with various crests and mottoes and 84 Richardsons in Burke (*Burke's General Armory—The General*

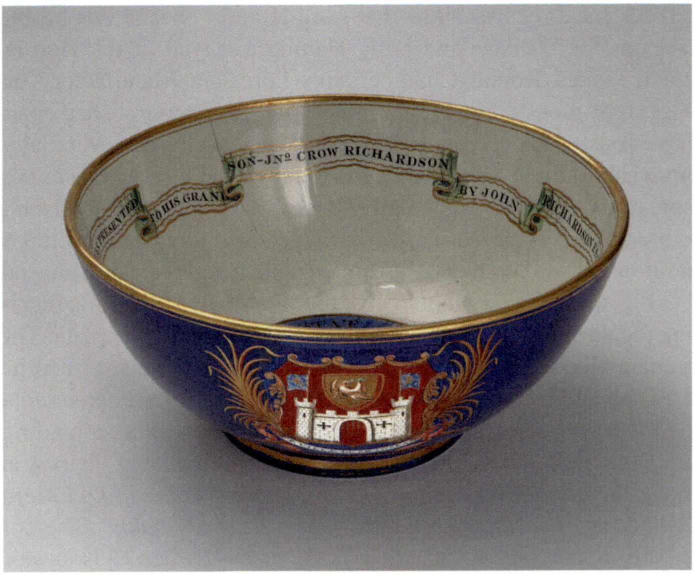

Fig. 4.16 The armorial punchbowl in Fig. 4.15 is dedicated to John Crow Richardson-Francis, his grandson, through the marriage of his daughter to Colonel George Grant Francis. Here the coat-of-arms of the town and borough of Swansea are shown: an embattlemented gateway of twin towers surmounted by an osprey devouring a fish. Underneath is enamelled a statement signifying it was made at Dillwyn's Pottery, Swansea, and "Painted by Stephen Dingley at the charge of Jno. Richardson Esqr. and rdesigned by Geo. Grant Francis FSA". *Copyright and reproduced with permission from the National Museum of Wales, Amgueddfa Cymru, Cardiff*

Armory of England, Scotland, Ireland and Wales: Comprising a Registry of Armorial Bearings from the Earliest to the Present Time, 1884) but the grandson, Richardson-Francis, mentioned on the bowl does not appear entitled to bear arms and is missing from these lists. Burke is more descriptive of the genealogy of Richardson, which definitively identifies him and his descendants as having settled in Glanbrydan Park, Manordeilo, co. Carmarthen. It is quite unusual to discover an armorial piece that is so descriptive of the entitled bearer of the arms and the reason for its commissioning. Fairbairn (Fairbairn, *Crests of the Families of Great Britain and Ireland and Their Kindred in Other Lands*, 1905) gives the Francis family crest as an eagle perched on a fructed tree stump—which appears very similar to another crest of a bird similarly perched and appearing on Swansea porcelain (Edwards, *Welsh Armorial Porcelain: Nantgarw and Swansea Crested China*, 2022b).

4.3.5.4 The *Horatia* Service

A 150 piece, 12-place setting, breakfast service of armorial porcelain was ordered by Admiral Lord Horatio Nelson from Robert Chamberlain's Worcester manufactory in August 1802; local records show that Lord Horatio Nelson, his brother the Rev William Nelson, Sir William and Lady Hamilton stayed at the Hop-Pole Inn in Worcester and visited Robert Chamberlain's Porcelain Manufactory on the 27th August, where they met staff including James Plant, an accomplished enameller who may well have assisted in decorating the service. The service was eventually delivered to Mrs Emma Hamilton at Merton Place in Surrey, where some pieces were used in house and the remainder taken on board *HMS Victory* for Admiral Nelson's use there at sea. This breakfast service was known as the *Horatia* service after Lord Nelson and Emma Hamilton's baby daughter. A specimen plate from a Nelson armorial service is illustrated in Fig. 4.17: readers should note the wealth of information that is available for checking the attribution of the heraldic achievements on this plate. The centre of the plate is enamelled with the full coat-of-arms of Horatio Nelson, incorporating the Order of the Bath and its motto of *Tria Juncta in Uno*, "*Three Joined into One*", and its two supporters of a sailor holding a pennant and a lion, the latter holding a tattered French flag. Both supporters hold a palm branch, symbolising victory, and a palm appears above the escutcheon with the motto below "*Palmam Qui Meruit Ferat*", which translates as "*Let him who has earned it bear the palm*". The coat-of-arms dates from Nelson's victory at Cape St Vincent in 1797 when Nelson was conferred with the Order of the Bath by King George III. The escutcheon comprises *a black cross flory on a shield or with a bend gules surmounted by another bend engrailed gold thereon with three bombs fired proper and surmounted by a baron's coronet.* He was created Baron Nelson of the Nile in August 1798 following the Battle of the Nile, where he annihilated the French fleet in Aboukir Bay. After the battle he was invested with a chelengk by the Sultan of Istanbul, which he adopted as his dexter crest, the sinister crest being the stern of the 112-gun battleship *San Jose* that he captured at St Vincent. In 1799 he was given the title of Duke of Bronte by King Ferdinand III of Sicily which he much appreciated and later adopted in his signature, Nelson & Bronte. After the Battle of Copenhagen in May 1801, Nelson was created Viscount Nelson of the Nile and Burnham Thorpe, and in August of the same year Viscount Nelson of the Nile and Hilborough. His escutcheon, now surmounted by his Viscount's coronet, was amended following the Battle of the Nile to include *a chief undulated argent and thereon a wave of the sea with a palm tree between a disabled ship on the dexter side and a destroyed battery in ruins on the sinister, all proper.* It is this latest coat-of-arms which appears on the *Horatia* service plate shown in Fig. 4.17. The border is richly decorated in a red and blue Imari type Japanese pattern including foliate panels and green birds. The plate is marked Chamberlain's Worcester pattern 240.

Reference to the order books at Chamberlain's Worcester factory reveals an interesting piece of information: Nelson's choice of the Imari "Fine Old Japan" pattern 240 was a costly and flamboyant one and he insisted that each piece bore his coat-of-arms, insignia and heraldic crest—he originally contemplated three armorial services, namely, a breakfast service, dessert service with ice pails and a dinner service. In the end only the breakfast service was completed and delivered, whilst it is believed that only specimen plates were prepared for the dessert and dinner services, and one of these is featured in Fig. 4.17. The breakfast service tableau was much simpler and

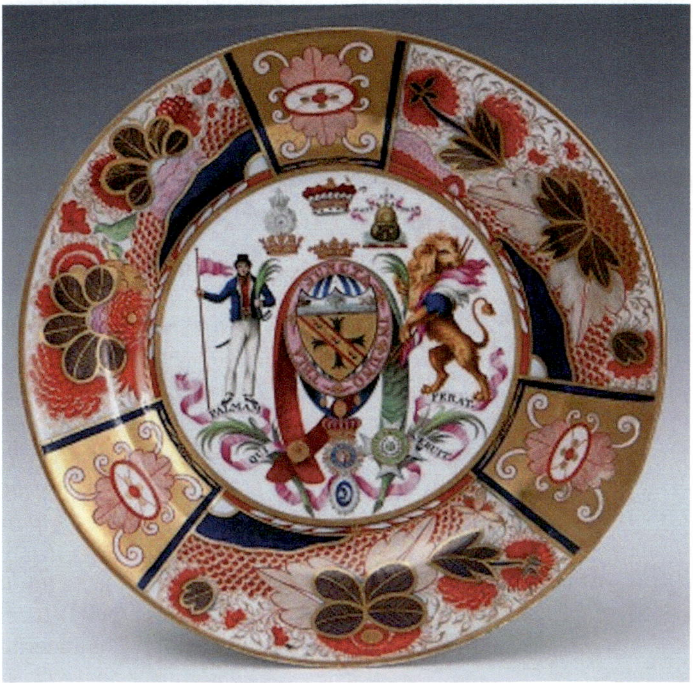

Fig. 4.17 Specimen plate from the proposed Admiral Lord Nelson armorial service in Chamberlain's Worcester porcelain, 1802, displaying his full coat-of-arms and investiture into the Order of the Bath. It appears that the service was never actually ordered by Nelson, although specimens of this type are still sometimes referred to in the literature as the *Lord Nelson* service or even mistakenly as the *Horatia* service. *Reproduced courtesy of Bonham's Antiques and Valuers, Bond Street, London*

as shown on the hot water dish depicted in Figs. 4.18 and 4.19, which features one of Nelson's crests and coronet only in three vignettes at the verge and only the two teapots in this service bore the full coat-of-arms as shown in Fig. 4.17. A similar hot water dish is displayed in the National Maritime Museum, Greenwich, London, and also the two teapots which bear the full coat-of-arms. Heraldic emblazoning on china was very expensive and workshop records at Chamberlain's factory indicate that the individual crests on the breakfast service were costed at between one shilling and two shillings and sixpence each in comparison to the four full coats-of-arms on the two teapots costed at £6. Hence each full coat-of-arms as shown on the specimen plate in Fig. 4.17 would have incurred an additional cost to Nelson of £1.10s-0d per item! At this stage Nelson was acutely aware of financial restrictions operating at his country estate of Merton Park and hence the order for the subsequent dessert and dinner services was never placed with the factory. The specimen plates it is believed were returned to the Worcester Porcelain Manufactory when the order was not proceeded with.

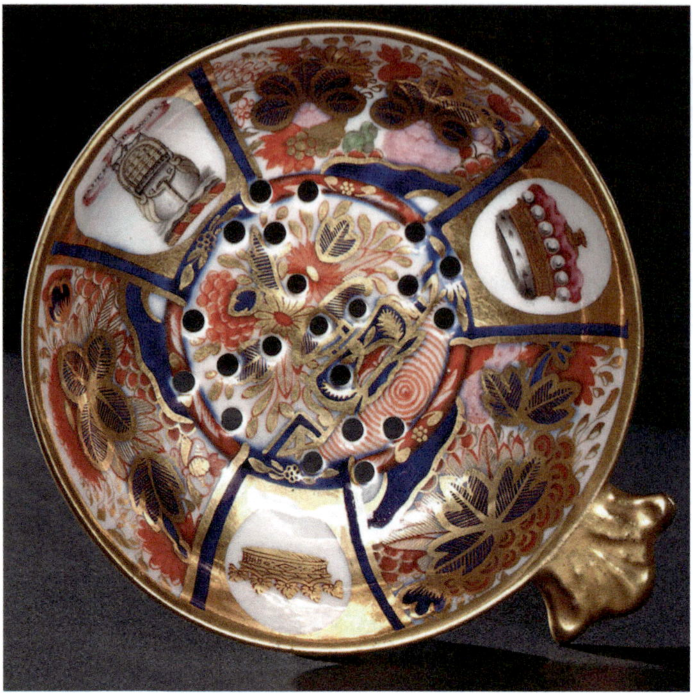

Fig. 4.18 Warm water strainer dish from the *Horatia* service, a breakfast service of 150 pieces commissioned by Admiral Lord Horatio Nelson from Robert Chamberlain's Worcester Porcelain Manufactory in August 1802 in the "Fine Old Japan" Imari pattern, marked 240 underneath in purple enamel. This features three panels in the verge which display the crest of the stern of the Spanish Battleship *San Jose,* captured by Nelson at the Battle of Cape St Vincent in 1797, a Viscount's coronet and a ducal coronet (Nelson had also been as the Duke of Bronte in Sicily). The remainder of the piece is decorated in the red, white and blue colours of the Imari pattern that was popular at that time. *Reproduced with permission from Martyn Downer Works of Art Ltd., Puttock's End, Anstey, Hertfordshire, UK*

Mrs. Emma Hamilton eventually received the invoice for the *Horatia* breakfast service from Chamberlain's Worcester factory, which was rather tactlessly presented to her during the week of Nelson's State funeral following his death at the Battle of Trafalgar (21st October 1805) in January 1806; the amount requested was £120 10s 6d, which today would be equivalent to approximately £12,000. An inventory of Emma Hamilton's assets shortly thereafter indicated that only 104 pieces of the *Horatia* service then remained and these were inherited by William, the first Earl Nelson and Horatio's brother, and eventually were dispersed amongst the family and at auction. As expected, when such provenanced historical armorial china becomes available at auction it commands premium bids from collectors and museums.

It would be appropriate here to mention a few sentences relating to Horatia Nelson, after whom this service is named. Horatia was born to Emma Hamilton in 23, Piccadilly, London, a house rented by Sir William Hamilton, and was the illegitimate

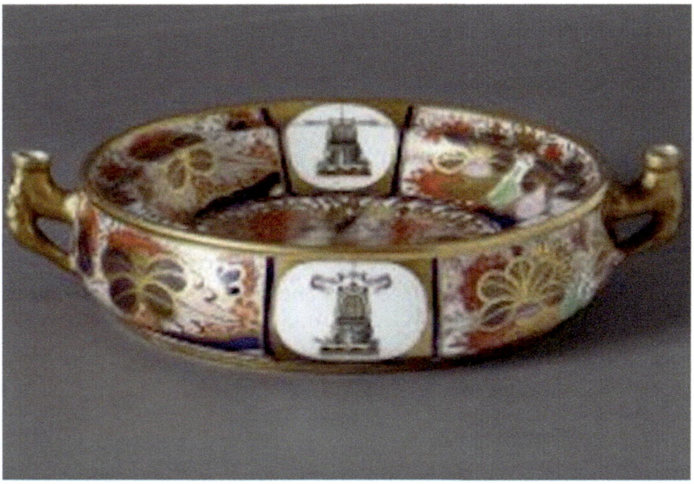

Fig. 4.19 Another view of the warm water strainer dish shown in Fig. 4.18 from the *Horatia* service, a breakfast service of 150 pieces commissioned by Admiral Lord Horatio Nelson from Robert Chamberlain's Worcester Porcelain Manufactory in August 1802 in the "Fine Old Japan" Imari pattern, marked 240 underneath in purple enamel. This features prominently the crest of the stern of the Spanish Battleship *San Jose,* captured by Nelson at the Battle of Cape St Vincent in 1797. The remainder of the piece is decorated in the red, white and blue colours of the Imari pattern that was popular at that time. *Reproduced with permission from Martyn Downer Works of Art Ltd., Puttock's End, Anstey, Hertfordshire, UK*

daughter of Horatio Nelson; her birth date was masked for some time to pretend that she was the adopted child of Sir William and Emma Hamilton following a liaison with another fictitious naval officer, after her being orphaned in Naples. However, it eventually emerged that she was born truly in January 1801. After Sir William had died in 1803, it was straightforward for Nelson and Emma to engage in their relationship and Nelson moved his new adopted family to Merton Park in Wandsworth, Surrey. Nelson adored Horatia and wrote to her a final letter as the Franco-Spanish fleet was coming out from Cadiz to do battle with him at Cape Trafalgar. Documentation surfacing in the mid-1800s confirmed that Horatio Nelson was Horatia's father. Emma Hamilton was rather neglected by the country after Trafalgar and experienced extreme poverty and debt, and she died in 1815 in Calais. The Rev William Nelson succeeded to the Viscountcy in 1805. The situation was partially rectified for Horatia, who married the Rev Philip Ward, vicar of Tenterden, in 1822 in the 1840s with the upsurge of interest generated by the erection of Nelson's Column in Trafalgar Square. She bore ten children between 1822 and 1836 and she died in 1881, her husband pre-deceasing her in 1859. Horatia never recognised Emma as her mother, but she was intensely proud of her father: her tombstone in Pinner Parish Cemetery reads, "Emma Ward, beloved daughter of Vice-Admiral Lord Nelson and widow of the Rev. Philip Ward".

4.4 Earliest English Porcelain Armorials

With such a palpable demonstration of the surge in the commissioning and use of Chinese armorial china services in the eighteenth century we need to consider the potential reasons underlying the hiatus that seems to exist in the apparent failure of the English manufactories to match the need of the armigerous clientele for a local equivalent. The earliest English porcelain armorial pieces that can be assigned to porcelain dinner services can be dated to around 1767/1768 from the Worcester manufactory: over the next decade, similar pieces are found from other factories such as Derby. The fundamental questions that may illuminate the reason for this chronological hiatus of some 50 or 60 years, during which the armorial porcelain provided for British families was exclusively originating in China, need now to be addressed:

- What was the reason for the failure of English manufactories to respond to the Chinese challenge for the decorating of armorial porcelain for local consumption ? Even in the 1760s, the prestigious porcelain decorating atelier of James Giles in Haymarket, London, which was well patronised by the aristocracy, only decorated Chinese porcelain that was bought "in the white": a good example of this procedure is the armorial teapot shown in Fig. 4.20, this is in Chinese porcelain which was decorated by James Giles, displaying an as yet unidentified and rather fanciful baroque escutcheon containing three lions heads sable and couped, and random sable pheons on a gold field and a crest of a golden hunting dog (wolf ?) on a gold and sable torse (Hanscombe, 'Armorial Porcelain Decorated in the London Workshop of James Giles', *Transactions of the English Ceramic Circle*, 2008).
- By the mid-1760s, several English manufactories were producing quality porcelains to rival those of the Chinese, yet none seem to have embarked upon an armorial theme, or at least most of their surviving pieces do not indicate that they had done so. Examples of the English manufactories that by this time had a following from clients are Bow, Chelsea, Limehouse, Vauxhall, Lowestoft, Liverpool, Worcester and Derby. An armorial Liverpool punchbowl is known which is dated 1758 but this is a noteworthy exception.
- It is perhaps not appreciated fully that a determining factor could well have been the time lapse incurred between the ordering and delivery of a porcelain service from China: this was typically about two to three years and was necessitated by the prevailing winds required for the access of sailing ships to the Far East ports of Canton and Nanking. The ships usually departed from London in January for South Africa, rounding the Cape of Good Hope and then proceeding on to India where they had to await favourable winds in the monsoon season to journey onward to China across the South China Sea. The order for the porcelain would then need to be placed with agents in Canton for its manufacture and decoration in Jingdezhen, some 1000 km distant. The return journey had to be accomplished similarly and the whole process repeated a year later to collect the completed porcelain orders. Did this factor generate the need for English contemporaries to compete successfully in the 1770s for Chinese armorial porcelain commissions?

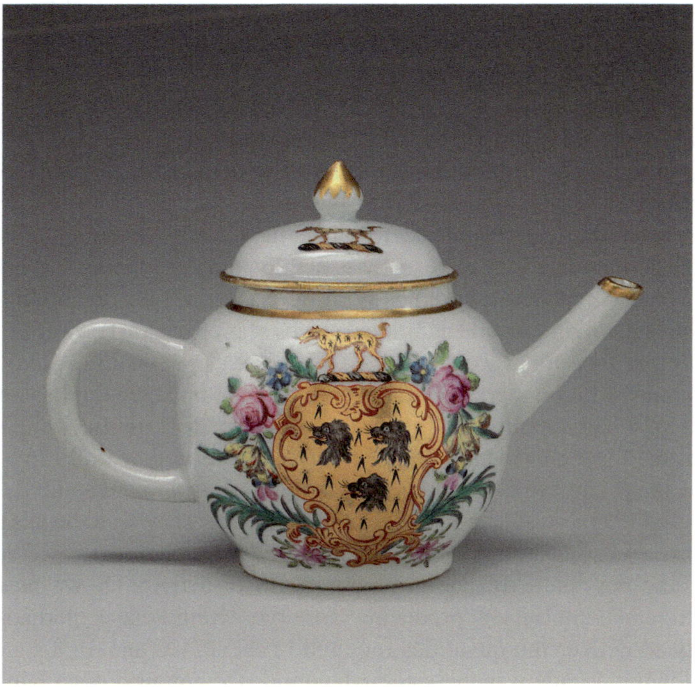

Fig. 4.20 A Chinese hard paste porcelain teapot decorated in the James Giles workshop in Haymarket, London, bearing an unidentified armorial escutcheon dating from the period 1760–1770, which is now in the Helena Woolworth McCann Collection of the Metropolitan Museum of Art, New York (Winfield Foundation; Accession Number 51.86.394 a, b). The escutcheon shape is of an unusual Italianate type and contains three wolf's heads couped armed sable on an erminois field, with the gilt crest of a wolf ambulant erminois on a sable and or torse. *Public domain*

- This time element could be significant, as if a porcelain armorial service was being commissioned for a special event or commemoration some forward planning was necessary: this was out of the question for commemorating the birth of a child, succession upon a death, a marriage or the award of an honour or a rank in the peerage!
- Could there be another reason for the reluctance of English manufactories to undertake armorial china manufacture initially? The size of the Chinese manufacturing operation was immense: observers have noted that the dragon kilns at Jingdezhen could hold up to 25,000 individual pieces of porcelain to be fired, so even after firing losses, significant quantities of china was available for making up into large services, and several hundred pieces in each of these commissions would not be abnormal. In contrast, the kilns at the English porcelain factories were perhaps two orders of magnitude smaller than their Chinese counterparts, accommodating perhaps just a few hundred pieces in one firing sequence. Even at the end of the eighteenth century, William Billingsley, an experienced porcelain manufacturer, wrote to his proprietor at the Pinxton porcelain manufactory

in 1796, John Coke, that a typical kiln charge was some 20 dozen items, and that was before kiln losses were incurred during and post-firing, that could amount to some 25% or more !

This particular theme is in contrast with the activities of the Meissen manufactory already surveyed above, where some very large porcelain services were created in the early to mid-eighteenth century: perhaps the explanation for this discrepancy in production capability lies in the fact that Meissen produced a genuine hard paste porcelain, exactly as did the Chinese manufactories, whereas the English manufactories were intent generally upon producing soft paste porcelains, which demanded a completely different set of kiln operating and firing procedures involving several additional stages of preparation (Edwards, *Porcelain Analysis and Its Role in the Attribution of Ceramic Specimens*, 2022a)?

- The obvious solution therefore, at first, would be to engage with local decorators who already had supplies of Chinese porcelain in stock in the white to order a service and this seems to have been carried out in the early to mid 1760s and what is more, the accuracy of the armorial decoration being applied could be inspected readily and interactively, and altered quickly if it was proving to be unsatisfactory or incorrect in some detail.
- Hence, the possible explanation of the dearth of activity shown by the major English manufactories in producing armorial china relates perhaps to their capacity to deliver the quantities required in the 1740s and 1750s and only in the following decade was there enough confidence amongst porcelain manufactory proprietors to embark upon the necessary volume of production required for armorial services?
- A related question on this theme is convolved with the reason that armorial china was required in the first place? What perceived need catalysed its inception around the turn of the seventeenth and eighteenth centuries away from the idea of it being just used for a diplomatic gift as seemed to be its *force majeure* at that time, whether armorial or not, in the seventeenth century? The authors believe that the use of armorial bearings such as crests on silver plate and especially table wares in the seventeenth century would give a natural impetus to its use as decoration on the dishes used at the same tables—in this, the use of silver engraved soup or sauce ladles, spoons and cutlery would certainly engender a desire to emulate the designs on the porcelain items with which they were being used. The advent of polychrome enamelling decoration in the early 1700s would only have enhanced this perception, when the full beauty of the armorial devices and achievements could now be attained, over and above that which could be demonstrated on the accompanying armorially engraved silver plate. This attitude was paralleled by the awareness of formal dining arrangements and an appreciation of classical and neo-classical themes for houses décor and design and social behaviour consequent upon younger members of aristocratic families returning home after making the European Grand Tour to Italy, Greece and Egypt in the mid-eighteenth century (Ferguson, '"Vase Madness": Vases and the Antique Taste in British Ceramics, 1765–1790', *A Taste of the Antique: The Neo-Classical Style and Ceramics in England*, 2018).

- Another avenue that requires a wider investigation than is appropriate here is the presence of armorial ceramics in European earthenwares, faience and majolica which have been found dating from the 15th and early sixteenth centuries: good exemplars of these are shown in Figs. 4.21 and 4.22, which are a basin in tin-glazed Manises (Valencia) earthenware with a brown lustre decoration which dates from about 1485 and bears the arms of Maria of Castile (Queen of Aragon and Naples and Princess of Asturias), and a similar period Italian tin-glazed maiolica dish from Nicola di Urbino bearing the arms of Isabella D-Este and Gianfranco Gonzaga dating to 1524, respectively. The latter is in an *istoriato* classical style on the theme of Hippolytus and Phaedra and includes a scene for Ovid, with a comment in Latin "*Without hope and without fear, you are defeated*".

This theme will be developed somewhat further below where several specific examples will be considered in some detail in the categories of non-porcelain European armorials in delftwares and stoneware from the 16th and 17th centuries.

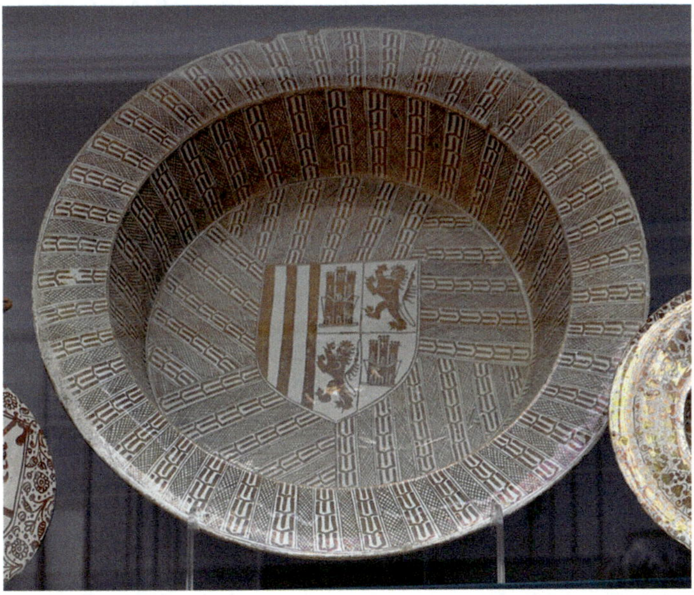

Fig. 4.21 Basin in tin glazed Manises (Valencia) earthenware with a brown lustre decoration which dates from about 1485 and bears the arms of Maria of Castile (Queen of Aragon and Naples and Princess of Asturias). Manises was famed for its lustre decoration which was achieved using a mixture of silver, copper, cinnabar and ochre ground in vinegar and fired in a reduction furnace. Victoria and Albert Museum. London, VA243-1853. *Public domain*

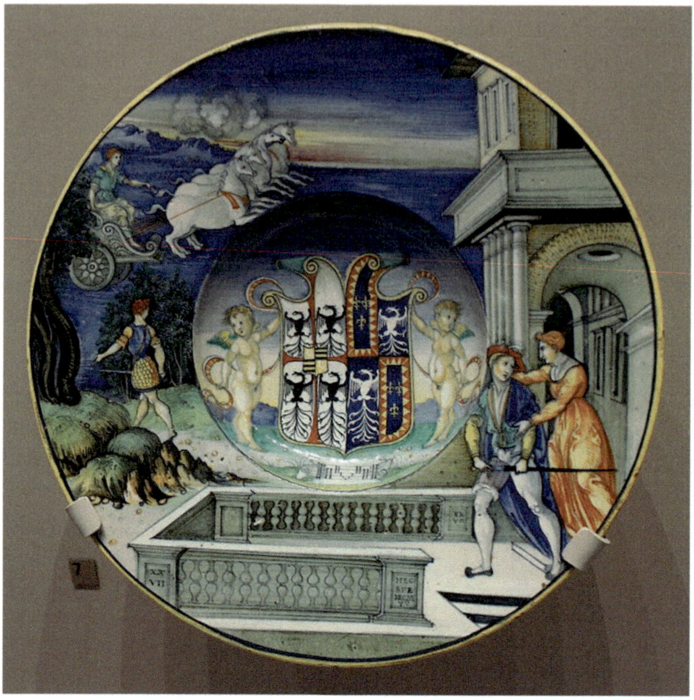

Fig. 4.22 Italian tin-glazed maiolica dish from Nicola di Urbino bearing the arms of Isabella D'Este and Gianfranco Gonzaga dating to 1524. This is decorated in an istoriato classical style on the theme of Hippolytus and Phaedra with a comment in Latin. *"Without hope and without fear, you are defeated". Salting bequest, Victoria and Albert Museum, London, C2229-1910, Public domain: credit Valerie McGlinchey*

4.4.1 An Early English Delftware Armorial Tankard

Recently, in December 2022, an early English dated tin-glazed earthenware (delftware) small tankard appeared at auction in Middle Claydon, Buckinghamshire: it was attributed to a London factory, possibly Southwark, and bears a bust length portrait of King Charles II in full armour and wearing a crown with a monogram CDR 2 and a date of 1660 (Fig. 4.23). His father, King Charles I, fought the English Civil War unsuccessfully against the Parliamentarians led by Oliver Cromwell between 1642 and 1646 and was executed in January 1649. His son, Charles, who had been declared King in Scotland, was defeated at the Battle of Worcester in 1651 and he exiled himself to France and The Netherlands. There followed the period of the Commonwealth under the Lord Protector Cromwell until he died in 1658: the Restoration of the Monarchy began in May 1660 when Charles was invited by Parliament to return to London on his 30th birthday, which was received with much public acclaim. It is believed that this tankard represented a celebration of the return of the Monarchy and can be considered an armorial because of the heraldic regal crown that it bears.

There are some sixteen surviving dated examples of such an armorial, most of which depict King Charles II with a crown in coronation robes bearing an orb and sceptre. His depiction wearing armour on such examples is rare and this tankard is one of only four examples known, three of which show the King in armour: the others are in The Museum of London (A.4946), Colonial Williamsburg, Virginia (1956-531) and one sold at Christie's in 2002 from the Margaret Cadman Collection. This tankard is now in the collection of E. & H. Manners Ltd, London, where it has been attributed to the Southwark factory based upon its stylistic similarity to other factory products (Lipski & Archer, *Dated English Delftware*, 1984; Britton, *London Delftware*, 1987; Archer, *Delftware: The Tin-glazed Earthenware of the British Isles*, 1997; Tyler et al., *London's Delftware Industry: The Tin-Glazed Pottery Industries of Southwark and Lambeth*, 2008).

An interesting feature of this particular tankard is that it was sent for testing to Oxford Authentication in October 2022 for potential analytical dating using the thermoluminescence technique, which involved the removal of two small samples

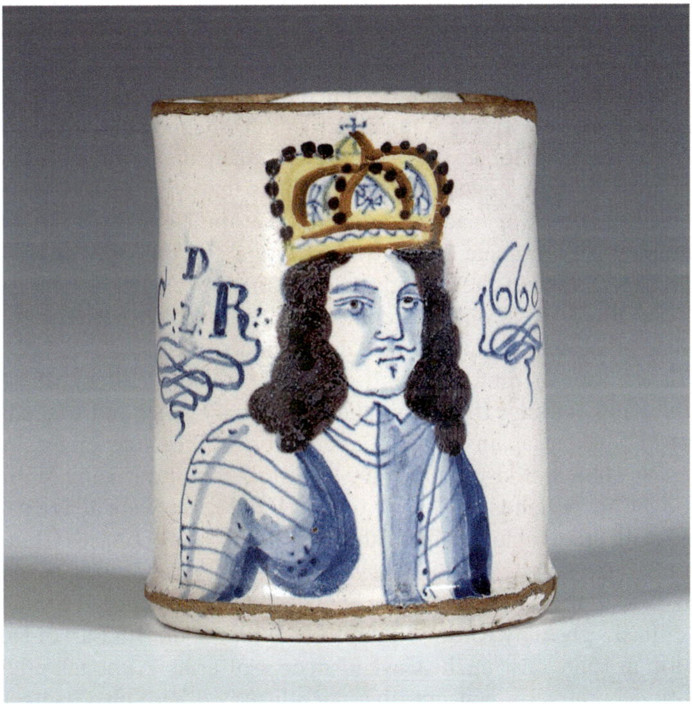

Fig. 4.23 English tin-glazed delftware armorial tankard, height 10 cm, diameter 7 cm, decorated in blue, ochre and purple pigments showing King Charles II in armour and bearing the monogram CDR 2, wearing the regal crown and dated 1660. Made in 1660 to celebrate the Restoration of the Monarchy. A rare example attributed to the Southwark earthenware manufactory, London. *Reproduced with permission from the Collection of E. and H. Manners Ltd., Church Street, Kensington, London*

using a drill from the base of the specimen: however, the report was disappointingly inconclusive because "the clay contained a spurious component which gave rise to an intense background signal making the analytical dating unreliable".

4.5 European Non-Porcelain Armorials: Four Specific Examples

4.5.1 A Delftware Diplomatic Gift

We have seen earlier that the diplomatic gifts of high-quality porcelain to cement alliances was highly esteemed in the sixteenth and seventeenth centuries and this has also been identified in other materials. A pair of delftware pilgrim flasks with screw lids have recently appeared with a clear diplomatic gift provenance (Errol Manners, personal communication). The pair of tin-glazed earthenware flasks are shown in Fig. 4.24, decorated in blue pigment on a white background and bearing the arms of Count Jacob Johann Hastfors, an influential member of the Swedish aristocracy in the Swedish Court: the flasks can be reliably dated to 1689–1690 and are marked AK for Adriaen Kocks of the De Griekse A manufactory in Delft and one with the monogram JW of Jacob Wemmersz Hoppestyn of the Het Moriaenshoofd manufactory in Delft. it is most unusual to find a matching pair of flasks that were made in two different manufactories: the first was made and decorated at the Het Moriaenshoofd factory but subsequent marks indicate that it was fired again at the De Griekse A factory some months later, whereas the second flask was made exclusively at the De Griekse A factory, and there are several differences to be observed between them in the organisational placement of the armorial decoration.

These flasks were a diplomatic gift from the Dutch Willem III of Orange, who became King William III of England in February 1689 and wished to secure the support of Sweden against French aggression in the Nine Years' War which commenced in 1688 and lasted until 1697. At this time both King William III of England and the Netherlands and King Louis XIV of France were actively seeking an alliance and cementing relationships with the Scandinavian countries (Hatton, *Gratification and Foreign Policy: Anglo-French Rivalry in Sweden During the Nine Years' War*, 1968). A second point of interest is revealed in the existence of a similar pair of Dutch silver flasks given by King William III to William, the 4th Earl of Devonshire, at Chatsworth in 1688, bearing the Cavendish coat-of-arms. Although pilgrim flasks were usually decorative in concept, the pair illustrated here also had a functional role in being used to store tea, which was an extremely valuable commodity in the seventeenth and eighteenth centuries. In his gift to Count Hastfors, King Willem III clearly wished to advance the appreciation and standing of Delftware as a diplomatic gift.

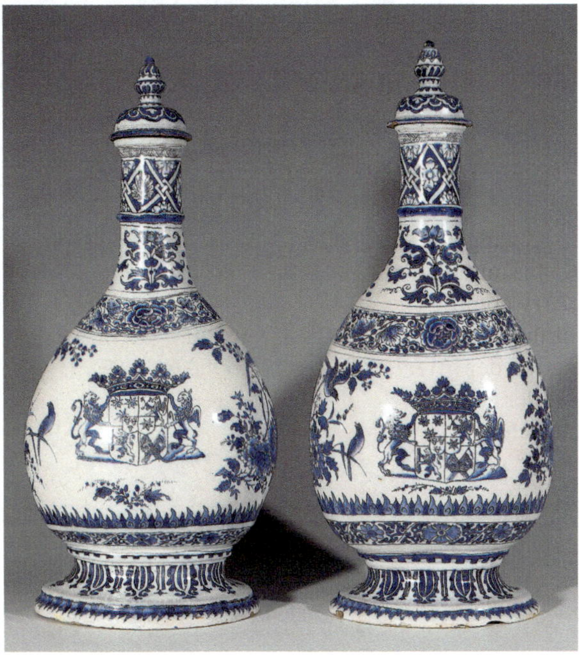

Fig. 4.24 Pair of tin glazed armorial earthenware pilgrim flasks, Delft, bearing the arms of Count Jacob Johann Hastfors of Sweden, 1689–1690, a diplomatic gift from King Willem III of the Netherlands. *Reproduced with permission from the Collection of E. and H. Manners Ltd., Church Street, Kensington, London*

4.5.2 A Spanish Pseudo-Armorial Dish

Figure 4.25 shows a Spanish tin-glazed armorial dish which displays an escutcheon beneath a stylised helm surrounded by dense polychrome foliage in yellow ochre, brown and manganese blue on a white background in the style of the Talavera *serie tricolor* wares and dating form *ca.* 1600–1625 (Ray, *Spanish Pottery, 1248–1898*, 2000). However, this item has been established to originate from Muel in Aragon (Errol Manners, personal communication) and in addition has a rather fanciful heraldry displayed on the escutcheon which firmly places it in the category of a pseudo-armorial, aligning it with the exemplars given earlier in Chapter 1.

4.5.3 A Portuguese Faience Pseudo-Armorial Jug

In Fig. 4.27 is shown an armorial Portuguese tin-glazed faience jug, *ca.* 1635–1640, made for export from Lisbon for the Northern European market in Germany and the Hanseatic States, in blue and yellow polychrome in imitation of the underglaze blue Ming porcelain exports that were so much admired and desirable in Europe at that time (Houkjaer, *Tin-Glazed Earthenwares from The Netherlands, France and Germany, 1600–1800*, 2016) and in competition with the Chinese kraak porcelain being imported by the Dutch East India Company (VOC). The remainder of the

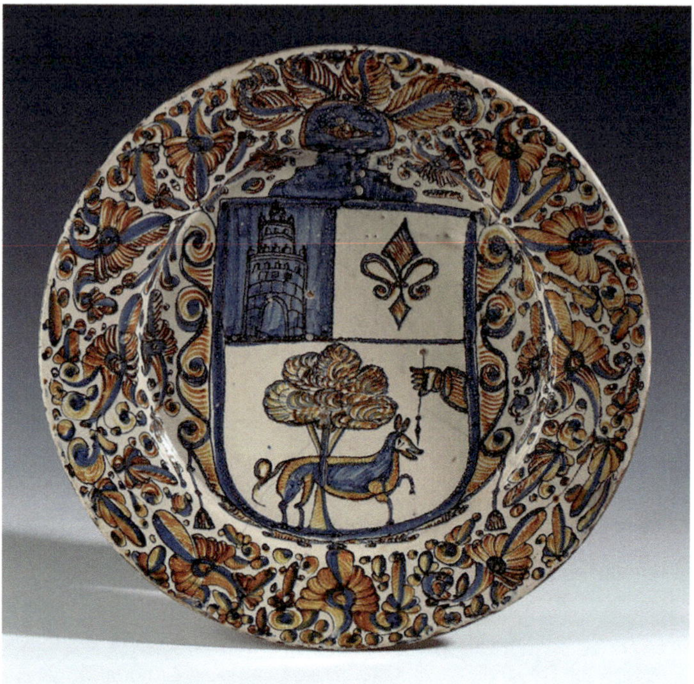

Fig. 4.25 A Spanish faux or pseudo-armorial dish, *ca.* 1600–1625, tin-glazed earthenware from Muel, Aragon. *Reproduced with permission from the Collection of E. and H. Manners Ltd., Church Street, Kensington, London*

decoration is a copy of the peony floral decoration that is found on contemporary Ming export porcelains. It appears to be genuinely armorial in concept with a recognisable helm, mantlings (lambrequins) a crest and torse and a twin-bouche baroque escutcheon but the escutcheon itself contains a floral motif spray which is a copy of the crest and clearly does not represent a specific heraldic achievement. So, again this item must be categorised as a pseudo-armorial representation. Originally, the jug would have been fitted with a pewter lid and mount and these are now missing but the marks of their attachment can still be seen.

4.5.4 A Cologne Stoneware Armorial Glazed Bellarmine Jug

A large salt-glazed stoneware jug, in Cologne ware, *ca.* 1560–1570, known as a *Bartmannkrug* from the presence of the bearded man on the lip is shown in Fig. 4.27. The bearded man was a popular representation of the "wild man" in European legend who bore a remarkable resemblance to Cardinal Roberto Bellarmino, an outspoken and vehemently anti-Protestant and opposer of the consumption of alcohol, who gave

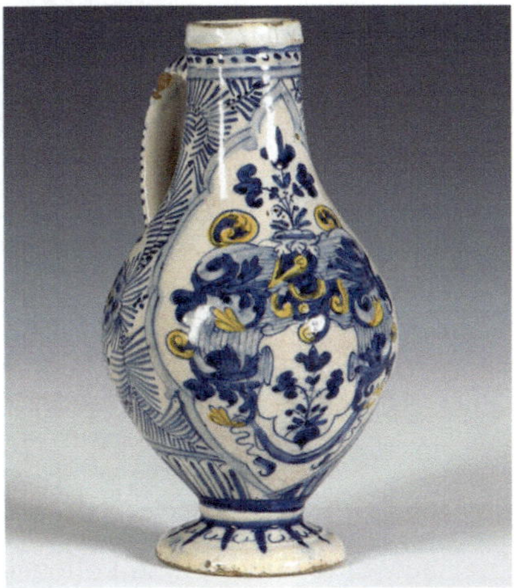

Fig. 4.26 A Portuguese tin-glazed faience pseudo-armorial jug, ca. 1635–1640, made for the Northern market in imitation of Ming porcelain analogues with a helm, crest, torse, lambrequins and escutcheon. The crest and escutcheon contain floral sprays rather than distinct heraldic achievements. The remainder of the decoration in blue enamel copies a peony design found on Ming porcelain. *Reproduced with permission from the Collection of E. and H. Manners Ltd., Church Street, Kensington, London*

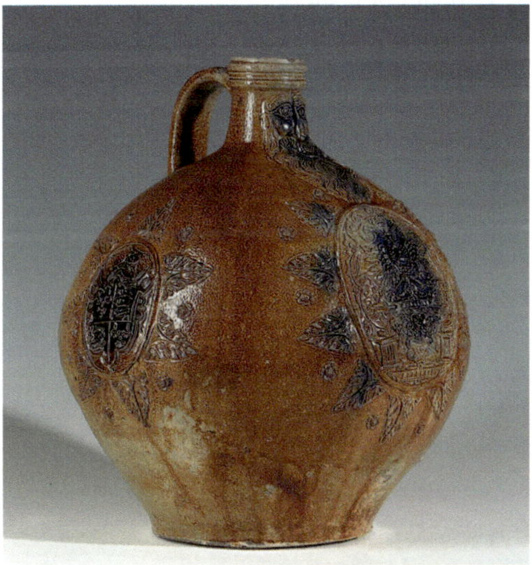

Fig. 4.27 An important Cologne stoneware Bellarmine jug displaying the Tudor coat-of-arms of Queen Elizabeth I of England and the arms of the City of Cologne, ca. 1560–1570. *Reproduced with permission from the Collection of E. and H. Manners Ltd., Church Street, Kensington, London*

his name to this type of jug in the sixteenth century amongst German Protestants. Made in Cologne or nearby Flechen, the quality and robustness of this stoneware was much admired in Europe and was a very desirable import into England (Gaimster, *German Stoneware, 1200–1900: Archaeology and Cultural History*, 1997; Reiniking

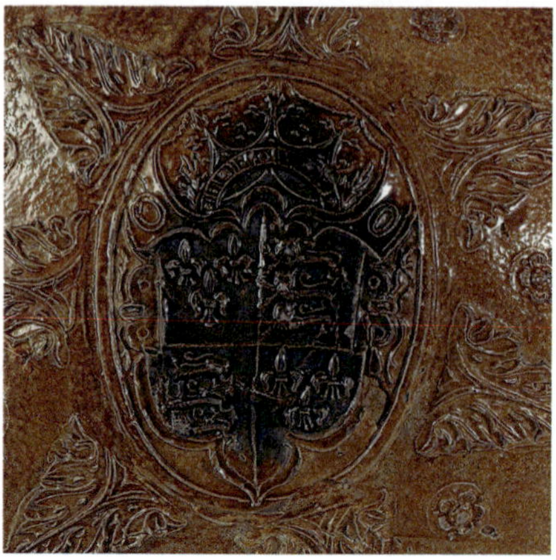

Fig. 4.28 Detail of the Tudor arms of Queen Elizabeth I of England on the Cologne ware jug shown in Fig. 4.27. *Reproduced with permission from the Collection of E. and H. Manners Ltd., Church Street, Kensington, London*

von Bock, *Kunstwerkenmuseum der Stadt Koln: Steinzeig*, 1986). This jug bears the Tudor arms of Queen Elizabeth I (shown in detail in Fig. 4.28) and the arms of the City of Cologne (shown in detail in Fig. 4.29), bearing the three crowns of the Magi (whose bones lay buried in a sarcophagus in Cologne Cathedral) and the ermine tails of St Ursula who was martyred by Atilla the Hun in Cologne in the fourth century, underneath the double-headed eagle of the Holy Roman Empire and flanked by rampant lion supporters.

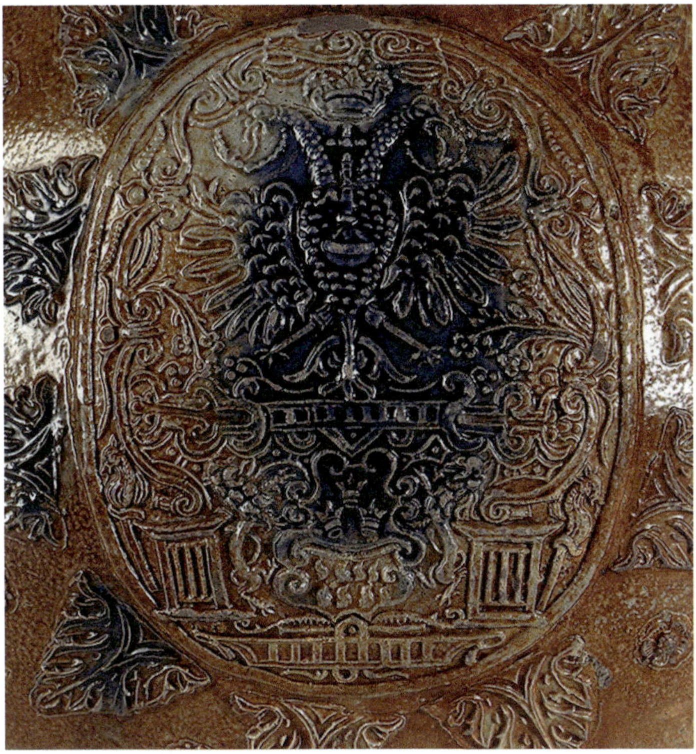

Fig. 4.29 Detail of the arms of the city of Cologne on the Cologne ware jug shown in Fig. 4.27. *Reproduced with permission from the Collection of E. and H. Manners Ltd., Church Street, Kensington, London*

References

Archer, M. (1997). *Delftware: The Tin-Glazed Earthenware of The British Isles.* A Catalogue of the Collection in the V&A Museum. HMSO/V&A Museum Publications.

Arnold, L. (1999). Princely Gifts and Papal Treasures: The Franciscan Mission to China and ITS INFluence on the Art of the West, 12501350. In M. Stephen (Ed.), *Princely Gifts and Papal Treasures; the Artistic Legacy of China in Europe* (pp. 119–134). Desiderata Press.

Bernard Burke, Sir J. (1884). *Burke's General Armory—The General Armory of England, Scotland, Ireland and Wales: Comprising a Registry of Armorial Bearings from the Earliest to the Present Time.* Harrison & Sons, Pall Mall.

Bowen Backus, I. (2014). *Asia Materialized: Perceptions of China in Renaissance Florence* (Ph.D. Thesis). University of Chicago.

Britton, F. (1987). *London Delftware.* Jonathan Horne.

Cassidy-Geiger, M. (2007). Porcelain and Prestige: Princely Gifts and *White Gold* from Meissen", Chapter 1. In M. Cassidy-Geiger (Ed.), *Fragile Diplomacy: Meissen Production for Foreign Courts, ca. 1710–1763* (pp. 2–23), Yale University Press.

Colomban, P., Milande, V., & Lucas, H. (2004). On-site Raman analysis of Medici porcelain. *Journal of Raman Spectroscopy, 35*(1), 68–72.

Cowell, M., & Zhang, F. K. (2001). *Analyses and Source of the Cobalt-Blue Pigments Employed on Chinese Ceramic, Ming Ceramic* (p. 604). The British Museum Press.
Daniels, P. (2007). *The Origin and Development of Bow Porcelain, 1730–1747, Including the Participation of the Royal Society, Andrew Duche and the American Contribution.* Resurgat Publishers.
de Plinval de Guillebon, R. (1995). *Faience et Porcelaine de Paris : XVIIIe -XIXe Siecles.* Faton.
Dawson, A. (1994). *French Porcelain—A Catalogue of the British Museum Collection.* British Museum Publications.
Ducret, S. (1962). *German Porcelain and Faience, with Vienna, Zurich and Nyon.* Oldbourne Press.
Edwards, H. G. M. (2022a). *Forensic Analysis and Its Role in the Attribution of Ceramic Specimens.* Springer-Nature Publishing.
Edwards, H. G. M. (2022b). *Welsh Armorial Porcelains: Nantgarw and Swansea Crested China.* Springer-Nature Publishing.
Fairbairn, J. (1986). *Crests of the Families of Great Britain and Ireland and Their Kindred in Other Lands*, Volumes I and II, T.C. and E.C. Jack, Edinburgh and London, 1905 and reprinted by New Orchard Publishing, London.
Ferguson, P. (2018). "Vase Madness": Vases and the Antique Taste in British Ceramics, 1765–1790, in *A Taste of the Antique: The Neo-Classical Style and Ceramics in England*, ca. 1770–1800, *Transactions of the English Ceramic Circle, 29*, 83–114.
Freeman, C. (2011). *Holy Bones, Holy Dust: How Relics Shaped the History of Mediaeval Europe.* Yale University Press.
Froissart, C. (2008), *Collection d'un Amateur, Rouen 1680–1740 : De la Premier Porcelaine de l'Age D'Or de la Faience.* Sotheby's Paris, 18th June.
Gaimster, D. (1997). *German Stoneware, 1200–1900: Archaeology and Cultural History.* British Museum Press.
Gerritsen, A. (2020). *The City of Blue and White.* Cambridge University Press.
Graca, J. (1977). The Portuguese Porcelain Trade with China. *Arts in Asia, 7*, 45–47.
Grandjean, G. (1999). The Porcelain of Rouen. In B. Rondot (Ed.), *Discovering the Secrets of Soft Paste Porcelain at the St Cloud Manufactory, ca. 1690–1696.* The Bard Graduate Center for the Studies in the Decorative Arts.
Green, C. (1999). John Dwight's Fulham Pottery: Excavations, 1971–1979. *Archaeological Report 6*, English Heritage, London.
Hanscombe, G. (2008). Armorial Porcelain Decorated in the London Workshop of James Giles. *Transactions of the English Ceramic Circle, 20*(Part 1), 124.
Haselgrove, D., & Murray, J. (1979). John Dwight's Fulham Pottery, 1672–1978: A Collection of Documentary Sources. *Journal of Ceramic History, 11*, Stoke-on-Trent City Museum.
Haslem, J. (1876). *The Old Derby China Factory.* George Bell.
Hatton, R. M. (1968). Gratification and Foreign Policy: Anglo-French Rivalry in Sweden During the Nine Years' War. In R. M. Hatton & J. S. Bromley (Eds.), *William III and Louis XIV Essays* (pp. 1680–1720). Liverpool University Press.
Honey, W. B. (1934). *Dresden China: An Introduction to the Study of Meissen Porcelain.* Faber & Faber.
Horvath, J. (1986). The Pedigree of Louis the Great's Ewer. In S. B. Vardy, G. Grosschmidt & L. S. Domonkos (Eds.), *Louis the Great, King of Hungary and Poland* (pp. 325–345).
Houkjaer, U. (2016). *Tin-Glazed Earthenwares from The Netherlands, France and Germany, 1600–1800* (pp. 106–107, 346–349). Designmuseum Danmark, Copenhagen.
Howard, D. (1974). *Chinese Armorial Porcelain* (Vol. 1). Faber & Faber.
Keating, M. (2011). *The Hunt for White Gold.* Hodder & Stoughton.
Keller, V. (2012). Accounting for Invention: Guido Pancirolli's Lost and Found Things and the Development of Desiderata. *Journal of the History of Ideas, 73*, 223–245.
Keller, V. (2015). *Knowledge and the Public Interest, 1575–1725.* Cambridge University Press.
Kerr, R., & Wood, N. (2004). *Science and Civilisation in China*, 5 Volumes, Volume 5, *Part 12: Ceramic Technology* (pp. 364–368). J. Needham (Ed.). Cambridge University Press.

Koninckx, C. (1978). The Maritime Routes of the Swedish East India Company During Its First and Second Charters, 1731–1766. *Scandinavian Economic History Review, 26*, 36–65.

Medley, M. (1989). *The Chinese Potter: A Practical History of Chinese Ceramics* (pp. 176–191). Phaidon Books.

Lacombe, C. (2006). *L'Apparition de la Porcelaine Tendré à Rouen Chez les Poterat : L'Hypothese Protestante?* (Sevres Museum Publications No. 15, Sevres).

Lane, A. (1954). *Italian Porcelain.* Faber & Faber.

Lane, A. (1961). The Gaignieres-Fonthill Vase—A Chinese Porcelain of About 1300. *The Burlington Magazine, 103*, 124–132.

Lipski, L., & Archer, M. (1984). *Dated English Delftware.* Sotheby Publications.

Lister, M. (1967). *Journey to Paris in the Year 1698.* R. P. Stearns (Ed.). Southern Illinois University Press.

Mack, R. E. (2002) *Bazaar to Piazza: Islamic Trade and Italian Art, 1300–1600* (pp. 109–111). University of California Press.

Mako, E., Kristof, J., Horvath E., & Vagvolgyi, V. (2013). Kaolinitye-Urea Complexes Obtained by Mechanochemical and Aqueous Suspension Methods. *Applied Clay Sciences, 83/84*, 24–31.

Mallett, J. V. G. (1995). Porcellana tenera del Seicento in Liguria? Due ciotole at Victoria & Albert. In *Atti XLV Courgno Internazionale della Ceramica Albisola* (pp. 337–345).

Mallett, J. V. G. (2008). The Earliest English Porcelains? Part 6: The Place of Buckingham Porcelain in Ceramic History. *Transactions of the English Ceramic Circle, 20*, 211–230.

Manners, E., Personal Communication.

Medley, M. (1974). *Yuan Porcelain and Stoneware.* Faber & Faber.

Morton Nance, E. (1942). *The Pottery and Porcelain of Swansea and Nantgarw.* Batsford.

Munger, J., & Sullivan, E. (2018). *European Porcelain in the Metropolitan Museum of Art: Highlights of the Collection.* Metropolitan Museum of Art.

Nicholson, J. (1715). *The History of Many Memorable Things in Use Among the Ancients, But Now Lost and an Account of Many Excellent Discoveries Made By and Now in Use Among the Moderns*, 3 volumes, London.

Pancirolli, G. (1599/1602). *Rerum Memorabilium, Iam Olim Deperditarium, Libro Duo* (2 volumes). Machaelis Forster.

Pietsch, U. (2000). *Schwanenservice – Meissner Porzellan fur Heinrich Graf von Buhl.* Edition Leipzig.

Plot, R. (1677). *The Natural History of Oxfordshire Being an Essay Towards the Natural History of England.* Printed at The Theatre.

Pollard, A. M. (2015). Letter from China: A History of the Origins of the Chemical Analysis of Ceramics. *Ambix, 62*, 50–71.

Polo, M. (1928). *Il Milione,* cap. CLVIII, dell edizione a cuva di L.F. Benedetto, translated 1928.

Ramsay, W. R. H., & Ramsay, E. G. (2017). *The Evolution and Compositional Development of English Porcelains from the 16th Century to Lund's Bristol c. 1750 and Worcester c. 1752–the Golden Chain.* Invercargill Press.

Ray, A. (2000). *Spanish Pottery, 1248-1898.* A Catalogue of the Collection in the V&A Museum. V&A Museum Publications.

Reiniking von Bock, G. (1986). *Kunstwerkenmuseum der Stadt Koln.* Steinzeig.

Roentgen, R. E. (1996). *The Book of Meissen* (2nd ed.). Schiffer Publishing.

Rui, L. (1990). Portugal Embusca de China: Imagens e Miragens (1498–1514). *Ler Historia, 9*, 33.

Schonfeld, M. (1998). Was There a Western Inventor of Porcelain? *Technology and Culture, 39*, 716–727.

Schulle, W., & Goder, W. (1982). Die erfinding des europanischen porzellans durch Bottger – eine systematische schopferisch entwicklung. *Keramische Zeitschrift, 34*, 598–600.

Schulle, W., & Ulrich, B. (1982). Ergebnisse gefugeanalytischer untersuchungen an Bottger-porzellan. *Silikatechnik, 33*, 44–47.

Seifi, S., Diatta-Dieme, M. T, Blanchart, P., Necomte-Nana, G. L., Kobor, D., & Petit, S. (2016). Kaolin Intercalated by Urea. Ceramic Applications. *Construction and Building Materials, 113*, 579–585.

Shaw, S. (1970). *History of the Staffordshire Potteries and the Rise and Progress of the Manufacture of Pottery and Porcelain*, Jackson, Hanley, Staffordshire, 1829; reprinted by David & Charles, Newton Abbot, Devon.

Solon, M. L. (1883). *Art of the Old English Potter* (pp. 32–35). Bemrose & Son.

Solon, M. L. (1905, May). The Rouen Porcelain. *The Burlington Magazine for Connoisseurs, 7*(26), 116–124.

Solon, M. L. (1906a). "The St. Cloud Porcelain", Part I. *The Burlington Magazine for Connoisseurs, 10*(43), 24–25, 27–28.

Solon, M. L. (1906b). "The St. Cloud Porcelain", Part II. *The Burlington Magazine for Connoisseurs, 10*(44), 89–93, 95–96.

Spallanzani, M. (1990, May). Medici Porcelain in the Collection of the Last Grand Duke. *The Burlington Magazine, 132*(1046), 316–320.

Spallanzani, M. (1994). *Ceramiche alla Corte dei Medici nel Cinquecento* (pp. 55–56). Franco Cosimo Panini.

Spataro, M., Bimson, M., & Meeks, N. (2008). Part 3: A Scientific Investigation of Three Pieces of Porcelain Presumed Seventeenth Century, from Burghley House. *Transactions of the English Ceramic Circle, 20*, 187–194.

Spataro, M., Meeks, N., Bimson, M., Dawson, A., & Ambers, J. (2009). Early Porcelain in 17th Century England: Non-destructive Examination of Two Jars from Burghley House. *British Museum Research Bulletin, 3*, 37–48.

Tite, M. S., Freestone, I. C., & Bimson, M. (1984). A Technological Study of Chinese Porcelain of the Yuan Dynasty. *Archaeometry, 26*, 139–154.

Tite, M. S., Bimson, M., & Freestone, I. C. (1986). A Technological Study of Fulham Stonewares. In *Proceedings of the 24th International Archaeology Symposium, Washington, DC* (pp. 95–104).

Tyler, K., Betts, I., & Stephenson, R. (2008). *London's Delftware Industry: The Tin-Glazed Pottery Industries of Southwark and Lambeth*. Museum of London Archaeology Publications.

Walton, G. (1992). The Era of Louis XIV: A Turning-Point in the History of Diplomatic Gifts. In H. C. Collison (Ed.), *Versailles: French Court Style and Its Influence*. University of Toronto Press.

Walton, G. (2001). Diplomacy and Ambassadorial Gifts in the 16th and 17th Centuries. In B. Shifman & G. Walton (Eds.), *Gifts to the Tsars, 1500–1700; Treasures from the Kremlin*. H.N. Abrams.

Wilson, T. (1993). *Western Decorative Arts, Part I*. National Gallery of Art (pp. 234-237).

Vasari, G. (1908). Le Vite de piu eccelenti pittori, scultori e architettori, Lorenzo Torrentino, Florence, 1568, transl. into English by E.L. Seeley.

Chapter 5
The Earliest English Armorials on Ceramics

Abstract This chapter looks at the earliest armorials produced on earthenware: delftware and slipware. It considers the role of the livery companies and the reasons why livery company officials placed armorial commissions. Manufacture of stoneware is discussed, including the development of key types of stoneware: white salt-glaze, redware, creamware, pearlware and the major manufacturers. Armorial decoration of each type is discussed.

Keywords Earthenware · Delftware · Slipware · Livery companies · Thomas Toft · Stoneware · White salt-glaze stoneware · Redware · Creamware · Pearlware Wedgwood · Leeds Pottery · Castleford Pottery · Cambrian Ware · Spode

5.1 Introduction

Ceramcs or pottery is a generic term for products made from clays formed from very fine particles of aluminocylicates less than 10 μm in diameter and a covers a wide variety of forms. The alumminocylicate particles when fired melt sufficiently to fuse forming a hard stable material. The stability and properties of ceramics are very dependent on the temperature at which they are fired; the higher the temperature, the more glassy and stronger the ceramic becomes. There are three major types of ceramics: earthenware, stoneware and porcelain (Vandiver, 'Ancient Glazes,' *Scientific American*, 1990).

Earthenware vessels have been used for eating and drinking since ancient times. Earthenware is baked at the lowest temperature; terracotta and faience are types of earthenware. Stoneware is a denser, harder and heavier material which is made by firing clay at a higher temperature. Porcelain is the hardest form of ceramic, produced at the highest temperature. Porcelain's key characteristics include being resistant to corrosive compounds, hard, non-porous, heat fracture resistant, translucent and strong enough to be used in the manufacture thin-walled vessels. All of these characteristics made porcelain a very desirable product, once it could be manufactured.

R. L. Denyer et al., *Armorial Porcelain*, https://doi.org/10.1007/978-3-031-63745-2_5

In comparison to porcelain, earthenware products are porus, meaning that liquids can seep through them. This is a problem that can be resolved by the application of a glaze formed from ground aluminosilicates mixed with fluxes such as potash (Vandiver, 'Ancient Glazes', *Scientific American*, 1990). The glaze is then applied to the item, and heated until the glaze fuses onto the clay and seals it so that it becomes non-porus. To all intents and purposes the firing of the glaze converts it into a glassy non-porous coating. However, the use of fluxes introduces impurities which alter the melting of the constituents allowing the glazes to develop specific characteristics in terms of viscosity and ability to bind to the underlying ceramics. These glazes cannot remain at high temperatures for prolonged periods of time because the underlying ceramic may also melt and collapse. The relatively short duration of the firing of glazes means that their melt rate can be manipulated by fluxes, and this can result in incomplete melting and defects in the glaze, which potentially gives different glazes very distinct finishes (Vandiver, 'Ancient Glazes' *Scientific American*, 1900). Glazes were also used in the production of stoneware and porcelains where they functioned by protecting underlying decoration or by giving a ceramic a specific finish, perhaps associated with colour transparency or texture.

The colours used to decorate ceramics must also be able to withstand heat, therefore they have to be minerals, not dyes. Colours can be applied under the glaze (under glaze), with the glaze or over the glaze (over glaze). The manufacture of ceramics and the application of glaze and colours involves a series of chemical processes, hence the role of chemical anlalysis in determining the origins of particular items of ceramics or the manufacturer.

In England in the Middle Ages, eating and drinking vessels were made from treen, leather, pewter or silver. The majority of people at this time would have had limited personal possessions; a mug, a knife and spoon for dining, and a shallow dish used as a plate and bowl. Treen became less popular in Tudor England, and ceramic items took over as the preferred material for eating vessels for the majority; only the wealthy had pewter or silver ware. Drinks were stored and consumed from pottery, pewter and silver vessels, but leather, treen and glass remained in use.

By the time of the Stuarts, a range of specific drinking wares was developing. Ale was drunk from glasses, mugs or tankards, typically in a public house. A mug was a smaller vessel than a tankard, which would have a lid. Beer, which was weaker than ale, was consumed from stoneware mugs or tankards made from leather, pewter or silver. Flagons were also used for drinks; these were taller tankards that were either drunk from directly or used to fill smaller drinking vessels. Silver tankards became very popular in the reign of Charles II, and they were usually engraved with armorials or initials.

Drinks were changing in the seventeenth century, mainly in response to the new products imported by the Dutch East India Company (VOC) or the English East India Company (EIC) which included rum, coffee, tea, spices and sugar. The traditional drinks of ale, beer, porter, mead and cider were supplemented by new fashions for drinking punch, coffee, hot chocolate, caudle and tea.

Punch was a social drink made from 5 ingredients: spirits, water, lemon juice, sugar and spice. Typically, rum and brandy were used in punch, which was prepared and served from a large punch bowl into smaller drinking vessels, usually small mugs. Punch became popular in the reign of Charles I and it was often drunk at social gatherings by groups, societies and clubs.

The first coffee house in England was set up in 1651 in Oxford, and the first coffee house in London was established in 1652 in St Michaels Alley off Cornhill London. Within 10 years there were 82 coffee shops in London and by 1703 there were over 500 (Edwards, *Porcelain Analysis and its role in the Forensic Attribution of Ceramic Specimens*, 2022). Coffee houses became popular as meeting places and were used for conducting business. Lloyds Coffee House, for example, was founded near the London docks in 1687 by Edward Lloyd and used to provide commercial information to generate business (Dale, *The First Crash, Lessons from the South Sea Bubble*, 2016).

Chocolate was imported into England from the seventeenth century onwards. Chocolate was a hot drink made with spices and sugar. This was a luxury drink. A watery version of hot chocolate could be obtained from a coffee house, but dedicated chocolate houses were established for wealthy clientele. Queen Mary had a dedicated chocolate kitchen installed as part of her renovations to Hampton Court Palace in 1690 (Otnes, *Royal Fever: The British Monarchy In Consumer Culture,* 2015).

Other drinks had dedicated drinking vessels; caudle was made from wine or other alcohol mixed with breadcrumbs, sugar and spices, whilst posset was made from wine or beer mixed with curdled milk, breadcrumbs and spices. Both drinks became popular after 1650 and were drunk from two handled cups that were usually pear-shaped with covers.

Tea was imported in significant quantities by the East India Company; initially from China and then from India. Tea bowls were originally stand-alone items, then tea services developed as tea-drinking became a fashionable way of entertaining friends in the home.

Food was also changing in the seventeenth century. In Tudor times, dinner was the main meal, and it was eaten at midday with savoury and sweet dishes mixed. Supper was eaten between 5 - 6pm. Breakfast, if eaten at all, would have been bread; with cheese and meat or fish for the wealthier. Food was normally taken from the central dishes onto a thick slice of bread (a trencher) or wooden board. Under the Stuarts, there was a greater variety of food. Country estates had kitchen gardens, complete with hot beds and glasshouses in which to grow fruits such as melons. Further down the social scale, meat was eaten more frequently and there were a series of agricultural improvements to land, including Fen drainage. Cook-shops appeared, providing hot dinners as a takeaway service.

Dining fashions changed from requiring only a spoon and a knife, to using a fork. The fashion for forks came from Italy. The very earliest known silver forks are hallmarked for 1632 and engraved with the armorial crests for Manners and Montagu of Broughton and are thought to have belong to John Manners 8th Earl of Rutland (Dickson Wright, *A History of English Food,* 1988).

All of these social changes combined to create a market for earthenware and stoneware products for dining, drinking or entertaining, as well as for porcelain. Earthenware and stoneware products continued to be manufactured once porcelain was available; they were decorated similarly to porcelains, and they also carried armorials.

5.2 Earthenwares

Delftware

Tin glaze pottery describes a form of earthenware finished with glaze that was high in lead (Pb) and tin (Sn) mixed with some form of silica. The glazing process involved two stages. Firstly a mixture of lead and tin was calcined by heating to give rise to a calx powder. This calx powder was then mixed with a frit consisting of ground aluminocylicates and fluxes such as potash before application to the body and firing. This mix could also be further modified by adding an alkaline to the frit such as calcium carbonate, thus giving rise to an alkaline glaze. On final firing, these coatings formed an opaque glaze, of differing thicknesses depending on the recipe, that when tapped with a fingernail gives a tinny ring thus the name. The first known tin glazed erathenwares were produced in the eighth century in Egypt and the Levant (Matin, Tite, and Watson, 'On the origins of tin-opacified ceramic glazes: New evidence from early Islamic Egypt, the Levant, Mesopotamia, Iran, and Central Asia', *Journal of Archiological Science,* 2018). The use of tinglazes, particularly the opague white glazes, steadily spread to Persia, Eastern Asia and westwards into Islamic Spain and Italy, and then into the rest of Europe. A form of tinglaze is still used today, but the lead and tin have been replaced with zirconium and titanium.

Tinglaze or delftware was first introduced into England during the early stages of the Dutch 80 year War of Independence with Spain by Huguenot refugees fleeing the Spanish inquisiton. Some of the refugees that arrived in Norwich continued their production in England, for example Henry VIII is known to have offered a certain Jasper Andries and Jacob Jansen of Antwerp the oportunity of producing delftware in England. They initially landed in Norwich and made paving stones around 1567, before moving to London in 1570. They were followed by the potter Jan Aryaensz van Hamme, who having declared bancrupcy in the Netherlands and fled to London to avoid his creditors, established a delftware pottery in Lambeth in 1571.

During the seventeenth century a new method was developed whereby polychrome decoration was applied to the finished delftware. This entailed using a low level of heat in a muffle kiln, the only limitation being that the colours used had to be able to withstand the firing process. The major problem with the finished delftware product was that the glaze could be very easily chipped and separated from the body of the ceramic. This is evidenced by surviving delftware pieces, the majority of which are missing sections of glaze which has flaked off after being chipped at some point since manufacture. The glazes could also be quite soft. According to Barber, Dutch delftware could be quite easily cut with a knife whilst with English delftware, the glaze was harder, less easily scratched and thinner, to the extent that sometimes the tint of the body showed through (Barber, Tin Enamelled Pottery: Majolica, Delft and other Stanniferous Faience, 1909) Clearly, shifting production to England also involved modifications to the glaze, possibly associated with availability of raw materials or due to experimentalism with the recipe.

The majority of English delftwares produced were of a practical form. Delftware items included drinking vessels such as mugs and jugs, dining wares such as plates, dishes, and practical household items such as candlesticks and storage items such as drug jars. However, it is unlikely that the delftwares saw much use, due to the fragile nature of the glaze. Delftware glaze can be easily chipped and damaged, which effectively prohibits the production of anything that will be subject to significant wear. A dinner service in delftware, for example, would be scratched and badly damaged after only a few uses.

English delftware was made in a number of locations across the UK, including London, Bristol and Liverpool. Lambath and Southwark became the main production centres for delftware in London, with multiple factories in each. Norfolk House and Glasshouse street factories were located in Lambeth, whilst Montague Close c1613 – 1755. Pickleherring, and Rotherhithe factories were all located in Southwark. Of these the Pickleherring Pottery was founded around 1618 by Christian Wilhelm, who ran the business until about 1630 when it passed to his son-in-law, Thomas Townsend. The business operated until around 1720 (Tyler, Betts, and Stephenson, *London's delftware industry: the tin-glazed pottery industries of Southwark and Lambeth,* 2008).

Delft was also made at Bristol from c1700 - 1780, when its products were superceded by Wedgwood. A prominent manufacturer of Bristol delftware was Redcliffe Bank (Jackson, Jackson and Price, *Bristol Potters and Potteries 1600–1800*, 1982). A range of goods were made, including punch bowls. Bristol delftware has a greenish tinge. There is also a decorative technique called *bianco sopra bianco* in which pure white is painted over the enamel, which is unique to Bristol. The bulk of Bristol delftware was destined for the America market. A second prominent Bristol delftware factory was owned by Joseph Flower, and it produced plates depicting local scenes along with 'marriage plates' marked with the couple's initials. Election plates were also produced, carrying the names of the candidates and the year of the election. Tiles were also made.

Another centre of delftware manufacture was Liverpool; the origins of this defltware production are unclear, but it was a well-established and important industry by 1716. Liverpool delftware has a blue-ish tinge and products produced included tiles and punch bowls. Liverpool delftware was decorated with transfer work; Sadler and Green pioneered its development and the initial colours used were red and black. As the transfer process developed, Sadler and Green decorated wares from all over the country: Wedgwood was a key client of theirs.

Delftware chargers were made, which were decorated with 5 colours: blue, green yellow, orange and a brownish purple. These were clearly decorative display pieces, and carry depictions of monarchs. The Stuart kings, William and Mary and Queen Anne are all depicted on delftware chargers believed to have been made between 1665 - 1710 (Lane, *A Style in Pottery,* 1973). These are likely to have been signs of allegiance, or to have marked coronations or celebrations.

5.2.1 Armorials on Delftware

Delftware mugs, plates and chargers may well have been intended for primarily decorative use, particularly when the decoration is considered. Many pieces of delftware carry initials and inscriptions; pieces were made in commemoration of Royal events, or to mark marriages or other significant occasions.

Delftware was decorated with armorials, and many of the surviving armorial delftware pieces appear to have been made for the livery companies or their officials. An early example is a Lambeth delftware caudle cup dated 1657, which carries the armorials of the Bakers Company (Hodgkin, *Examples of Early English Pottery*, 1891). Another instance is a mug dated for 1660 which carries the Leather Sellers Guild armorials (Moor, Hudson, N *Delftware, Dutch and English*, 1908). Another armorial delftware is a puzzle posset cup dated for 1674, which carries the armorials for the Merchant Taylors Company (Lipski and Archer, *Dated English Delftware*, 1984). These pieces were doubtless only used on special occasions, if at all. Judging by how much delftware from the seventeenth and eighteenth centuries has survived, it is likely that the majority of English delftware, despite the apparent practicality, was mostly used for decorative display purposes in the home.

Ornamental delftwares became very popular towards the end of the seventeenth century, as evidenced by the evolution of delftware racks and dressers developed specifically for their display in the home, which created a flourishing market for armorial delftwares.

5.2.2 Slipware

Slipware was the most common form of ceramic produced in England in the late Middle Ages. Slipware tended to be made from a common clay with a high iron oxide content. After being formed slipware was often coated with a white slip by diping or pooring and then further decorated with slips of different colours. The slips were formed by mixing clays of different colours with water, they could be diluted to different constutuences so that they could be aplied to the surface by painting or piping, They could also be removed to reveal the underlying clay. Once decorated the wares were fired at temperatures of about 1150 °C and then glazed with powdered lead via a second firing. This glazing technique tended to give any underlying white slip a yellow brown honeyed colouration (Dean, 1994, *A Slipware Dish by Samuel Malkin: An Analysis of Vernacular Design, 1994*). The process of manufacturing slipware made the application of fine coloured detail difficult, but master craftsmen could acheive highly decorative results. Slipware was produced in the Staffordshire potteries by makers such as Thomas Toft and Ralph Simpson, who produced chargers. There was also a tradition of slipware manufacture in North Devon (Atterbury, *English Pottery and Porcelain, A Historical Survey,* 1978). Despite the technical difficulties, slipware has been used to produce armorial ceramics, an excellent example of which appeared in a Bonhams Fine British Pottery and Porcelain Auction in New Bond Street London in 2014. This particular example was manufactured by Thomas Toft around 1670 - 1690 and is decorated with a trellis border around a central shield ornamented with foliate scrolls containg the arms of the Worshipful Company of Cordwainers, consisting of a chevron and three goats heads. The decorative design also incorporates a panel to the base marked Thomas Toft (Fig. 5.1).

Thomas Toft was a Staffordshire maker who specialised in slipware dishes. It is unknown if he was single worker or owned a small workshop, but products of this type were typically made in potworks with less than 6 workers (Elliott, *Aspects of Ceramic History,* 2006).

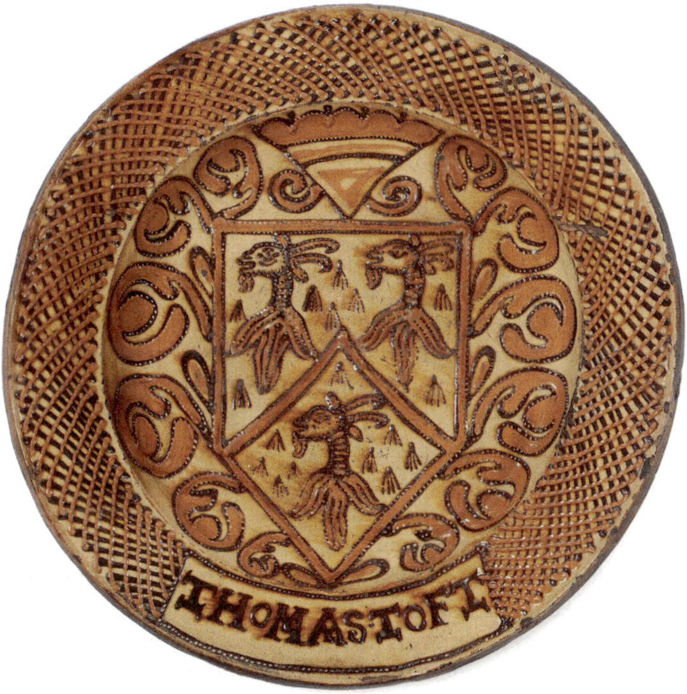

Fig. 5.1 Thomas Toft Slipware plate c1670-1690, decorated with the arms of Worshipful Company of Cordwainers consisting of *a chevron and three goats heads. Reproduced with the Courtesy of Bonhams Auctioneers and Valuers, Bond Street, London*

5.3 British Stoneware

Stoneware describes an eartherware pottery fired to a high temerature between 1100 and 1200 °C in which the silica fuses to a higher degree producing a harder less porous ceramic. Stonewares were first introduced into England by the Germans who made brown stoneware household goods such as jugs, bottles and dishes some of these were impressed with armorials. The Staffordshire potters adopted these higher temperature techiques to generate a range of stonewares including redwares, salt glazed stonewares, creamwares and pearlwares.

5.3.1 Redware and Salt Glaze Stoneware

The manufacture of redwares and salt glaze wares share a common history. The technique of manufacture arrived in England via immigrants such as the Dutch Elers brothers, John Philip Elers and David Elers, who established a small pottery at Bradwell about 2 miles away from Burslem in Staffordshire. They initially produced a fine red unglazed stoneware formed from iron rich clays fired at high temperature to give a fine hard impervious red ceramic comparable in comosition and physical properties to the red Chinese Boccaro ware (Barber, 'So-Called 'Red Porcelain,' or Boccaro Ware of the Chinese, and Its Imitations', *Bulletin of the Pennsylvania Museum*, 1911). These wares were often finished by lathe - turning. Their wares included fine delicately formed teawares such as teapots, cups, mugs, jugs and bowls decorated in relief by the attachment of moulded clay designs; an example is given in Fig. 5.2. Their techniques and designs were adopted by a number of neighbouring potteries including Twyford and Astbury.

The Elers brothers have also been credited with introducing salt glazing to England, in around 1690. Acording to the writings of Wedgwood in 1765, he learned from discusions with an 84 year old workman called Steele, who witnessed the event, what happened when the Elers brothers first glazed their wares. Apparently, a huge plume of black smoke suddenly poured out of the chimney and Steele, like other workers, ran to the pottery to see what had happened. They were no doubt expecting there to have been an awful accidental explosion; but they found that the Elers brothers had errected scaffolding around their furnace, and as it reached its peak temperature during the firing of their wares, they had workers wrapped in wet clothes throwing considerable quantaties of salt through vents onto the white hot ceramics beneath (Hobson, 'Early Staffordshire Wares Illustrated by Pieces in the British Museum', Article III *Salt - Glazed Wares,* 1904). The workers were salt glazing their stoneware; the vapourised salt attacked the silica of the stonewares, giving rise to sodium cilicate or soda glass glaze. None of the wares that were salt glazed by the Elers brothers are thought to have survived, but they were probably redwares. However, the potter John Astbury, having already acquired the technique of manufacturing red stoneware from the Elers brothers, also learned the salt glaze technique. He used it in his own

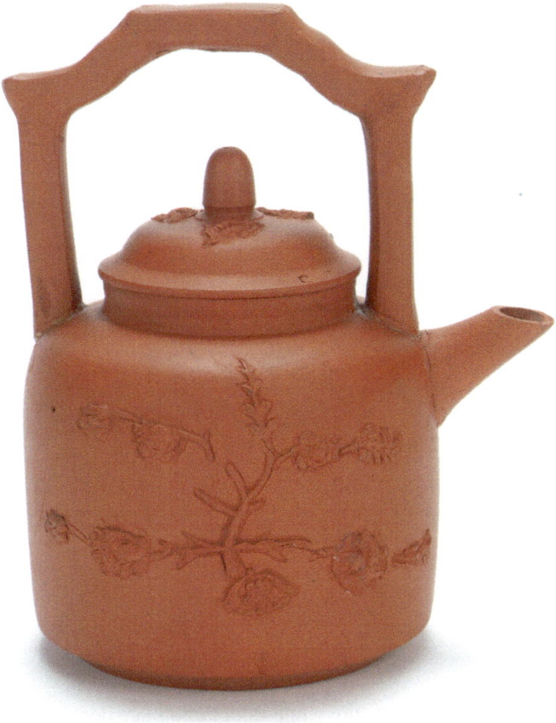

Fig. 5.2 A rare Elers Brothers redware teapot decorated with an applied botanical relief and finished with a domed lid decorated in relief with leaves. Modelled on a Yixing ware original. *Reproduced with the Courtesy of Bonhams Auctioneers and Valuers, Bond Street, London*

pottery where he later developed a paste consisting of white clay from Devon and Dorset combined with a white powder formed from calcined flint instead of sand. As a result of this, Astbury in 1698 was able to produce a white bodied stoneware with what was described as a brilliant white paste, with a pitted glaze. The glaze was later improved by mixing the salt with red lead (Pb_3O_4). This new ceramic had a thin, durable body with a translucency similar to that porcelain, which gave it an advantage over delftware and other stonewares of the time. White salt glazed stonewares became very popular and could be moulded into exquisite forms. White salt glazed stonewares were decorated with armorials; these were applied as a transfer print in red, an example of which is given in Fig. 5.3.

By the 1750s, the manufacture of salt glazed wares had become mechanised and associated with the use of soft plaster of Paris moulds, the edges of which became bluntened with continued use which then gave rise to products with less well defined features. Needless to say, the popularity of salt glaze wares waned, and manuafactures such as Wedgwood ceased its production by around 1779 (Hobson, *Early Staffordshire Wares Illustrated by Pieces in the British Museum*, 1904).

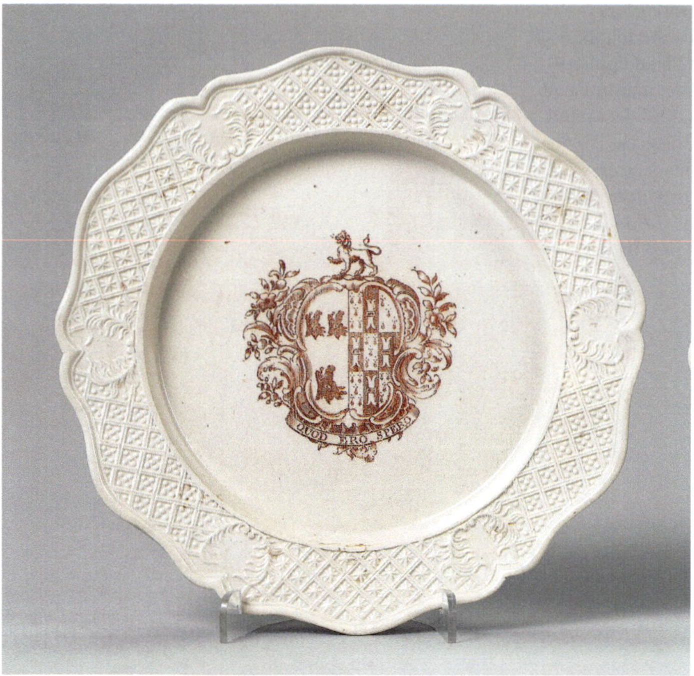

Fig. 5.3 A Staffordshire white salt glazed plate decorated with a transfer printed armorial in red. *Reproduced with the Courtesy of Bonhams Auctioneers and Valuers, Bond Street, London*

5.3.2 Creamware

Creamware is another derivative from stoneware developed in Staffordshire by Wedgwood. The company was founded in 1759 by Josiah Wedgwood (1730 - 1795). He developed a slightly off-white stoneware ceramic known as creamware. This was a cream coloured earthenware that was a cheaper alternative to porcelain. His creamware surpassed the durability of developing English soft paste porcelain, and in design and decoration, creamware was similar to porcelain, but it was versatile and affordable; two key features for its commercial success. A creamware body could be moulded and decorated with colour through the glazing or applied under the glaze.

Creamware caught the attention of Queen Charlotte, who ordered an entire tea service in the June of 1765. Queen Charlotte later allowed Wedgwood to call his new wares Queensware. Queensware, carrying the endorsement of the Queen, became very fashionable very quickly, which caused Wedgewood a manufacturing problem; his factory couldn't make creamware fast enough to meet the demand. Accordingly, he used other local Staffordshire manufacturers to produce his creamware to his specifications. The popularity of his creamware was thought to have been one the factors that reduced the demand for Chinese ceramics (Copeland, *Spode's Willow Pattern and Other Designs After the Chinese,* 1980).

Josiah Wedgwood was apparently reluctant to undertake armorial commissions, because he believed that it would be dificult sell any armorial decorated pieces that were not perfect, but he did produce creamware armorials in the 1770s. Following his death, the company began to use bone china, lustre and some of the company's first underglaze blue printed patterns.

The Leeds Pottery was first formed in 1760. According Jewitt, the Leeds Pottery was founded by two brothers called Green. Those brothers were most likely Joshua Green of Middleton and John Green of Hunslet, both of whom represented the company of Humble and Green and Co. In 1775, they signed an agreement with Messrs Hutchinson and Evers to build a water wheel for grinding flint at their mill in Hunslet (Jewitt, *The Ceramic Art of Great Britain,* 1883). By 1783, the Leeds Pottery, now under the name of Hartley, Green and Company, published a book of "designs" in which designs of earthenware, Queenswares, and creamwares were described. The book of designs also indicated that these wares could be decorated or ornamented with any pattern or coat-of-arms required. The quality of the creamware produced at Leeds Pottery was comparable to that of Wedgewood, and this was attributed to a local vien of fine white clay that retained its whiteness following firing.

Another creamware manufacturer was Castleford Pottery, established in Castleford about 19 km from Leeds in the 1790s by David Dunderdale. The Castleford Pottery made a range of earthenwares, but mostly creamwares decorated with applied neoclassical mouldings. The company ran under the direction of Dunderdale and his partners until it closed in 1820. The factory then reopened in 1821 and continued to make ceramics under a number of partnerships until 1871 and the death of Thomas Hartley of Nicholson & Co. After then, the factory became a glass bottle manufacturer under Hugh McDowell Clokie and John Masterman of the company Clokie and Masterman (Jewitt, *The Ceramic Art of Great Britain*, 1883). The Castleford factory was best known for its teawares, particularly the crenelated creamware Castleford teapots, an example of which is shown in Fig. 5.4. Castleford is not thought to have made armorial services.

The Cambrian Pottery was another manufacturer of creamware. The factory was established in an old copper-works in 1764 by William Coles, who then took on a partner, George Haynes. In 1783 the pottery was offered for sale and bought by George Haynes. He established the company of Haynes & Co and signifiantly increased the size of the factory, such that by 1800 it was said to have been modelled on the factory of Josiah Wegwood in Staffordshire. Soon after this in 1802, Haynes sold his interest in the factory to Mr Lewis Weston Dillwyn who changed the name of the factory to Haynes Dillwyn and Co.

The factory under Coles and Haynes initially produced earthenware but, under the direction of Haynes, production moved towards producing a fine white creamware-like ceramic. Haynes is thought to have attempted the production of a porcelain-like ceramic. It seems that on purchasing the factory Dillwyn aimed to improve upon this opaque ceramic, and with the help of William Billingsley produced the first Swansea porcelain in around 1814 to 1815. Dillwyn continued at the pottery until 1817, when a Mr Bevington, an ex manager of works, took it over. Bevington sold the porcelain moulds to John Rose of the Coalport works around 1823, but continued making

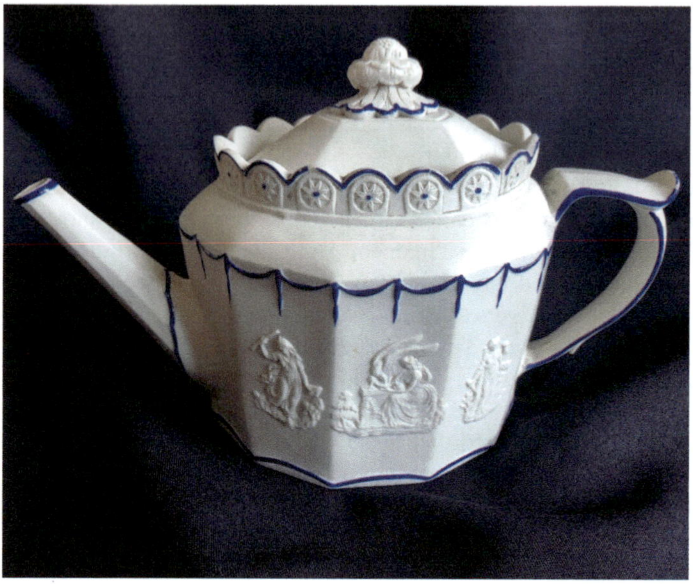

Fig. 5.4 Creamware Castleford crenelated teapot decorated with blue detail and moulded classical scenes and fitted with a sliding lid. Private collection

earthenwares under the company name of Bevington and Roby and then Bevington, Roby & Co. Dillwyn then re-acquired control of the works for a period, before his son Lewis Lewelleyn Dillwyn, M.P. for Swansea took over. The company continued in business under variuos owners until closing in 1869.

The Cambrian Pottery did undertake commisions to decorate its wares with armorials, and an example of which was illustrated by Nance (Nance, *The Pottery & Porcelain of Swansea and Nantgarw,* 1942).

Pearlware

Despite the popularity of creamwares, the Chinese blue and white porcelain still remained relatively popular. This, however, didn't mean that there was any respite to the steady erosion of the Chinese porcelain trade via the sale of earthenwares, and another new form of earthenware, pearlware, was developed in the mid-late 1770s.

Pearlware is a derivative of stoneware. Flint and Cornish china stone were added to the clay, which increased both the strength and whiteness of the product (Atterbury, *English Pottery and Porcelain, A Historical Survey,* 1978). Pearlware was essentially a creamware whitened via a glaze containing a small amount of cobalt, giving it a very faint blueish tint, which made it look white. Pearlwares could be decorated with underglaze blue transfer printed Chinoiserie designs, then covered in a cobalt containing glaze, giving the finish a pearl-like whiteness.

This technique was attributed to Josiah Wedgwood in 1779, but this view has been challenged by Reilley, who asserts that the first forms of pearlware, also known as China glaze, were produced by Staffordshire potters around 1774 (Reilley, *Wedgwood,* 1989).

A number of Staffordshire manufacturers produced pearlware, and one of the largest producers was Josiah Spode. He owned a small independent pottery manufactory in Staffordshire but expanded by taking over the Stoke works of Turner, and Banks in 1770. He then went into partnership with William Copeland and they established a warehouse in London in 1778. By 1784, Spode had introduced transfer printing into the Stoke factory, which produced a range of wares including creamwares and pearlwares. It was the later pearlware he manufactured to great effect in combination with transfer printing, to produce large quantities of blue and white Chinese style ceramics. Spode did not mark the majority of his pottery, however, so much of it is unattributed.

Pearlware was used to produce armorial ceramics, as evidenced by a pearlware punch bowl decorated for J. J. E. Pepperday around 1800. The punch bowl carries the Taylors Guild armorials. It has been suggested by Lewis (*A Collector's History of English Pottery*, 1985) that the piece was an electioneering bowl that depicted a Member of Parliament, presumably Pepperday himself, approaching London, however no record of a Member of Parliament named Pepperday around this date has been found.

5.4 The Impact of Earthenwares and Stonewares on the Chinese Porcelain Trade

It could be assumed that earthenwares and stonewares were less expensive alternatives to imported Chinese porcelains, but this is an over-simplification. It can also be said that earthenwares such as the delftwares, slipwares, redwares, creamwares and pearlwares were produced in imitation of Chinese ceramics, and this is true of all except the slipwares. However, all were valued in their own right. An example of this was King William III and Queen Mary II. When they first moved into Hampton Court having acceeded to the throne in 1688, Queen Mary brought her collection of of Chinese Kraak porcelain with her, and also her collection of Dutch delftware, which was displayed along with her porcelain. Clearly, both were valued by her. The constituents of delftware, particularly the glaze, the lead and the tin, made delftware relatively expensive to produce, as did the blue mineral cobalt used to decorate them.

The key advantage that earthenwares and stonewares had over porcelain in relation to the commission of armorials was availability. All of the earthenwares and stonewares could be made by the home market, rather than having to be shipped from China. Given that bespoke Chinese porcelain commissions took up to 3 or 4 years to be delivered, this was a significant consideration when a patron wanted an armorial piece. Furthermore more, colour could be incorporated into the design making armorial designs more accurate. For example, one of the earliest surviving armorial English delftware pieces decorated in polychrome, the Markham plate, which appeared at Bonhams Auctioneers in 2008 (Fig. 5.1). This is believed to have been manufactured by the Pickleherring Pottery and has a very distinctive form. Three other plates of an identical form and no doubt manufactured using the same mould are known. One is located in the Fitzwilliam Museum (Glaisher Collection no. 1408–1928), it is dated 1649 and decorated with polychrome grotesques, the second is dated 1651 and decorated in blue with the Master's initials along with the arms of the Grocers' Company and the third, dated to c1649 the 'Clerks' dish, carries the arms of the Worshipful Company of Parish Clerks (Lipski and Archer, *Dated English Delftware*, 1984). The Markham plate (Fig. 5.5) carries the arms of Markham impaling Faringe. The Markhams originated from Nottinghamshire and although no record of a marriage between a Markham and a Faringe has been traced, there are records of Markhams living in the London parish of St Dionis, Backchurch who were association with the Fishmongers and the Vintners Companies established in 1625 (Bonhams, *Fine British Pottery and Porcelain Including Welsh Ceramics,* 2008).

There are repeated instances of armorial delftwares, slipwares, redwares, salt glaze wares, creamwares and pearlwares in the seventeenth and eighteenth centuries that carry armorials of livery companies. Armorials had originally been used as a means of identification on the battlefield for noble knights. With a land-owning nobility, as

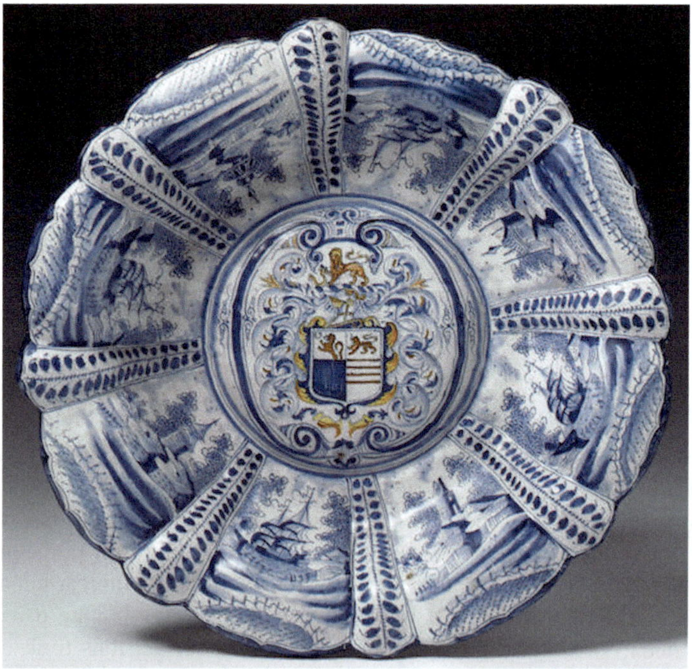

Fig. 5.5 The Markham plate. A rare example of an English Armorial Delftware plate manufactured by the Pickleherring Pottery. The plate caries the arms of Markham impaling Faringe, c 1649, and is believed to have been associated with the Fishmongers and the Vintners Companies established in 1625. *Reproduced with the Courtesy of Bonhams Auctioneers and Valuers, Bond Street, London*

land passed from generation to generation, so did the armorials, and in this respect, armorials functioned as a means of tracing landed estates ownership; coats-of-arms enabled subsequent landowners to show their familial link to the earlier holders of their estates. Armorials also served as a sign of kinship, belonging and allegiance; servants wore the livery of their masters, and the armorial image, therefore, served as a practical sign to indicate who belonged in a magnate's retinue or following. When the livery companies were founded, they followed suit in registering their own armorials. This meant that a liveryman could use the armorial badge of the Guild to demonstrate his membership to it. Family members often worked in the same trade, and businesses were handed down within families. The use of livery company armorials on these early ceramics needs to be seen within this tradition of armorial display.

Livery companies were trade guilds that were granted Royal Charters from the Late Middle Ages onwards. Their function was to manage and regulate trade, and to set trade specific related standards. Iliteracy was still comonplace in the seventeenth century, so these livery coats-of-arms were used as images to represent the livery companies, and were emblazoned on the guildhalls and displayed by their members.

Although delftware was produced in London from 1570 onwards and many delftware pieces had a functional form such as mugs plates and chargers, the fragile nature of the material made delftware an impractical option for every day use. It is probable that most if not all of the surviving delftware was treated as ornamental and this is supported by the high number of pieces that are commemorative and carry initials dates inscriptions and armorials. This decorative nature of some of the delftwares could also apply to the slipwares, the decoration of which could be easily damaged.

Truly functional ceramics only first became available in England in the early seventeenth century when stonewares were introduced from Germany. These brown ceramics were used for producing cups, jugs and storage vessels and could be manufactured to incorporate moulded designs. The Germans incorporated armorials into their stonewares, however, these very utilitarian ceramics did not really fit with the aesthetic of the day in England. Thus no hard-wearing decorative ceramics were available until the very late seventeenth century and early eighteenth century, when larger quantities of Chinese porcelain started to arrive in England via the English East India Company (EIC).

The ongoing development of English ceramics throughout the 1700s had a significant impact on the trade in Chinese porcelain. During this period there were a number of improvements in the stoneware products available. First, red stonewares were produced, these were later glazed in salt glazes and soon after that white salt glaze stoneware were developed. These advances in ceramics were then followed by the production of creamware and pearlware. All of these materials were used to create practical household items as well as dining and drinking vessels. They increased in popularity throughout the eighteenth century and were a cheaper alternative to porcelain. Armorial decorations were also commissioned on all of these product types. The growing popularity of these English stonewares, along with the development of the British porcelain industry, combined to sound the death knoll for the Chinese porcelains. Finally, porcelain imports via the EIC ceased in the 1798s when increases in import taxes finaly delivered the fatal blow, making the Chinese trade unprofitable.

The next chapter looks at the development of the British porcelain industry.

References

Atterbury, P. (1978). *English Pottery and Porcelain: A Historical Survey*. Universe Books.

Barber, E. A. (1907). *Tin Enamelled Pottery Majolica, Delft and Other Stanniferous Faience* (p. 99). Doubleday, Page & Company.

Barber, E. (1911). So-Called 'Red Porcelain', or Boccaro Ware of the Chinese, and Its Imitations. *Bulletin of the Pennsylvania Museum, 9*(34), 17–23.

Braudel, F. (1985). *The Wheels of Commerce*. Fontana Press.

Bonhams, (2008, December 10). *Fine British Pottery and Porcelain Including Welsh Ceramics, Auction Catalogue*.

Cooper, J. K. D. (1966). The 'Hot Well' Plate by Joseph Flower (City Art Gallery, Bristol). *The Burlington Magazine, 108*(761).

Copeland, R. (1980). *Spode's Willow Pattern and Other Designs After the Chinese*. Studio Vista.

Dale, R. (2016). *The First Crash: Lessons from the South Sea Bubble*. Princeton University Press.

Dawson, G. J. (1976). Montague Close Excavations 1969–1973, Pt. I, a General Survey. *Research Volume of the Surrey Archaeological Society No. 3* (pp. 37–58).

Dean, D. (1994). A Slipware Dish by Samuel Malkin: An Analysis of Vernacular Design. *Journal of Design History, 7*(3), 153–167.

Dickson, W. C. (1988). *A History of English Food*. Random House Books.

Elliot, G. (2006). *Aspects of Ceramic History: A Series of Papers Focusing on the Ceramic Artifact as Evidence of Cultural and Technical Developments,* 1. G W E Publications.

Hobson, R. L. (1904). Early Staffordshire Wares Illustrated by Pieces in the British Museum. Article III. Salt-Glazed Wares. *The Burlington Magazine for Connoisseurs, 4*(10), 65–72.

Hodgkin, J. E. (1981). *Examples of Early English Pottery*. Castle & Co.

Jackson, R., Jackson, P., & Price, R. (1982). Bristol Potters and Potteries 1600–1800. *Journal of Ceramic History No.1982* Stoke-on-Trent (pp. 18–21).

Jewitt. (1883). *The Ceramic Art of Great Britain*, J. S. Virtue & Co.

Lane, A. (1973). *Style in Pottery*. Faber.

Lewis, G. (1985). *A Collector's History of English Pottery*. Antique Collectors' Club.

Lipski, L. L., & Archer, M. (1984). *1984*. Dated English Delftware. Sotheby's Publications.

Matin, M., Tite, M., & Watson, O. (2018). On the Origins of Tin-opacified Ceramic Glazes: New Evidence from Early Islamic Egypt, the Levant, Mesopotamia, Iran, and Central Asia. *Journal of Archaeological Science, 97*, 42–66.

Moor Hudson, N. (1908). *Delftware, Dutch and English*. Hodder & Stoughton.

Nance, E. M. (1947). *The Pottery and Porcelain of Swansea Nantgarw, B. T*. Batsford Ltd.

Otnes, C. C. (2015). *Royal Fever: The British Monarchy in Consumer Culture*. University of California Press.

Reddaway, T. F. (1951). *The Rebuilding of London After the Great Fire*. Butler & Tanner Ltd.

Reilley, R. (1989). *Wedgwood* Volume II Stockton Press (p. 325).

Tyler, K., Betts, I., & Stephenson, R. (2008). *London's Delftware Industry: The Tin-Glazed Pottery Industries of Southwark and Lambeth*.

Vandiver, P. B. (1990). Ancient Glazes. *Scientific American, 262*(4), 106–113.

Chapter 6
Catalyst and Historical Beginnings of British Porcelain Armorials

Abstract This chapter looks at the genesis of armorial porcelain in Britain. It considers the establishment of the first porcelain manufactories in the 1740s and considers why porcelain factories were not established earlier in Britain. Ponoma, Chelsea, Bow, Limehouse and Vauxhall are examined. The establishment of Worcester and the enamelling workshops of James Giles are discussed. The difficulties in establishing a commercially viable porcelain manufactory are reviewed, with reference to William Littler, William Reid, and Richard Champion as key entrepreneurs. Other manufactories are considered, including Lowestoft, Richard Chaffers, Philip Christian, the Pennington brothers, New Hall, Caughley, Pinxton Swansea and Nantgarw. Those factories that did and did not produce armorial porcelain are discussed, and the reasons for not doing so are considered.

Keywords Ponoma · Chelsea · Bow · Limehouse · Vauxhall · William Littler · William Reid · William Ball · William Cookworthy · Richard Champion · Richard Chaffers · Philip Christian · Pennington · New Hall · Caughley · Pinxton · Swansea · Nantgarw

6.1 Introduction

Ceramics have been used for eating, drinking storage and ornamental purposes since the Neolithic period. There are 3 main types of ceramics: earthenware, stoneware and porcelain.

Earthenware is the earliest form of ceramic. In its simplest forms, earthenware was a loose grained sandy ceramic fired at temperatures of around 700 °C. By about 1600 BCE in China during the late Xia period the first proto-porcelains were starting to appear. These proto-porcelains were fired at temperatures of 1000–1200 °C and were akin to modern day stonewares, being harder, significantly less porous and much more resistant to heat fracture, making them much more durable. However, these proto-porcelains lacked translucency.

R. L. Denyer et al., *Armorial Porcelain*, https://doi.org/10.1007/978-3-031-63745-2_6

Porcelain is formed from a more refined clay and is fired at an even higher temperature of between 1200–1400 °C (Li et al., *Ceramics in China*, 2008) This enables the production of a ceramic in which the silicaceous components fuse giving rise to a glassier body. Porcelains are even less porous than stonewares and can be used to manufacture fine thin translucent ceramics. The manufacturing process for porcelain was developed in China as early as the Tang Dynasty (618–906 CE). By the late tenth century, Chinese porcelain was appearing in Islamic Spain (Gutiérrez et al., 'The earliest Chinese ceramics in Europe?' *Antiquity*, 2021), but it didn't really appear in the rest of Western Europe until the late thirteenth to fourteenth century. The very first Western European attempt at the manufacture of porcelain was undertaken in the 1560s, when Barnardo Buontalenti experimented with porcelain manufacture in Florence albeit without success (Lane, *Italian Porcelain*, 1954).

The first commercial production of porcelain in Europe is generally accepted to have been that of the Meissen factory; a hard-paste porcelain appeared in 1713 (Schulle & Ulrich, 'Ergebnisse gefugeanalytischer untersuchungen an Bottger-porzellan', *Silikatechnik*, 1982). By this time Japanese porcelains rather than Chinese porcelains had started to appear on the market. This was because the production of Chinese porcelain was temporarily interrupted during the transition from the Ming to the Qing dynasties by the destruction of the kilns in Jingdezhen. During this time the Dutch were ejected from Formosa (Taiwan), their base near Guangzhou (Canton). This resulted in a switch from trade in Chinese porcelain from Canton to Japanese porcelain from Arita. Western European porcelain factories took the Arita porcelain as the ideal product, and by the 1730s, it took connoisseurship to distinguish Arita porcelain from the European imitations. In terms of desirability, the Kakiemon style was ranked the highest; these products have a milky-white body and an asymmetrical design. From 1670 to 1690 the Arita porcelain was produced in this style, and the finest of these pieces had a brown iron oxide on the rim, which framed the decorative design. Arita porcelain was made mainly for export, but it was nevertheless a rare and very expensive product. Even as late as the 1720s, Arita porcelain could only be acquired by complex social networks, second hand or via specialist retailers (Aslet, *The Story of the Country House: A History of Places and People, 2023*).

From 1685, Chinese porcelain started re-emerged on the market; the Qing Emperor Kangxi, had by this time rebuilt the kilns at Jingdezhen and reopened the port of Canton to trade with any nation. He was on a mission to relaunch China's trading activities. By the early 1700s his potters had learned the secrets of overglaze enamels from French Jesuit priests sent as emissaries from Louis XIV of France. In exchange, the Jesuits learnt the secrets of hard-paste porcelain manufacture, which were communicated back to the court of Louis XIV by the Jesuit priest Father Francois Xavier d'Entrecolles via his superior in Paris, Pere Orry. This trade in technology between the French Jesuits and the Jingdezhen potters allowed the Chinese to start producing polychrome porcelains so that they could once more compete successfully with the Japanese porcelain producers. It also allowed the French to establish their own state sponsored production of porcelain. However, the production of hard-paste porcelain in France was delayed because it took until 1768 to find the first French deposits of kaolin.

The manufacture of British commercial porcelain did not begin until the 1740s, some 30 years after Meissen. This chapter looks at the early British porcelain manufacturers; considers why the manufactories were set up in the 1740s, and then traces the genesis of armorial porcelain manufacture in Britain.

6.2 The Genesis of Porcelain Manufacture in Britain

The very earliest known attempts at porcelain manufacture in Britain relate to John Dwight's Fulham Pottery, which operated between 1670 and 1703. There are no known, complete, pieces from the Fulham factory, but sherds from the site have been analysed and confirmed as porcelain (Mallet, 'The earliest English Porcelains? Part 6: The place of Buckingham Porcelain in Ceramic History', *Transactions of the English Ceramic* Circle, 2008). Dwight's experimentation with porcelain manufacture was supported by the Royal Society who had an interest in the scientific processes associated with porcelain production, but not its commercial production. Thus, the Fulham site is thought to have been a non-commercial operation (Edwards, *Porcelain Analysis and its Role in the Forensic Attribution of Ceramic Specimens*, 2022). There is no evidence indicating that armorial porcelain was made at John Dwight's Fulham pottery.

Dwight died in 1703, and the next known attempt at porcelain manufacture was not until the 1740s. This represents a 40-year hiatus between Dwight and the English porcelain factories that were set up in the 1740s. By this time, Meissen was fully established, along with the French soft-paste porcelain factory Manufacture de Vincennes, which under the patronage of Louis XV of France and his mistress Madam de Pompadour became Manufacture Royale de Porcelain de France located in Sevres (Haywood, 'Sevres Porcelain', *The Decorator and Furnisher*, 1889).

It is possible that there were some unrecorded attempts to manufacture porcelain during this time that were simply not successful and thus no record exists today. The secret of how to make a hard-paste porcelain just like the porcelain imported from China, was realised by Bow in the 1740s, who acquire a patent for the process, however the product was not commercially available until 1768 when Cookworthy succeeded in replicating the Chinese manufacturing process using the kaolin rich china clay deposits from Cornwall. Interestingly, Cookworthy's discovery occurred at about the same time as did a similar discovery in Limoges in France. This later discovery allowed Louis XV's Royal Factory of Sevres to start producing the first true French hard-paste porcelain in 1769. It seems both the English and French ceramics manufacturers had the same problem; although they knew they needed a kaolin rich china clay, they struggled to locate a supply. As a result of this, like the French, the first porcelain factories established in Britain made soft-paste porcelains.

Progress in the manufacture of porcelain was very different in Britain when compared with other Western European countries. Those in Europe had Royal patronage. Meissen was established under the auspices of King Augustus; he provided the necessary funding to ensure the successful development of the product and was himself a keen collector of porcelain. Sevres also had royal support; King Louis XV and Madam de Pompadour, were instrumental in its establishment, and the King became its sole owner in around 1759–1760. This level of royal support and endorsement did not occur in Britain, where porcelain manufacture was seen as a commercial enterprise, and thus had to be commercially viable.

In the 1740s, a number of English porcelain factories were set up in swift succession: Pomona in Newcastle Under Lyme, along with Chelsea, Bow and Limehouse in London These were soon followed by the London Vauxhall factory, Lowestoft, and several Liverpool factories in the 1750s, including Richard Chaffers and Samuel Gilbody. Porcelain manufacture was also started in Worcester and Derby in the 1750s by Dr John Wall and William Davis along with another 13 partners for Worcester, and William Duesbury and his partners for Derby. All of these factories were founded by potters and experimental ceramicists, often in partnership with businessmen or traders, and funded by local investors. The English market for porcelain was dominated by Chinese porcelain imported via the English East India Company (EIC), although European styles, particularly French, were steadily gaining popularity. Thus, the newly forming English factories copied the designs and decoration styles of both the Chinese and European porcelains, in an effort to acquire market share.

At this time in England the only porcelains decorated with armorial designs were Chinese. Armorials were applied to earthenwares such as delftware and slipware, but such ceramics were not very durable. Armorials were also applied to stonewares, for example armorials were applied to salt glazed stonewares, but stonewares only really became fashionable following the success of Wedgwood and his Queensware. Some of the early English porcelain manufacturers obviously perceived the manufacture of armorial porcelain as an opportunity. After all, the delay between ordering a Chinese armorial service and its delivery could be as long as 3 to 4 years, and sometimes on delivery Chinese services contained significant armorial errors. Both of these problems could be circumvented by enterprising English manufacturers. However, some of the English manufacturers were a bit more cautious and were reluctant to make armorial porcelain.

6.3 Commercial Soft-Paste Porcelain Manufacture in Britain: 1740–1770

6.3.1 Pomona, Newcastle Under Lyme

The Pomona factory is sometimes overlooked in early English porcelains because of the lack of recognizable, surviving pieces. Nevertheless, it was established in 1744, which makes it contemporary with Chelsea and Bow, and as such it has a key place in an account of the genesis of porcelain manufacture in Britain. The factory was founded by William Steers and operated from 1744 to 1747. The factory is known to have produced tea wares; sherds from a mug and teapot have been found at the factory's waste site along with tankards, mugs, jugs, a mustard pot and a spittoon (Edwards, *Porcelain Analysis and Its Role in the Forensic Attribution of Ceramic Specimens*, 2022). The pieces are decorated in an underglaze blue in the Chinese style. It is highly likely that the factory was attempting to capitalise on the Chinoiserie trend, which was at its peak in England in the 1750s.

Given the short-lived nature of the factory, it is thought that it did not produce sufficient porcelain to be a commercial success. No sherds with armorial decoration have been discovered.

6.3.2 Chelsea

The precise date for the establishment of Chelsea is uncertain. It is known to have been producing high-quality soft-paste porcelain wares by 1745, due to surviving decorated pieces that carry the date. Newspaper advertisements in the *General Advertise* of 1750 refers to a Mr Charles Gouyn as late proprietor and chief manager of the Chelsea House and N. Sprimont as the manager of the Chelsea factory. Nicholas Sprimont was a Huguenot silversmith, being registered at Goldsmiths Hall in 1742. His background in silver is evident from some of the shapes found in early Chelsea porcelain, which resemble silverware of the same date.

Chelsea is not thought to have produced any armorial porcelain commissions. Based on its location, it may have been thought that Chelsea would have produced armorial porcelain. The Chelsea factory was located between Justice Walk and Lawrence Street; there is also evidence of works in the Cheyne Row West area (Church, *English Porcelain*, 1886). Following the Great Fire of 1666, when four-fifths of the City of London were destroyed in the blaze, nobles who had townhouses in London had generally moved west towards Whitehall; Covent Garden, Bedford Street, Henrietta Street and King Street were all sought-after locations. By the 1740s, Chelsea was a very affluent area, therefore there would have been any number of potential armigerous clients in the immediate vicinity of the Chelsea works. Nicholas Sprimont was a silversmith by background, well accustomed to creating luxury products for wealthy clientele; as such, he must have had contact with armigerous customers. A practical consideration for Chelsea could have been that accepting a single commission to produce an armorial porcelain dinner service, which would have entailed decorating hundreds of pieces of porcelain with the patron's heraldic device, may have superseded the factory's capacity. On a practical level, the hard-paste Chinese porcelain may have been, or perceived to have been, a more robust product than Chelsea's soft-paste porcelain, particularly for serving hot drinks and hot food. Chelsea's clientele may have bought Chelsea porcelain figurines for their novel decorative value, but not armorial Chinese porcelain dinner services.

6.3.3 Bow

The Bow factory was established around 1744–1745 by Edward Heylyn, a merchant, and Thomas Frye, a painter. They experimented with porcelain production, and together they acquired two patents for porcelain manufacture. The first patent was granted in 1744 and involved the use of a calcium rich frit and kaolinitic clay they referred to as *uneka* imported by Andrew Duché from South Carolina. This clay was known as *uneka*, meaning white, by the Cherokee native Americans and used by them to produce a white ceramic. Experimental re-creation of the ceramic described by this first Bow patent using Cherokee clay was found to give rise to a hard-paste porcelain with chemical properties comparable to a very rare eighteenth century English porcelain produced by a previously unknown manufacturer referred to as the 'A marked group' (AMG) of porcelains because the base is marked with an A. This is very interesting because it has been suggested that the AMG porcelain was actually produced by Bow (Ramsay, Hill, & Ramsay, *Re-Creation of the 1744 Heylyn and Frye Ceramic Patent Wares Using Cherokee Clay: Implications for Raw Materials Kiln Conditions,* 2004); (Owen & Mills, *Earliest English Porcelain Production,* 2003); (Owen & Mills, 'From London to Liverpool: Evidence for a Limehouse - Reid Porcelain Connection Based on the Analysis of Sherds from the Brownlow Hill (ca. 1755–1767) Factory Site', *Geoarchaeology: An International Journal*, 2003). It also indicates that Bow initially experimented with the formulation and production of a silicious, aluminous and calcic hard-paste porcelain using Cherokee clay long before Cookworthy discovered china clay in Cornwall. A recent paper on the analysis of an A-marked snuff box has appeared which confirms clearly the attribution of A-marked porcelain to the Bow manufactory (Ramsay, Edwards, Manners, and Howkins, 'Geochemical Investigation of a Ceramic Snuff Box: Body, Glaze, and Enamels: 'A'- Mark English Porcelain Attribution Confirmed'. *Ceramics in America*, 2023).

In 1748, Heylyn and Frye moved to an area between Sugar House Lane and Marshgate Lane, after which they acquired their second patent in 1749 associated with the production of the bone china for which they are well known. Their move placed them near Three Mills Island, which they used to source animal bones. By adding bone ash to their soft-paste, they produced a milky-white phosphatic porcelain. Their factory was named New Canton, and it produced Blanc de Chine, which was based on the Dehua porcelain from Fujian, China. The factory was actually based in West Ham, Essex, despite the Bow name. It produced sufficient porcelain to warrant employing 300 workers, 90 of whom were porcelain decorators. Nevertheless, their products were more prone to imperfections than those of Chelsea; Bow porcelain often had an uneven finish and appeared ivory in colour. The Bow factory also produced a significant number of drinking vessels and decorative wares aimed at the middling class. By 1750, ownership had moved to John Crowther and Weatherby, although Frye was still involved as a manager.

In contrast to Chelsea, the Bow factory did make armorial porcelains. The earliest known examples are dated to 1757 if not earlier. Bow made a number of blue and white underglaze decorated mugs. One carries armorials for the Worshipful Company of Tylers and Bricklayers, and the second for the Worshipful Company of Merchant Taylors. It is significant that these very early armorial porcelains were decorated with the livery company armorials rather than those of armigerous individuals; this points to the East India Company's armorial Chinese porcelain still being the preferred option for wealthy armigerous individuals who wanted to commission armorial porcelains in the 1750s. Although Bow had the technical capability to make armorial porcelain commissions, it was producing armorial products that were small, and one-off items rather than services. From a commercial perspective, this may have been better business: an order from a livery company official for a single armorial mug was one thing but taking on an order from a titled aristocrat who wanted a 400-piece armorial dinner service would have been quite a different proposition. A large commission could occupy a workforce for a considerable period of time during which they still had to be paid, whilst the factory owner had to wait to be paid until after the large commission was completed. This could have a significant impact on cashflow, potentially leading to bankruptcy.

6.3.4 Limehouse

The Limehouse factory was a short-lived venture: it was established in 1745 and ran until 1748. Limehouse was established by William Ball, John Wilson and Joseph Wilson, and was located in Fore Street, London, on the North bank of the Thames. During its time in operation, the factory experimented with the porcelain manufacturing process, and trialled several recipes for producing porcelain; a soft-paste porcelain was made initially, then a magnesium porcelain was manufactured in 1747 (Edwards, *Porcelain Analysis and Its Role in the Forensic Attribution of Ceramic Specimens,* 2022) however its product was inferior to Chelsea and it is thought the factory was simply unable to produce sufficient saleable products to be commercially viable.

The factory site has been excavated; Limehouse is known to have produced tea wares plus tableware items such sauce boats and pickle dishes. It also made decorative vases (Drakard, *Limehouse Revealed*, 1993). It is not thought to have produced any armorial porcelain.

6.3.5 Vauxhall

The Vauxhall factory was established around 1751 by Nicholas Crisp, a jeweller, and John Sanders, a Delft potter from Lambeth. Vauxhall porcelain does not have a factory mark, which potentially means there could be a high level of misattribution for surviving Vauxhall pieces. Crisp and Sanders produced a form of soapstone soft-paste porcelain with additional calcium, which gives a greenish hue when viewed through light. They are known to have produced underglazed blue and polychrome pieces. Products include tea bowls, saucers, coffee cans, jugs and teapots as well as some figurines and decorative vases. It seems that the business was set up to compete with Bow and Chelsea in terms of product, and a number of Vauxhall products followed Sevres designs. However the business was relatively short-lived, and the eventual bankruptcy of Crisp indicates it was not commercially successful whilst operating,

The factory was active throughout the 1750s, until the death of Sanders in 1758. Crisp continued the business until 1763 when he was declared bankrupt, and the business ceased the following year.

Vauxhall did produce armorial porcelains; there is a surviving blue and white jug which carries the Axtell armorials, which is discussed in Chapter 8.

6.3.6 *Bovey Tracey, Devon*

Lord William Courtenay, Earl of Devon, acquired the china clay and lignite mines near Bovey Tracey in Devon in 1750. He encouraged the establishment of pottery factories, but there were issues getting the kilns hot enough to fire porcelain when using the local coal. After the failure of Vauxhall, Nicholas Crisp moved to Bovey Tracey, with a number of experienced pottery workers. He set up Indeo Pottery at Indeo House as a porcelain manufacturing business from 1766 onwards, however, the venture was short-lived as he was made bankrupt again in 1768 (Edwards, *18th and 19th Century Porcelain Analysis: A Forensic Provenancing Assessment*, 2020). Other potters continued making stoneware and pearlware at Bovey Tracey; William Ellis operated the Indeo Pottery from 1774 until 1836. There is no apparent evidence of armorial porcelain having been manufactured at Bovey Tracey before Crisp's second bankruptcy.

6.3.7 *Worcester*

There were three Worcester major producers of porcelain in Worcester, the Worcester Warmstry House factory founded by Dr John Wall and William Davis in 1751, the Chamberlain Worcester Diglis factory founded by Robert Chamberlain in 1783 and the St Martins Gate factory founded by Thomas Grainger and John Wood in 1801.

The Worcester factory (Warmstry House) was established in 1751. Its founders were Dr John Wall, William Davis and 13 other partners. John Wall was a physician, who sought to make a porcelain product that could withstand hot drinks. William Davis was an apothecary, and he undertook experiments with John Wall that gave rise to a new porcelain within 9 months. The recipe then formed the basis allowing them to attract the support of their additional partners and the establishment of the factory. Analysis of sherds from the lowest levels in the waste tips of the factory reveal that the first formulation developed by Wall and Davis was phosphatic, however, with the acquisition of Lund's Bristol factory in 1752, this paste changed. Lund's paste contained magnesium rich soapstone and low phosphorus calcined bone ash. The acquisition of Lund's factory allowed the Worcester paste to evolve into the low phosphorus, magnesium soapstone paste that is well recognised as forming the body of the Wall period porcelains.

In terms of armorial production, Worcester dominated the market throughout the periods of ownership by Dr Wall, Flight Barr and Barr Periods. 1750s–1840. The earliest Worcester armorials seem to have dated from around 1757. A knife and fork with porcelain handles was decorated with the armorials for Stoughton of Bally-Horgan, Kerry around that date (Gardiner Museum, object G05.18.6a-b). A coffee service was produced for Marke of Liskeard 1758–1760. This service cartouche was transfer printed, rather than hand-decorated (Spero & Sandon, *Worcester Porcelain 1751–1790: The Zorensky Collection*, 1996).

Another example of a Worcester armorial is the Gavin-Hersey dinner service produced around 1767. This service is the first known surviving armorial dinner service produced by an English porcelain manufacturer. Its commissioning patrons were David Gavin and Christina Hersey. Gavin was a successful merchant who generated sufficient wealth to acquire a substantial property in 1765, and he rebuilt the Manor House of Langton. The commissioning of this service is discussed further in Chapter 8; however, the Gavin-Hersey dinner service was decorated by the Giles workshop (Coke, *In Search of James Giles*, 1983). James Giles (1718–1780) owned a London atelier or enameling workshop, which undertook armorial decoration for several of the porcelain manufacturers. Worcester was a large client for James Giles, with many non-armorial services also being decorated there (Hamscombe, *James Giles Chine and Glass Painter*, 2005).

There were multiple ateliers working in London from the 1750s onwards. John Sims of Pimlico, and Robbins and Randall of Islington are two such enameling workshops who employed specialist artists and enamellers such as Moses Webster and James Plant. Robbins and Randall employed over 40 people at their atelier in Spa Fields Islington. As well as decorating English porcelains, the London ateliers also enamelled Chinese porcelain blanks, which they painted with European style birds, flowers and landscapes (Peck, *Consumer Splendour: Society and Culture in Seventeenth Century England*, 2005).

James Giles Jr (1718–1780) was one of the best known of the ateliers. His workshop was in Cockspur Street, London. James Giles decorated porcelains for Worcester, Derby, Bow and Chelsea in considerable quantities. In the 1770s he had some 50,000 pieces of Worcester in his workshop (Edwards, *Porcelain Analysis and Its Role in the Forensic Attribution of Ceramic Specimens*, 2022).

Using the London ateliers of James Giles enabled Worcester to outsource large quantities of porcelain for decoration; this included the enamel painting of the patterns as well as gilding. The outsourcing arrangement enabled Worcester to undertake significant volumes of armorial porcelain commissions; Worcester was one of the largest manufacturers of armorial porcelain, if not the largest in the 1760s. James

Giles' relationship with Worcester broke down in the 1770s, however, and having lost Worcester as a client, he was declared bankrupt in 1774.

Robert Chamberlain had been an apprentice at the Warmstry House factory. In 1783 he established himself as an independent decorator of whiteware at what became the site of the later Royal Worcester factory. From 1791 onwards, Chamberlain started to manufacture his own ceramics; initially he produced what was described as a hybrid hard-paste porcelain which was high in aluminocsilicates fluxed with sodium oxide and potash. After 1811, Chamberlain changed his porcelain to a white glassy body referred to as Regent China, which in turn was replaced by bone china in the 1830s. Chamberlain Worcester made a large number of armorial services including a breakfast, dinner and dessert service commissioned by Admiral Lord Nelson in 1802 in the Fine Old Japan pattern, although only the breakfast component of the service was delivered before his death at the Battle of Trafalgar in 1805. The remainder was apparently delivered to Emma Hamilton one week after Nelson's state funeral along with a bill for the sum of £120 and 10 shillings. The teapot belonging to this service is now held in the Museum of Royal Worcester. Chamberlain sold his porcelains via his shop in Worcester and his showroom in Piccadilly before moving to New Bond Street in 1816.

Thomas Grainger had been an apprentice at the Chamberlain factory. He moved on from Chamberlains to establish Grainger and Wood at St Martins Gate Porcelain Factory in Worcester with his partner John Wood in 1801. They produced a bone china and a true porcelain. The company continued trading for 88 years and was eventually incorporated into Royal Worcester in 1889. Grainger's factory, like Warmstry House Worcester and Chamberlain Worcester, undertook armorial commissions.

6.3.8 *William Duesbury and the Old Derby Works*

The Derby factory was founded in 1756. There were three founding partners: William Duesbury, an enameller, Andre Planchè, whose father had worked at Meissen as a china maker, and John Heath, a financier. The factory was established in Nottingham Road, Derby. The Derby works quickly established itself as a producer of high quality soft-paste porcelain and had engaged a London agent by 1757.

William Duesbury was the driving force behind the partnership. He had served an apprenticeship from 1742 in the London enameling workshops, where he had decorated Chelsea porcelain. In 1754, Duesbury moved back to Staffordshire, where he worked at Longton Hill for William Littler. In 1770, Duesbury acquired the Chelsea Works, and from then on, Derby porcelain was decorated at the Chelsea site in London. In 1774, Duesbury established a London warehouse in Covent Gardens, and in 1775 he bought the Bow China works. He then bought the Longton Hall factory. Several key porcelain manufactories were thus united under the one owner, and this gave Duesbury a very experienced and skilled workforce, able to produce and decorate Derby porcelain. In particular, it enabled the Derby factory to carry out the decorative enameling work, which was highly profitable. This approach was financially successful as it enabled the Derby factory to trade until 1845.

By the early nineteenth century, the production of armorial porcelain at Derby had become a major income stream. Twitchett suggests that at this time Thomas Soar, Derby's chief painter in gold probably undertook the majority of the gold decoration on armorial porcelain at Derby (Twitchett, *Derby Porcelain 1748–1757 An Illustrated Guide*, 2002). Soar himself certainly perceived that armorial decoration generated sufficient business for him to leave Derby in 1810 to set up his own workshop decorating armorial porcelain as evidenced by his advertisement in the *Derby Mercury* under the title 'Enamelled China' in which he stated: -

> Thomas Soare with the greatest respect, begs leave to inform the Nobility, Gentry and Public at Large, that he enamels Dessert, Breakfast and Tea Sets with arm, Crests, Cyphers, &c., in the most elegant manner, and the most liberal terms.

Thomas Soar is known to have gilded a number of named services including the Pendock Barry service (Fig. 1.7), which was also most likely decorated by William Billingsley (Denyer, Denyer, & Edwards, *The Pendock Barry Porcelain Service: A Forensic Evaluation*, 2023). The manufacture of armorial porcelain extended into the Bloor period with services such as the Kemeys-Tynte service being commissioned in around 1820 (see Chapter 8).

6.3.9 William Littler (1724–1784): Longton Hall & West Pans

William Littler was the son of William Littler of Brownhills near Burslem, who was a potter. He established the Longton Hall Pottery, near Stoke on Trent, around 1749 and it operated until around 1760. Not all Longton Hall products had a factory mark, however those that did, carry crossed L's and three dots in blue.

Longton Hall porcelain was a soft-paste product produced in a range of colours including a dark blue, red, pink, and green. Littler experimented with porcelain recipes to improve his product however it was frit based and thus the porcelain was unable to withstand sudden temperatures changes (Nightingale, *Contribution Towards the History of Early English Porcelain from Contemporary Sources*, 1881).

A range of tableware was manufactured and decorated in a cobalt blue which was unique to this factory and named *Littler's Blue.* Sauce boats and leaf shaped dishes were made as well as teapots and mugs. The manufacturer also produced some decorative wares, including a series of figurines. Littler employed his own decorators, William Duesbury, founder of the Derby China Works who later acquired the Chelsea manufactory, worked at Longton Hall as an enameller. Longton Hall also used Sadler of Liverpool for transfer prints on his wares. John Sadler invented a means of transfer printing on pottery, and he used this technique to decorate Longton Hall's armorial mugs.

The Longton Hall porcelain venture was not a commercial success and William Littler was declared bankrupt in 1760. Following this, he moved to West Pans, near

Musselburgh in Scotland, where his bankruptcy would not be impactful. There was already a pottery factory at West Pans, producing redware, but it began producing soft-paste porcelain once Littler took over in 1764. The West Pans porcelain was very similar to Longton Hall in that it was comprised of frits and the products were again decorated in *Littler's Blue*. The products made during this period included porcelain mugs, tea services and coffee services. Littler also manufactured practical items such as inkstands.

A number of armorial pieces were also produced at West Pans, including an ink stand. Littler continued to produced porcelain until around 1777 at which point Littler himself returned to Shenton in Staffordshire where he again produced a soft-paste porcelain product.

6.3.10 Lowestoft

The porcelain factory was founded around 1757 and continued in operation until 1802. The factory was situated on Crown Street in Lowestoft, Suffolk, and manufactured a range of goods including tea services. The factory was originally founded by Hewlin Luson however it did not survive for long. A partnership group then took over the factory, which included Philip Walker, Obed Aldred, James Richman, and Robert Browne. The owners experimented with the porcelain recipe used to manufacture their goods; the very earliest examples of Lowestoft porcelain had a pinkish hue, however once kaolin was introduced into the paste, this was no longer the case. The factory developed an underglaze blue for decoration. Later, as production developed at the site a larger range of porcelain items were produced. Polychrome decoration was also established at the factory and the designs rivaled those of Bow and Worcester, both of which Lowestoft copied extensively. The factory adopted transfer printing techniques to its porcelains in around 1770. The company's location meant that it had to ship the kaolin to the factory site, which added to its overall manufacturing costs. It also suffered significant losses when Napoleon's army sacked Rotterdam and destroyed the Lowestoft porcelain stored in the warehouses there; for many years Lowestoft porcelain had been exported to Rotterdam and held in warehouses before being sold on. Following the destruction of these stocks, the company ceased trading in 1803.

Lowestoft did make armorial porcelain; an armorial tea and coffee service was made for the Rev Potter, the vicar of Lowestoft parish in 1790 (Godden, *Lowestoft Porcelains*, 1985).

6.3.11 William Reid and Co (1756–61)

William Reid opened his factory on Brownlow Hill, Liverpool in 1756. The factory on Brownlow Hill was purpose built and William Reid's operations were big enough for there to have been a branch factory at Shelton, Stoke on Trent.

Before setting up in Liverpool, William Reid was associated with the Bow porcelain factory. His business partner in the Liverpool venture was William Ball, who had previously been a proprietor of Pomona and Limehouse. This movement of people between the different porcelain factories is significant as it explains the similarity between the chemical signatures revealed by the analysis of porcelain sherds from the different early factories. Setting up a porcelain business was a risky business undertaking, demonstrated by Reid and Co going into liquidation in 1761, and the business closing, despite all their technical expertise in porcelain manufactory. The Brownlow Hill site was taken over by William Bull and then by James Pennington in 1770. A recent paper on William Reid and his interaction with William Ball has been published (Ramsay and Edwards, New insights into the porcelains of William Reid & Co., Liverpool, and their prior links to London, 1744–1748' *Northern Ceramics Society Journal*, 2024): in this paper, the variety of different porcelain compositions produced by Bow in this period is explored. William Reid produced armorial porcelain; an armorial punch bowl dated to c.1758 is known (Edwards, *Welsh Armorial Porcelain*, 2022).

6.3.12 Richard Chaffers and Company

Richard Chaffers and Company was another of the early porcelain factories in Liverpool. The business was established at Shaw's Brow before 1756. The earliest porcelain produced was a phosphatic porcelain, but by 1756 the company had started to make a soap rock porcelain. Philip Christian (1765–1776) was his business partner and he took over the business following the death of Chaffers in 1765. Christian then developed an under graze blue printing technique. He produced a range of tea wares and copied Worcester patterns. He continued to run the business until 1777 when he sold it to Seth Pennington and John Part. Seth Pennington managed the business from 1778 until 1804 he also used the blue on the glaze printing on his porcelain items.

The company is known to have produced armorial porcelain during Philip Christian's tenure; an armorial teapot dated to 1769 is known which carries the armorials of Brougham impaling Lamplugh (Hillis, *Liverpool Porcelain*, 2011).

6.3.13 John Pennington

John Pennington is thought to have established his business around 1768, and it is believed to have been located in Copperas Hill until 1779 when it moved to Folly Lane. He also advertised at that time that he would take armorial commissions. He ran the factory until his death in 1786, after which his widow Jane ran the business until it closed down in 1794.

There was a third brother, James Pennington, who had established his porcelain factory at Brownlow Hill Liverpool in 1763; he subsequently moved his factory to Park Lane where he continued producing porcelain until 1773. Little is known about the porcelains produced by James Pennington (Hillis, *Liverpool Porcelain*, 2011).

6.3.14 Samuel Gilbody

Samuel Gilbody also had a porcelain factory on Shaw's Brow, where he was a neighbour of the Chaffers factory. His father had an earthenware factory, which he took over and converted to a porcelain manufactory in around 1754. He produced a phosphatic porcelain However, he was unable to make a commercial success of this business; he slipped into bankruptcy in 1760 and the business shut in 1761.

Gilbody's porcelain was decorated with enamels, and he was also involved in the early development of transfer printing onto ceramics.

6.3.15 Spode

The Spode factory was established at what had been the works of Turner, or Turner and Banks in Stoke on Trent in 1770 by Josiah Spode (1733–1797). At this factory Spode produced earthenwares, creamware similar to that of Wedgwood and china glazed stoneware decorated in underglaze blue chinoiserie designs. In the mid 1770s Spode formed a partnership with William Copeland who by all accounts was associated with the tea trade, resulting in them establishing a warehouse in Fore Street, Cripplegate, London, from which Spode's goods could be sold. The venture was very successful, and in 1779 William Copeland became a full partner in the company. After becoming a partner, Copeland established a large outlet for selling Spode goods in London by buying 37 Lincoln's Inn Fields; the partnership was such that Copeland was responsible for sales in London and Spode was responsible for the manufacture of the ceramics. At about this time, Spode introduced the technique of transfer printing into Stoke. Spode died in 1797 and his son Josiah Spode the younger (Josiah Spode II) took over control of running the factory. By 1800, Spode the younger had started to use a new bone china paste and was apparently the first to use feldspar mixed with china clay to produce the opaque ceramic ironstone china. In

1805, Spode was appointed potter to the Prince of Wales, which reflected quality of the factory's output. Spode did undertake armorial commissions, but the majority of the factory's early output was unmarked and as such is unattributed. However, there are numerous examples of Copeland Spode armorial ceramics dating to the mid to late nineteenth century, an example being the service commissioned by William Amcotts Ingilby c.1825 which is discussed in Chapter 8.

6.3.16 *Caughley*

The Caughley porcelain factory had been founded in Caughley, Shropshire before 1775 by Thomas Turner (born 1749). The Caughley factory produced a range of porcelain items including tea services and tableware items such as dinner services and other household items that would appeal to the middling class. The earliest porcelain produced at Caughley was a soft-paste, however the factory moved to manufacturing a hard-paste porcelain around 1790.

Much of Caughley's porcelain was decorated in underglaze blue although polychrome enamels and gilding were used. In 1799, Turner sold Caughley to the Coalport manufacturers John Rose, Edward Blakeway, and Richard Rose; thereafter manufacture continued as the Coalport factory. The Caughley factory is known to have manufactured armorials, for example, a Caughley muffin dish decorated with the arms for Braithwaite of Yorkshire and Westmoreland (Halls Fine Art Auctions & Valuations, *The Autumn Auction*, 2020).

6.4 Hard-Paste Porcelain & the Genesis of British Hard-Paste Armorials: William Cookworthy & Plymouth

William Cookworthy was born in Kingsbridge near Plymouth in 1705. By trade he was an apothecary, which gave him an understanding of chemical properties. He built a laboratory at his business Notte Street Plymouth, where he experimented with the manufacture of porcelain. In 1745 Cookworthy apparently met with Andrew Duché from South Carolina who showed him samples of china clay and samples of Chinese type ceramics made with it. Duché was apparently willing to sell Cookworthy clay imported from Carolina at £13 a ton; it was the same Andrew Duché who approached Edward Heylyn and Thomas Frye of the Bow factory with the same offer. The latter took him up on it, which gave rise to the first true genesis of hard-paste porcelain at Bow, and Bow's first patent c1745. It seems that Duché was doing the rounds of the porcelain manufacturers because he also approached Josiah Wedgwood with the offer of supplying him with the white clay of the Cherokees (Pounds, 'The Discovery of China Clay', *The Economic History Review*, 1948).

Cookworthy's encounter with Duché may have given him an insight into what was needed to make porcelain; a local source of china clay, as he identified his own source just outside the hamlet of Godolpin on Tregonning Hill c.1748. He also found china clay in the tower of St Columb Church. It turns out that the entire tower was made of china stone. Cookworthy used the china clay from Treggonin Hill to experiment on the production of porcelain, and by 1768 he acquired a patent with Lord Camelford covering the manufacture of porcelain. He then established a factory at Coxside Plymouth to produce hard-paste porcelain, and this continued to operate until 1774 (Pounds, 'The Discovery of China Clay', *The Economic History Review*, 1948). Porcelain manufactured by Cookworthy at Plymouth carries the factory mark of the alchemical symbol for tin and the astrological sign of Jupiter.

The first hard-paste porcelain armorial mug manufactured by Cookworthy in Plymouth, now in the British Museum, is decorated in underglaze blue and white with the Plymouth City arms and the initials CF and dated for three days before the patent was issued (Hurlbutt, *Bristol Porcelain*, 1928).

6.4.1 Richard Champion & the Hard-Paste Armorial Porcelains of Bristol

The Bristol works were founded by Richard Champion in 1768 where he is believed to have worked under license from Cookworthy to manufacture hard-paste porcelain until 1773 when he purchased the patent in 1775; he later extended the patent for another five years.

His business produced Meissen style porcelain. He manufactured a number of products such as tea wares. He also produced armorials; Champion was himself armigerous, he used the armorials for Sir Richard Champion Lord Mayor of London from the Berkshire Champion family. The two armorial tea services that are best known from this factory are the *Burke* service and the *Smith* service; these are discussed in more detail in Chapter 8.

Champion also produced decorative ware in the form of plaques. These plaques were usually floral; however, he also made a small number of armorial plaques which are believed to have been decorated by a French artist called Monsieur Le Quoi. One of the armorial plaques was given to Joseph Smith, whilst another was made for Isaac Elton who married the daughter of George Tierney. A third plaque is decorated with the armorials for Hartford and Lloyd; these plaques are discussed in more detail in Chapter 8. These plaques are believed to have been gifts which he presented to influential patrons as well as friends; the high manufacturing cost of these items indicates that they would not have been commercially successful as products.

Although the quality of his products was very high and favourably compared to Meissen, the patent protecting the manufacture of porcelain was due to expire in 1782. In 1775 he went to Parliament to extend his right to use Cookworthy's patent for another 14 years beyond that date, to 1796. He then tried but failed to establish a partnership with Wedgwood to manufacture porcelain in Staffordshire, before he sold the patent in 1781 to a new Staffordshire porcelain manufacturer located at the 'New Hall' at Shelton.

6.4.2 New Hall & Keeling

The group of Stafford potters who came together to form the New Hall company were John Turner, Anthony Keeling, Samuel Hollins, Jacob and Peter Warburton and John Daniel (Hollgate, *New Hall and Its Imitators*, 1971).

Soon after the purchase of the patent, John Turner and Anthony Keeling left the partnership to establish their own porcelain factory at the Tunstall works. This later fell under the sole ownership of Anthony Keeling, who went into partnership with his son. The company was only attributed the name 'A E Keeling' in the 1960s; a cup marked to its base 'A E Keeling' appeared in an auction sale (Barratt, *A. E Keeling,* 2009).

The establishment of the New Hall & Keeling companies differs from other, earlier porcelain manufactories in that the founders were starting off with a proven porcelain recipe, rather than experimenting to create their own product. However, as a result of Champion failing to fully instruct the potters at New Hall on the production of porcelain because of taking up the lucrative position of Deputy Paymaster General, they did change the manufacturing process to a two-stage firing process which reduced costs by using a lower temperature.

Although New Hall and Keeling were separate competing companies, their products were remarkably similar. Both produced porcelain wares principally for drinking and eating; tea and coffee services predominated, but New Hall and Keeling also produced dessert services, spoon trays and tea canisters. Both used decorative styles that were often simple; cobalt blue underglaze patterns or floral overglaze designs. Patterns were often shared between the two companies and that probably arose because they poached one another's decorators. The Keeling factory ceased trading in 1804 but New Hall continued, and started to use overglaze printed patterns. One of the drawbacks of the New Hall and Keeling hard-paste porcelain was the slightly greyish colouration of the body. To circumvent this, New Hall started to manufacture a white bone china from around 1811. The simplicity of the designs paid off for New Hall which continued producing porcelain until around 1835.

There is no evidence of either New Hall or Keeling having produced any armorial porcelain.

6.5 The Factories and Decorating Workshops Associated with William Billingsley

William Billingsley deserves a particular mention in this text because he represents the end of the era of the itinerant ceramicist. This era spanned from c.1740 to c.1820 and involved potters such as William Littler, William Ball, Robert Podmore, Benjamin Lund, and William Billingsley, all of whom moved around the country in search of work and opportunity, thus spreading knowledge of porcelain manufacturing techniques. This led to commonalities in formulation appearing in quite disparate and geographically separate manufacturers of ceramics (Owen & Mills, 'From London to Liverpool Evidence for a Limehouse—Reid Porcelain Connection Based on the Analysis of Sherds from the Brownlow Hill (ca. 1755–1767) Factory Site', *Geoarchaeology: An International Journal*, 2003).

William Billingsley started his career in 1774 in Derby aged 16 as an apprenticed flower decorator, where he worked under the guidance of Edward Withers, the Derby chief flower decorator, for five years. He excelled at enamelling and developed a new, more realistic method of floral painting referred to as the wash out or wipe out technique. In 1790, Billingsley was appointed as the chief flower decorator at Derby. However, Billingsley wanted to do more than just paint flowers for Derby; he had developed a wish to manufacture his own porcelain. This wish was fuelled by experiments he and Zachariah Boreman undertook in order to develop a new porcelain paste. Billingsley was fortunate enough to meet with a young John Coke, then aged 19, who had just completed his educational Grand Tour which encompassed a stay in Meissen. He had returned with the wish to use coal from his father's coal mine to make porcelain. As a result of their mutual interest in the development of a porcelain manufactory, they formed a partnership to establish the Pinxton porcelain factory in 1796.

The Pinxton factory was built at Pinxton Wharf, right by the canal, which was very convenient for the delivery of ingredients and fuel for the kilns, and equally convenient for the shipment of finished wares. Billingsley worked as the factory manager for two years without pay, but by 1799 he found that he was running out of money and in April of that year left the factory. Pinxton during this time produced a translucent but slightly grainy porcelain that was described as having a brilliant glaze.

The Pinxton factory continued until 1813, but is not thought to have produced armorial porcelains, however the decorators employed there did have the enamelling skills to produce armorial porcelain. They included several decorators with extensive experience, such as William Slater, who worked at Pinxton, Derby, the London enamelling workshop Robbins and Randal, and Davenport (Haslem, *The Old Derby China Factory: The Workmen and their Productions*, 1876).

Billingsley moved from Pinxton to establish a decorators workshop at Mansfield, where he continued to decorate whiteware from Pinxton for Coke. Business boomed for Billingsley at Mansfield, and to meet this demand he employed a number of decorators including Joseph Tatlow, a gilder who later worked at Derby, George

Hancock, his nephew William Wheeldon and probably his daughter Sarah Billingsley who undertook the burnishing of the gilding. Despite the activity at Mansfield there is no evidence of Billingsley having undertaken any armorial commissions.

Whilst at Mansfield Billingsley developed a good client base; his services were in high demand; however, he still had a wish to make his own porcelain. At this point Billingsley acquired the financial support of Henry Banks to establish another new factory at Brampton in Torksey in 1802. There are no records indicating that Billingsley manufactured any porcelain at Torksey, however, he did continue undertaking commissions to decorate whiteware from a number of factories including Derby and he is the most probable decorator of the *Pendock Barry* service (Fig. 1. 7) in c.1805 (Denyer, Denyer, & Edwards, *The Pendock Barry Porcelain Service: A Forensic Evaluation*, 2023). However, post 1805 the commissions started to dry up as the recession associated with the European blockade of Britain during the Napoleonic wars took hold. As a result of this, the virtually destitute Billingsley along with his two daughters Lavinia and Sarah and Sarah's future husband, Samuel Walker, left Torksey in late 1807 to seek employment elsewhere. After walking to Worcester, they moved on to Wales before walking back to Worcester because of news regarding an opportunity of employment with Martin Barr of Barr of Barr Flight and Barr.

At Worcester Billingsley and Samuel Walker worked on improving the Worcester paste and in November 1812, Martin Barr of Worcester paid Billingsley the sum of £200 (John, *William Billingsley 1758–1828*, 1968) for a secret porcelain recipe, on the basis that Billingsley would not share the secrete with any other manufacturer. Billingsley, his now son-in-law Samuel Walker and his two daughters suddenly left Worcester for Wales in November 1813, where Billingsley rented Nantgarw House (Tyla Gwyn) under the name of William Beeley with the aim of establishing yet another new porcelain factory.

Nantgarw House was very well located close to the Glamorgan canal, which meant the property had excellent transport links with a ready supply of fuel from the coal mines of South Wales. He established a business partnership with William Weston Young in the spring of 1814, and by August of that year, the output of the factory had reached 300 pieces of white unfinished porcelain per week, although his production experienced a huge kiln wastage of around 90%. By this time, William Weston Young had forwarded Billingsley £600, but funds were running low. They sought support from the government but were turned down, and the Nantgarw works looked doomed.

Lewis Weston Dillwyn visited the Nantgarw towards the end of September in 1814. He already had a viable ceramics business in Swansea that mostly produced creamware-like eathenwares, but Dilwyn also wanted to expand his business into the production of porcelain. He noted the huge kiln losses at the Nantgarw works, but Billingsley assured him that those lossses could be reduced, given the right conditions. This would make the production of Billingley's porcelain viable at the Swansea works, so Billingsley and Samuel Walker moved to Swansea. By doing this, however, Billingsley and Walker had broken their agreement with Martin Barr. Flight Barr & Barr wrote to Billingsley and Walker in November 1814, remonstrating with them for the sudden departure from Worcester and stating their intention to seek damages. The company also wrote to Dillwyn in a similar vein, however he appears to have ignored this letter and employed both Billingsley and Walker at the Swansea works, where Samuel Walker was responsible for building new kilns. These were completed in 1815 at a cost of £340, and the first duck egg Swansea porcelain was produced soon after. It was found to be of an excellent quality and very quickly became popular with the London retailers.

Swansea undertook armorial commissions, and a number of armorial services are known from this period (Edwards 2022, *Welsh Armorial Porcelain: Nantgarw and Swansea Crested China*, 2022) including the *Clarke Impaling Parkinson* service, which is discussed in Chapter 8.

Although kiln losses improved at Swansea, they were still very high, and in 1817 Dillwyn changed the porcelain paste to one containing soaprock and steatite. This new trident ware was much more resilient, therefore kiln losses dropped, but it didn't sell; the translucency became murky, and the glazed finish was granular and described as being pigskin-like (Edwards, *Swansea and Nantgarw Porcelains*, 2017).

Billingsley's return to Nantgarw in 1817 was supported by William Weston Young who provided £1100 worth of funds that year, and a further £1000 in 1818. The pure white porcelain produced at Nantgarw was as popular as the Swansea porcelain had been; the vast majority of it was snapped up by the London ateliers. Nantgarw porcelain was used by the London ateliers to produce a number of armorial services including the *Viscount Weymouth* service discussed in Chapter 2. Production of porcelain at Nantgarw was not commercially viable, however, and in 1820 Billingsley and Walker left Nantgarw for the last time. Billingsley moved on to work for John Rose at Coalport, and Walker eventually emigrated to North America where he set up his own pottery in West Troy in New York by 1849.

6.6 Conclusion

This chapter has considered a number of British porcelain manufacturers that were established before 1820. John Dwight's factory at Fulham is the earliest known attempt to manufacture porcelain in England, although it was primarily for scientific experimentation rather than commercial output. Following his death in 1703, there was a hiatus in England porcelain manufacturing until the 1740s, when four factories were opened in quick succession: Pomona in Newcastle-under-Lyme along with Chelsea, Bow and Limehouse, which were sited in London.

The Pomona factory operated for three years from 1744 to 1747 and is not known to have been commercially successful. Chelsea was founded before 1745 and it was successful in establishing itself as a high-quality porcelain manufacturer. It continued trading until it was purchased by Duesbury of Derby in 1770. Despite its success in producing a high-quality product, Chelsea did not move into the armorial porcelain market. Given its location and product, this must have been a commercial decision; perhaps the factory owners deemed it not commercially viable to do so, given the volumes of armorial Chinese porcelain being imported into London at that time. The

East India Company was still importing significant quantities of Chinese porcelain throughout the 1740s and 50 s, when the fashion for Chinoiserie was at its peak, and those who wanted armorial Chinese porcelain dinner services were acquiring it via the East India Company. Another consideration could have been that goods were typically supplied on credit, and it was quite standard for aristocratic patrons to take their time in settling their accounts. Executing a single very large order then waiting for payment could have caused serious cashflow problems for a porcelain manufacturer.

The Bow factory took a different path; its porcelains were aimed at a more middling clientele. Bow did produce armorial porcelains; however, it is noticeable that these were on smaller pieces such as mugs for the livery companies. Bow was meeting the needs of a different customer to those commissioning the large and expensive Chinese armorial porcelain dinner services; its products were not competing with the imports. The Limehouse porcelain manufactory, meanwhile, lasted only three years from 1745 to 1748. It seems to have suffered the same fate as Pomona in that it did not succeed commercially. Thus, of the earliest five commercial porcelain manufactories, only one, Bow, produced armorials—and the earliest of those were not until the 1750s, meaning there was a gap of at least 5 years between Bow being established and the genesis of British armorial porcelain.

In the 1750s, a number of porcelain manufactories were established, including Vauxhall, Langton Hall, Worcester, Lowestoft, William Reid, Richard Chaffers, and Simon Gilbody amongst others.

Vauxhall only traded for 12 years from 1751 to 1763. It had two founding partners; Sander died in 1758 and Crisp was declared bankrupt in 1763. Its contemporary, Worcester, fared better. Worcester porcelain was specifically designed to withstand boiling water temperatures. Given the new fashion for hot drinks, it enjoyed significant success with drink wares and dining wares. Worcester was able to produce enough volume of porcelain for it to be profitable for the company to outsource the decorative enameling work to the London atelier workshops. Volume was critical for its success, because there was less profit to be made by selling the undecorated whitewares. Worcester was shipped in great quantities to the London enameling workshops in the 1750s and 1760s. Worcester porcelain was sold to customers through the fashionable retailers, who then passed the orders to the ateliers who applied the decoration to meet the commissions. This process also enabled armorial commissions to be taken, as they were executed by the specialist London enameling ateliers. Worcester was one of the most prolific, if not the most prolific, producer of armorial English porcelain commissions. As with Bow, however, there seems to have been a gap between the factory being established and armorial commissions being made; one of the earliest known armorial commissions from Worcester was made around 1758 for the Marke family. This indicates there was a gap of 7 years between the factory being established and armorial porcelains being produced.

William Littler, owner of Longton Hall, established the company around 1749 and was declared bankrupt by 1760. Undeterred, he moved to Scotland in order to set up again at West Pans in Scotland. Lowestoft was established in 1757. Being located on the Suffolk coast meant the manufactory had to sail kaolin round the coast from

Cornwall; a costly undertaking, but Lowestoft was able to continue trading until 1803 when the loss of a warehouse full of porcelain at the hands of Napoleonic troops in Rotterdam forced its closure. Lowestoft is known to have made armorial porcelains. In Liverpool, Richard Chaffers is thought to have established the first porcelain factory he traded successfully from around 1756 until 1777 when he sold his business to Seth Pennington. This compares favorably to Samuel Gilbody, who set up his business in Liverpool 1757 and had closed by 1761 having gone bankrupt.

The success rate of these porcelain manufactories illustrates what a risky venture it was to set up a porcelain factory at this time; many of these businesses failed within a short period. This was a time when porcelain manufacture was still an uncertain process, and kiln losses could be considerable.

All of these early porcelain companies focused on making various forms of soft-paste porcelain when they set up. True hard-paste porcelain was not made commercially in England until the William Cookworthy patented in 1768, despite Bow's earlier experiments which had resulted in a patent by 1745. William Cookworthy was the first person to manufacture a British armorial hard-paste piece of porcelain, a mug carrying the Plymouth city arms in 1768.

Cookworthy sold his patent rights to Richard Champion of Bristol. He made a beautiful, high quality porcelain comparable to Meissen, but was commercially unsuccessful—in a large part, due to the cost of the patent rights. In his short period of production, however, Richard Champion made the first armorial English hard-paste porcelain services. He manufactured two, the *Burke service* and the *Smith service.* He also produced floral decorative plaques and had a small number made with armorial decoration. These are thought to have been made as gifts as the high costs of manufacture would have rendered them commercially unviable. Unlike Bow and Worcester, Champion made these armorial pieces within 4 years of having established his manufactory, however, armorial porcelains for Champion appear to have been a means of networking rather than commercially viable products.

When Champion closed his business in 1781, he sold the hard-paste porcelain manufacturing rights to the consortium of owners of the New Hall factory. They were careful not to repeat Champions mistakes; they manufactured predominantly tea ware and coffee ware which were simply decorated and reasonably priced, so that they appealed to the growing middling class of tea and coffee consumers. New Hall did not make armorials.

Derby was able to scale its production by absorbing its rival porcelain makers. Duesbury purchased Chelsea, then Bow, then Langdon Hall. This gave Derby a large and experienced workforce with decorative enameling expertise. Derby also established its own retail shops in London, having a London agent in place by 1777. This meant that the majority of Derby commissions were manufactured and decorated at the Chelsea and Derby factories by their own skilled workforce, although there is evidence that Derby had some services decorated by ex-Derby decorators on a commission basis. This approach led to commercial success for Derby, which enabled the company to continued trading until 1845.

By the 1770s, prominent Georgian towns such as Brighton, Bath and Harrogate had developed as places for the fashionable nobility and gentry visit, and fashionable shopping enclaves had been built, to tempt the wealthy visitors to purchase items during their stay. Shopping for decorative items was a fashionable leisure activity, and there is no doubt that shops enabled porcelain manufacturers to connect with their wealthy clients, far from their factory premises which were often selected for reasons other than nearness to potential clients. However, even as late as the 1790s, setting up a new porcelain factory was still a risky venture, as seen by the fate of Pinxton, Swansea and Nantgarw. John Coke established Pinxton in Derbyshire in 1796 with William Billingsley, and it had ceased trading by 1813. In part, this was due to the economic difficulties of the 1790s, and the career of William Billingsley is illustrative of these difficult trading times. Born in Derby, Billingsley did his apprenticeship at the Derby China Works where he became the chief flower painter. He then moved to Pinxton to set up in business with John Coke and made his first attempt to manufacture porcelain. He left Pinxton in 1803 and moved to Mansfield, where he decorated porcelain, and then to Brampton In Torksey before he moved to Worcester. There he created a novel porcelain body with Samuel Walker, before they moved on again, first to Nantgarw and then to Swansea, where he made high quality porcelain however each venture was commercially unsuccessful and lasted only a short period of time.

The number of factories which were established, only to fail within a short period, is indicative of how difficult the porcelain manufacturing process was in the period 1740–1820. Multiple businesses ended in bankruptcy because the costs outweighed the factory's profitability, or because of cashflow issues. The number of factories also indicates how many varieties of porcelain were in production, and variations in output from the factories demonstrates the repeated efforts to refine and improve the porcelains.

The reason for the porcelain factories being established in the 1740s and not before then, may be as simple as the climate was right for porcelain manufacture to be commercially viable. By 1740, porcelain manufacture in Western Europe had spread to St Cloud, and to Rouen. There was greater knowledge of the manufacturing process, more workers with knowledge of porcelain making, and itinerant migration saw skilled workers settling in London. Some of the early manufactories were set up specifically to make porcelain, others originated as pottery works that moved into porcelain manufacture.

Although the manufacture of porcelain in the 1740s may feel a little 'late' compared to Western Europe, it still pre-dates the Industrial Revolution, and it was still 25 years before the patent for making the true hard-paste porcelain used in China was filed. It also pre-dated the establishment of the canal networks, which was a crucial transport system that under-pinned many industrial factories by enabling the shipping of bulky goods across the country; in fact, it was Josiah Wedgwood who pushed for canals to link his porcelain works in Staffordshire to Birmingham and Manchester in the 1770s.

Porcelain was, after all, a fragile material that required great heat in its manufacturing process, which in turn mandated supplies of coal and wood for fuel, as well as the heavy raw materials used to manufacture it, therefore setting up a commercially viable manufactory was still a risky undertaking as late as the 1790s and into the Regency period.

References

Aslet, C. (2023). *The Story of the Country House*. Yale University Press.

Barratt, J. (2009). *A & E Keeling*. Gomer Press.

Church, A. H. (1886). *English Porcelain, a Handbook to the China Made in England During the Eighteenth Century as Illustrated by Specimens Chiefly in the National Collections, Part 1*. Committee of Council on Education, Chapman and Hall.

Coke, G. (1983). *In Search of James Giles*. Micawber Publications.

Denyer, R., Denyer, M. C. T., & Edwards, H. G. M. (2023). *The Pendock Barry Porcelain Service: A Forensic Evaluation*. Springer Nature.

Drakard, D. (Ed.). (1993). *Limehouse Revealed*. English Ceramics Circle.

Edwards, H. G. M. (2017). *Swansea and Nantgarw Porcelains*. Springer Nature.

Edwards, H. G. M. (2020). *18th and 19th Century Porcelain Analysis: A Forensic Provenancing Assessment*. Springer Nature.

Edwards, H. G. M. (2022a). *Porcelain Analysis and Its Role In the Forensic Attribution of Ceramic Specimens*. Springer Nature.

Edwards, H. G. M. (2022b). *Welsh Armorial Porcelain: Nantgarw and Swansea Crested China*. Springer Nature.

Godden, G. A. (1985). *Lowestoft Porcelains*. Antique Collectors' Club.

Gutiérrez, A., Gerrard, C., Zhang, R., & Guangyao, W. (2021). The Earliest Chinese Ceramics in Europe? *Antiquity, 95*(383), 1213–1230. https://doi.org/10.15184/aqy.2021.95

Hamscombe, S. (2005). *James Giles Chine and Glass Painter*. Stockspring Antiques.

Haslem, J. (1876). *The Old Derby China Factory: The Workmen and Their Productions*. G Bell.

Haywood, M. (1889). Sevres Porcelain. *The Decorator and Furnisher, 14*(2), 41–43.

Hillis, M. (2011). *Liverpool Porcelain 1756–1804* (p. 247, pl. 6.60).

Hollgate, D. (1971). *New Hall and Its Imitators*. Faber & Faber Ltd.

Hume, I. N. (1977). *Early English Delftware from London and Virginia*. Colonial Williamsburg Foundation.

Hurlbutt, F. (1928). *Bristol Porcelain*. Medici Society.

Lane, A. (1954). *Italian Porcelain*. Faber & Faber.

Mallett, J. V. G. (2008). The Earliest English Porcelains? Part 6: The Place of Buckingham Porcelain in Ceramic History. *Transactions of the English Ceramic Circle, 20*, 211–230.

Nightingale, J. E. (1881). *Contribution Towards the History of Early English Porcelain from Contemporary Sources*. Bennet Brothers.

Owen, J. V., & Mills, M. (2003). From London to Liverpool Evidence for a Limehouse—Reid Porcelain Connection Based on the Analysis of Sherds from the Brownlow Hill (ca. 1755–1767) Factory Site. *Geoarchaeology: An International Journal, 18*(8), 851–882.

Owen, J. V. (2003). The Geochemistry of Worcester Porcelain from Dr Wall to Royal Worcester: 150 Years of Innovation. *Historical Archaeology, 37*(4), 84–96.

Pounds, N. J. G. (1948). The Discovery of China Clay. *The Economic History Review, 1*(1), 20–33.

Ramsay, W. R. H., Edwards, H. G. M., Manners, E., & Howkins, A. (2023). Geochemical Investigation of a Ceramic Snuff Box—Body, Glaze, and Enamels: 'A'- Mark English Porcelain Attribution Confirmed. *Ceramics in America* (pp. 91–112). Chipstone Foundation.

Ramsay W. R. H., & Edwards, H. G. M. (2024). New Insights into the Porcelains of William Reid & Co., Liverpool, and Their Prior Links to London, 1744–1748. *Northern Ceramics Society Journal*, in press, 2024

Ramsay, W. R. H., Hill, G. R., & Ramsay, E. G. (2004). Re-Creation of the 1744 Heylyn and Frye Ceramic Patent Wares Using Cherokee Clay: Implications for Raw Materials Kiln Conditions, and the Earliest English Porcelain Production. *Geoarchaeology: An International Journal, 19*(7), 635–655.

Schulle, W., & Ulrich, B. (1982). Ergebnisse gefugeanalytischer untersuchungen an Bottger-porzellan. *Silikatechnik, 33*, 44–47.

Spero S., & Sandon, J. (1996). *Worcester Porcelain 1751–1790: The Zorensky Collection.* Antique Collectors' Club.

Twitchett, J. (2002). *Derby Porcelain: An Illustrated Guide, 1748–1848.* Antique Collectors Club.

Online Sources

Gardiner Museum. *Knife and Fork Handles Painted with the Arms of Stoughton of Ballyhorgan, Worcester, England.* Gardiner Museum, Toronto, Canada. Object Reference: G05.18.6a-b. Available online at https://emuseum.gardinermuseum.com [accessed 30 Jan 2024].

Halls Fine Art Auctions & Valuations. Caughley Armorial Muffin Dish, c.1790. Lot 187 in *The Autumn Auction.* 16 September 2020. Available online at https://fineart.hallsgb.com [accessed 30 Jan 2024].

Chapter 7
The Commissioning Patrons of Armorial Chinese Porcelain

Abstract This chapter looks at the demand for armorial porcelain. It considers the people placing the commissions for armorial porcelain, and their reasons for ordering it. It discusses the first known armorial Chinese porcelain commission by Henry Johnson and Martha Lovelace, and the first known armorial Chinese porcelain dinner service ordered by Thomas Pitt, and then the first known armorial Chinese porcelain coffee service ordered by Henry Gough. The chapter discusses the social background of those ordering armorial porcelain from China and considers specific services including the commissions of Robert Child, Anthony Duncombe and the order Robert Blatchford placed on behalf of James Brydges Duke of Chandos. The chapter examines on the kinship networks of those ordering porcelain and looks at the armorial porcelain commissions of the Pitt family, the Gough Hinde families and finally the Walpole family. The chapter then considers the practice of replacing armorial Chinese porcelain service pieces with English armorial porcelain pieces, and considers the armorial services of Mary Bootle, Dukes of Hamilton and the Talbot family and Pulteney family.

Keywords Gifts · Court · Diplomats · Collectors · Decorative · Garniture · William and Mary · Practical · Tableware · East India Company · Sir Thomas Smythe · Henry Johnson and Martha Lovelace · Thomas Pitt · Henry Gough · Robert Child · Anthony Duncombe · Robert Blatchford · James Brydges Duke of Chandos · Humphrey Parsons · Sarah Crowley · Thomas Pitt 1st Earl of Londonderry · Eldred Lancelot Lee · Robert Walpole · Rachel Cavendish · Thomas Bootle · Mary Bootle Wilbraham · Duke of Hamilton · William Talbot · Exon Sayer · Charles Talbot · William Pulteney

R. L. Denyer et al., *Armorial Porcelain*, https://doi.org/10.1007/978-3-031-63745-2_7

7.1 Introduction

7.1.1 Early Porcelain in England

Porcelain has been known in England since the fifteenth century; Henry VIII owned a piece of porcelain, as did Archbishop Warham (d. 1532) (Lach, *Asia in the Making of Europe, Volume II*, 2010). These early porcelain items were often acquired as diplomatic gifts. Elizabeth I had received porcelain ewers and bowls by the 1580s, and she in turn gave porcelain items to her principal courtiers. The Walsingham Bowl in the Burghley Collection, for example, was presented by Elizabeth I to Elizabeth Walsingham, the daughter of Sir Francis Walsingham (1532–1590) to mark her christening in 1585 (Peck, *Consumer Splendour: Society and Culture in Seventeenth Century England*, 2005). Elizabeth I's principal courtier William Cecil Lord Burghley (1520–1598) who served as Secretary of State as well as Lord High Treasurer in a long career of service to the Crown, acquired 4 pieces of porcelain dating to the Wanli period, which were set in English silver-gilt mounts (Avery, 'Chinese Porcelain in English Mounts' *The Metropolitan Museum of Art Bulletin*, 1944).

In the seventeenth century, collecting developed as an aristocratic pastime which was led by the Stuart monarchs. Art was the primary focus of many aristocratic collections, particularly the works of the Italian Renaissance artists, however their interests also extended to collections of tapestries, statues and antiquities as well as luxury imported furniture such as lacquered cabinets and screens. The earliest Chinese porcelain pieces were ornamental in nature, and the rarity and expense of porcelain items at this time, meant that they found their way into collections as well. Chinese porcelains at this time consisted of vases, pots, and other decorative display items. The blue and white items were greatly favoured, particularly by the Dutch.

Notable aristocratic courtly collectors included Henry Howard 1st Earl of Northampton (1540–1614), Robert Carr 1st Earl of Somerset (1587–1645) George Villiers 1st Duke of Buckingham (1592–1628) and Thomas Howard 2nd Earl of Arundel (1585–1646).

Robert Carr had a collection of porcelain by the 1620s, as did George Villiers (Peck, *Consumer Splendour: Society and Culture in Seventeenth Century England*, 2005). Thomas Howard, known as the Collector Earl, married Althea Talbot (1581–1654) who was the daughter of Gilbert Talbot 7th Earl of Shrewsbury (1553–1616) and Mary Cavendish (1556–1632). In 1633, the Arundels bought Tart Hall, near Whitehall Palace. The entire property was remodelled, and a new Dutch Banqueting House was built, which the Countess filled with an extensive collection of porcelain. An inventory taken in 1641 shows that the Banqueting House contained decorative porcelain items such as a figure of a woman, a man, a frog and a lion in white 'purcelin' which were 'for ornament' but also contained items such as dishes, pots, bottles, flowerpots, ewers and basins (Claxton, 'The Countess of Arundel's Dutch Pranketing Room: 'An Inventory of all the Parcells or Purselin, glasses and other Goods now remayning in the Pranketing Roome at Tart Hall, 8th Sept 1641', *Journal of the History of Collections*, 2010). These were clearly items intended for use within households and demonstrates that porcelain had begun to transition from being purely decorative pieces to practical household pieces by the late 1630s.

The court and nobility were not alone in their admiration for porcelain. Prominent merchant families also owned porcelain. Lady Noel, for example, wife of London merchant Martin Noel (1614–1665) a member of the Scrivener Company, is said to have had a collection of porcelain by the 1660s (Peck, *Consumer Splendour: Society and Culture in Seventeenth Century England*, 2005). Nor was Lady Noel the only prominent society hostess who owned porcelain. Porcelain for the dining table was still very rare, but diarist John Evelyn recorded a supper held at Lady Gerard's London home in 1652 at which all the vessels were made from porcelain. Lady Gerard née Digby (1637–1692) was married to Charles Gerard 4th Baron Gerard (1634–1667). She was the daughter and heiress of George Digby of Sandon Hall in Stafford. Evelyn credits Lady Gerard with having "the most ample and richest collection of that curiosity in England" (Bray, *Memoirs of John Evelyn, Comprising His Diary from 1641 to 1705–06*, 1827).

It is probable that Evelyn was referring to a supper served in the new fashion of *service à la française*. The traditional banquet dinner of a central serving dish and few courses had begun to give way to multiple courses and individual servings. Dining services began to come into vogue in the mid-seventeenth century for the wealthy, as the approach towards dining had changed. Silver and pewterware were used for dining vessels, as was earthenware pottery.

John Evelyn also records a visit to Belsize House in June 1676, where he comments on the furniture, which included Indian cabinets, porcelain and other solid and noble moveables. (Evelyn et al., *Diary and Correspondence of John Evelyn*, 1859).

Belsize was owned by Charles Henry Kirkhoven, Lord Wotton (1643–1683), who had inherited the property from his mother Catherine Wotton (1609–1667), Countess of Chesterfield. Her first husband was Henry Stanhope, Lord Stanhope (1605–1634), and as Lady Stanhope, she had been Governess to Princess Mary, Princess Royal (1631–1660). She accompanied Princess Mary to Holland at the time of her marriage to William II of Orange, and it was during her time in Holland that she married her second husband Johannes Polyander van den Kerchoven (1595–1660). Their son Charles grew up in Holland at the royal court and was a favourite of Princess Mary until her death in 1660 (Stephen, *Dictionary of National Biography*, vol 11, 1922).

Chinese and Japanese porcelain had reached the Dutch court via the Dutch East India Company, which had dominated European trade with Asia. The very first Dutch porcelain and lacquer rooms had been created by the grandmother of the Princess Mary's husband, Amalia van Solms Braunfel (1602–1675) at her palaces in Noordeinde and Huis ten Bosch in the 1630s and 1640s. Her 4 daughters had also created similar porcelain and lacquer rooms in their own homes after they had married. Princess Mary (1662–1694) absorbed this taste and acquired her own porcelain collection, which she brought with her when she returned to England as monarch.

King William III and Queen Mary II were joint sovereigns, first cousins, who assumed their crowns following the flight and abdication of James II in 1688. They had married in 1677 and lived in the Netherlands. When William and Mary acceded to the throne in 1688, Queen Mary's extensive collection of blue and white porcelain was shipped to London. With their ascension to the throne, the cultural influence of the Orange-Nassau royal family became aligned with the Stuart court. The new monarchs undertook a considerable expansion and redecoration program of the royal palaces during their reign. On moving to London, they bought Nottingham House in Hyde Park, which would become Kensington Palace. Christopher Wren was engaged to update Hampton Court, which was to be used as their summer palace. Both of these renovations were major projects, which Queen Mary oversaw. Her Water Gallery was the first in England of its kind, and it demonstrated Mary's passion for interiors. The gallery contained all manner of decorative objects, with picture frames and furniture painted blue and white. Blue satin and a damask fabric with blue silk fringing was used to cover the furniture and for curtain material, so that they matched the blue and white porcelain. Daniel Defoe (c. 1660–1731) visited the Hampton Court Palace, and recorded seeing china piled 'upon the Tops of Cabinets, Scrutores [desks], and every Chymney-Piece, to the Tops of the Ceilings'. This was the 'garniture de cheminée' decorative style; porcelain objects, typically vases in groups of three, five or seven, were grouped together in a decorative display on the chimneypiece. These were themed by colour, pattern, or size—and outsized porcelain pieces were frequently used to create a decorative effect.

Queen Mary filled several rooms with porcelain. One room alone contained 144 pieces of porcelain from China and Japan, which she had acquired whilst living in Amsterdam. Much of the porcelain stood in Japanese lacquer cabinets, but pieces were displayed on mantelpieces and wall brackets. Even more porcelain was displayed at Kensington Palace, where hundreds of pieces were stacked on mantelpieces, door lintels, shelves, brackets and specially carved wooden pedestals. The porcelains were placed symmetrically on the walls as well as on the chimneypieces. This display attracted great interest. In 1695, diarist John Evelyn visited the Kings House at Kensington, where he saw the gallery with the royal art collection, a great collection of porcelain and a library (Evelyn, *Diary and Correspondence of John Evelyn*, 1859). An avid porcelain collector, Queen Mary owned some 800 pieces of porcelain at the time of her death (Peck, *Consumer Splendour: Society and Culture in Seventeenth Century England*, 2005).

7.1.2 The East India Company

The East India Company had been founded in the reign of James I (1566–1625) and it was closely associated with the London livery companies. Most of the Company's members were City merchants and aldermen, who invested in the early voyages, however there were always close links with the court and the aristocratic families around the king, due to the diplomatic aspects of the East India Company's trade. Trade and diplomacy went hand in glove for a company seeking new trade agreements with other countries. The twin considerations of trade and diplomacy ensured that the monarch and the aristocracy at court were always involved with the affairs of the East India Company.

There were no fixed definitions for social groups, therefore reference to groups such as ‘the aristocracy’ or ‘the merchants’ can be a little misleading. Marriages would generally be contracted between families of broadly similar social standing and wealth, but this was not a hard and fast rule. The East India Company’s first chairman, Sir Thomas Smythe (1558–1625), illustrates this point. He was the grandson of Sir Andrew Judd (1492–1558), a founding member of the Muscovy Company, who had been an Alderman and Lord Mayor of London in 1550. Andrew Judd’s grandmother Philippa Chichele was the niece of Henry Chichele (1364–1443) Archbishop of Canterbury and founder of All Souls College of Oxford (Buckler, *Stemmata Chicheleana: Or A Genealogical Account of Some of the Families Derived from Thomas Chichele of Higham Ferrers*, 1764).

Thomas Smythe was a member of both the Skinners and Haberdashers livery companies by 1580, and he served as an Alderman and Sherriff of London. He had been Member of Parliament for Dunwich in 1604, but in the general election of 1614, he was nominated for Sandwich by Henry Howard, Earl of Northampton (Thursh & Ferris, *The History of Parliament: The House of Commons 1604–1629*, 2010), a Privy Councillor and one of the noted collectors at court. Thomas Smythe amassed sufficient wealth to buy Otford Park in Kent, as well as the manor of Sutton At Hone in Kent. He married three times, and his third wife was Sarah Blount (1580–1656). Their son Sir John Smythe (1599–1640) married Isabella Rich (1595–1632) who was the daughter of Robert Rich, 1st Earl of Warwick (1559–1619), and the sister of Henry, 1st Earl of Holland (1590–1649). After Thomas’ own death in 1625 his widow Sarah married Robert Sidney, 1st Earl of Leicester (1563–1626). Nor was this the only marriage between the Smythe and Sidney families; Thomas’ nephew Thomas Smythe, 1st Viscount Strangford (1599–1635), married Barbara Sydney (1598–1652), who was the daughter of Robert Sidney, 1st Earl of Leicester, and Barbara Gamage (1563–1621). This illustrates that the acquisition of great wealth through the East India Company could lead to aristocratic patronage as well as marriage into aristocratic families: social mobility in action.

7.1.3 The Acquisition of Armorials

Armorials were granted by the monarch to the noble knights who served on the battlefield, and the original purpose of an armorial was to enable the identification of knights in armour. Over time, the use of armorials evolved from a battlefield identification; later monarchs still awarded armorials to those who served them in battle, but also bestowed them on those who served the crown in its executive functions. Elizabeth I granted armorials to courtiers such as Thomas Neale (1521–1601) who served as an Auditor in the Exchequer, for example. James I sought to raise funds in the seventeenth century by creating baronetcies, a new rank of armorial-bearing hereditary dignity which could be purchased for 1000 guineas (Stone, 'The Inflation of Honours 1558–1641', *Past & Present*, vol 14, 1958).

By the seventeenth century, the wider right to use an armorial is evident. The *London Visitations* of 1633, 1634 and 1635 by Sir Henry St George, Richmond Herald, demonstrates the migration to London that had occurred by that time, as well as the social mobility that was taking place. Multiple armorial-bearing goldsmiths are listed; John Acton, goldsmith, descendant of the Acton family of Acton Hall in Worcester; Francis Hall, goldsmith, the son of Francis Hall of Newsam, Bishopric of Durham; Simon Gibbon, goldsmith, a grandson of Richard Gibbon of Budsdale Suffolk. In 1611, a grant of armorials was made to Richard Arnold, Haberdasher of London, and there are pedigrees for members of the 12 great livery companies (St. George & St. George, *The Visitation of London Anno Domini 1633, 1634 and 1635*, 1880). It should be borne in mind that armorials could be used by those in the direct line of descent of the original holder; further discussion of armorial inheritance practice is found in Chapter 2. By the end of the seventeenth century, those claiming descent from armigerous ancestors included the baronets, minor gentry families, merchants, clerics, medics, naval and army officers as well as the traditional landed gentry and aristocratic families. As the number of families with armigerous antecedents grew, the role of validating and verifying claims became more problematic. Henry VIII had instigated the process of Heralds from the College of Arms making periodic Visitations to the counties, to record and update the College records of armigerous families. It is no coincidence that the role of the Heralds changed at this time, with the final Visitations being carried out in 1687. After that, the Heralds no longer inspected armigerous claims as a routine process; although the College of Arms retained its role as the arbitrator of who was and was not entitled to bear arms, there was no effective process to punish or sanction those who failed to follow the rules.

7.2 1695–1710: The Genesis of Armorial Chinese Porcelain

7.2.1 Henry Johnson (1659–1719) and Martha Lovelace (1667–1745)

The earliest known commission for Chinese porcelain displaying an English armorial was made by Henry Johnson of Blackwall, Middlesex (1659–1719) and Martha Lovelace (1667–1745), who ordered a set of jardinières c. 1695, which were decorated in blue and white underglaze with a Chinese pattern, to which his armorials were added (Pierson, *Collectors, Collections and Museums—the field of Chinese ceramics in Britain 1560–1969*, 2007).

Henry Johnson came from a long-established Norfolk family. His great-grandfather John Johnson of Wood Norton, Norfolk, married Anne Thimblethorpe of Foulsham, Norfolk. Their son George married Elizabeth Thimblethorpe of Foulsham, likely to have been a cousin. Their son Henry Johnson (b. 1574) acquired Friston Hall and Aldeburgh in Suffolk. Henry was the father of Francis Johnson (1595–1636) who married Mary Pett (b. c. 1588). She came from a long line of shipbuilders; her grandfather Sir Phineas Pett (d. 1544) had been Surveyor General of the Navy for Elizabeth I (Burke & Burke, *A Genealogical and Heraldic Dictionary of the Landed Gentry*, vol 2, 1850). Her father Peter Pett of Deptford, Kent (d. 1589) was granted armorials in 1583 of *or on a fess gules between 3 ogresses, a lion passant of the field and the crest, out of a ducal coronet, a demi-pelican with wings expanded.*

Francis Johnson and Mary Pett had a son Henry Johnson (1623–1683) who served an apprenticeship with Phineas Pett to become a shipbuilder himself. By 1651, Henry Johnson was building merchant ships at Deptford, and in 1655 he bought the East India Company's shipyards and docks at Blackwall. He became a Director of the East India Company, and held shares in several ships (Lubbock, *The Blackwall Frigates*, 2015). Henry Johnson acquired great wealth through the East India Company, and he was knighted by Charles II at Blackwall in 1679 (Le Neve, *Pedigrees of the Knights Made by King Charles II, King James II, King William III and Queen Mary, King William Alone, and Queen Anne*, 1873). Henry married Dorothy Lord, of Melton, Kent, and they had two sons: Henry (1659–1719) and William (b. c. 1665).

The bulk of Henry's wealth, including the family home, The Gate House, Blackwall, and the dockyards, were inherited by his eldest son Henry. He too, was knighted by James II at Blackwall in 1685.

Henry served as Member of Parliament for Aldeburgh from 1689 until 1719. Henry retained his interests in shipbuilding, having shares in 38 vessels at one point, however he had become a stout opponent of the East India Company by 1690, following a dispute with Sir Josiah Child who was a strong advocate of the East India Company's trading monopoly. In 1691 Henry served on the parliamentary subcommittee appointed to break the Company's monopoly, but after the Company's charter was renewed in 1693, he abandoned hostilities and re-subscribed to the Company.

Henry was twice married: he had married Ann Smithson in 1682, and after her death, he married Martha Lovelace (1667–1745) in 1693.

Martha was the great granddaughter of Thomas Wentworth, 1st Earl of Cleveland (d. 1667). Her grandmother Anne Wentworth (1623–1697) married John Lovelace, 2nd Baron Lovelace of Hurley (1616–1670). Martha inherited the Lovelace estate of Bradenham and Toddington (Rutton, *Three Branches of the Family of Wentworth. I Wentworth of Nettlestead, Suffolk. II Wentworth of Gosfield, Essex. III Wentworth of Lillingstone Lovell, Oxfordshire*, 1891). She also inherited the title of Baroness Wentworth of Nettlestead on the death of her grandmother in 1697.

Henry Johnson commissioned the first known armorial porcelains from China: a set of armorial jardinières. The jardinières are decorated in blue and white underglaze. The armorial consists of eight coats-of-arms, four each for Henry Johnson and Martha Lovelace (Fig. 1.1). Henry's armorials include Johnson (*or a water bouget sa on a chief of the second three bezants*) quartered with Lord (*ar on a fess between three cinquefoils az two pheons of the field*) and one other. Martha's include Lovelace (*gu on a chief indented sa three martlets ar*) with Pye (*or on a pile az three escallops of the first)* Wentworth (*sa a chev between leopards faces or)* and one other (Burke & Burke, *Encyclopaedia of Heraldry or General Armory of England Scotland and Ireland* 1844): there is further discussion on the armorial decoration of this commission in Chapter 1 and see Fig. 1.1.

The couple had multiple properties but appear to have lived primarily at The Gate House, Blackwall, as well as at Bradenham, Buckingham, which Martha Lovelace had inherited from her maternal grandfather Sir Edmund Pye (1607–1673). It can be said that this first known armorial Chinese commission represented a significant investment by Henry Johnson and Martha Lovelace, and it is highly likely that the purpose of the commission was to decorate their home in the garniture style and the jardinières would have been on prominent display in the public rooms of their home. The novelty of armorials on Chinese porcelain jardinières would have been significant, however having 8 ancestral coats-of-arms decorated onto the jardinières was a very clear signal as to the social standing of the commissioning patrons. Not only were they members of the social elite, with armorial connections to multiple landowning gentry families, but they also had the connections and wealth to commission highly personalised versions of luxury porcelain decorative items.

In 1711, Henry's only child Ann (1684–1754) married Sir Thomas Wentworth, 1st Earl of Strafford (1672–1739), son of Sir William Wentworth, 1st Baron of Wentworth Woodhouse, a distant connection of her stepmother Martha, Baroness Wentworth of Nettlestead in her own right; both were descended from Sir Philip Wentworth of Nettlestead (1423–1464), and both Wentworth families shared the same armorial displayed on the porcelain jardinières of *sa a chev between leopards faces or* (Burke & Burke, *Encyclopaedia of Heraldry or General Armory of England Scotland and Ireland*, 1844).

7.2.2 *Thomas Pitt (1653–1726): Chinese Armorial Porcelain Dinner Service Commission C. 1702*

The first known dinner service in Chinese porcelain with an English armorial decoration was commissioned by Sir Thomas Pitt (1653–1726) Governor of the Honourable East India Company at Fort St George, Madras (see Fig. 1.3).

Thomas was born in Blandford St Mary, Dorset, where his family had been settled for several generations. His great grandfather John Pitt (d. 1602) had been a mercer, and it was he who had been awarded the Pitt family armorial *sa. a fesse, chequy, ar. and az betw. three bezants* (Robson, *The British Herald; Or, Cabinet of Armorial Bearings of the Nobility & Gentry of Great Britain & Ireland, from the Earliest to the Present Time*, 1830). Thomas Pitt was the second son of the Rev John Pitt (1610–1672). In 1673 he went out to India on the *Lancaster*, an East Indiaman, but he left the ship in the Bay of Bengal to join the East India Company under Matthew Vincent (1645–1687). He took advantage of the opportunities his position gave to undertake private trading, as did Vincent. Thomas Pitt married Vincent's niece in 1674, Jane Innes

(d. 1727). Both he and Matthew Vincent attracted attention, criticism, and censure for their private trading activities; Pitt was fined on his return to England in 1683. He had, however, obtained significant wealth from his activities in India, and used this to become a propertied landowner. He already had a town house in London. In 1688 he bought the manor of Stratford from James Cecil, Earl of Salisbury, which he enlarged. He also bought the manor of Blandford St Mary, where he pulled down the original house to build a new house in its place. He became a member of Parliament for Old Sarum and sat in two Parliaments before he returned again to India.

In 1698 Thomas Pitt was appointed as President for the Company's affairs on the Coast of Coromandel and Orissa, and Governor of Fort St George and Fort St David. As Governor, Pitt lived in great style. His personal lodgings within the inner fort took one third of the building. At dinners, the Company's plate was used until 1703 when it was melted down. Old and battered, this plate was valued at less than £100. Pitt had brought a set of plate with him, said to weigh 2240 ounces and valued at £700. This service consisted of 55 plates, 12 dishes, 7 covers, 4 large candlesticks, salvers, salts, porringers, a teapot and flatware (Dalton, *The Life of Thomas Pitt*, 1915).

The size and scale of the plate gives an indication of Pitt's lifestyle and the level of entertaining that Pitt undertook at the time he placed his commission for an armorial Chinese dinner service. The service he ordered is of the Kangxi, Qing Dynasty. Decorated in underglaze blue and overglaze enamel design, the pattern is known as Chinese Imari, a Chinese version of Japanese Imari ware (see Chapter 1, Fig. 1. 3). The blue and orange colours are enhanced by gold gilding, creating an opulent look. The plates have the Pitt armorial displayed at their centre. The family crest, a stork, also appears on the plates, and its position is significant. In heraldry, the crest appears above the armorial, however that is not the case here. The stork is painted at the bottom of the plate, where it forms part of the overall decorative scheme; the stork appears to be wading amongst the flowers. In designing the decoration, the technical accuracy of heraldic imagery appears to have been rejected in favour of the aesthetic decorative appeal. The service would have been used by Thomas Pitt during his tenure as Governor until 1709, where it doubtless augmented the silver plate he already used at Fort St George. This armorial commission met the practical needs of entertaining and dining with visiting dignitaries and senior East India Company officials, in addition to being a clear signal as to the host's social standing.

7.2.3 Armorial Chinese Porcelain Coffee Service Commission of Henry Gough (1681–1751)

The Gough family were established in Staffordshire by the 1500s, when Thomas Gough (d. 1532) had settled in Wolverhampton. Henry Gough (d. 1655) held the Old Fallings estate near Wolverhampton, where his son John (1606–1665) was born. John married Bridget Astley (b. 1633). Their eldest son Henry (1649–1724) married

Mary Littleton (1653–1722) eldest daughter of Sir Edward Littleton, 2nd Baronet (1632–1709), of Pillaton Hall.

Their younger son Richard Gough (1659–1728) went to London where he was apprenticed to James Houblon (1629–1700) in the Dyers Company, who had ties with the East India Company. Houblon was an Alderman of London, and he was knighted in 1692. He became a Member of Parliament for London and served as Deputy Governor of the Bank of England. His brother Sir John Houblon (d. 1712) was the first Governor of the Bank of England, Lord of the Admiralty in 1694, and Lord Mayor of London in 1695.

Richard Gough became a merchant in his own right. He had begun trading with China by 1692. He travelled to China several times, and served as a super-cargoer on the *Sarah*, an East India Company ship. He is known to have travelled to China and Fort St George around 1702 (Reeves, '*Sir Richard Gough and his broadside collection*', *Harvard Library Bulletin* X (1), 1956). Whilst at St George it is highly likely he would have called on Thomas Pitt, then Governor.

Henry or Harry Gough (1681–1751), the son of Henry Gough and Mary Littleton, joined Richard Gough, his uncle, in his business and travelled with him to China. Harry Gough joined the East India Company and served as Captain on the *Streatham* from 1707 until 1715 (Stephen, *Dictionary of National Biography*, 1922). It is during this time he is thought to have commissioned the armorial coffee service carrying the Gough armorials.

This commission differs from those of Henry Johnson and Thomas Pitt in that the coat-of-arms for Gough—*gu on a fess arg between three boars heads couped or, a lion pass az*—is not depicted. Instead, only the crest is shown—*a boar's head couped ar devouring a broken spear gu* (Burke & Burke, *Encyclopædia of Heraldry Or General Armory of England, Scotland, and Ireland*, 1844).

In 1717, Henry Gough married Elizabeth Hynde (1698–1774) in London. Elizabeth was the daughter of Morgan Hynde. Another service was commissioned around 1720 with the arms of Gough *gu on a fess arg between three boars heads couped or, a lion pass az* impaling Hynde *gu a chev between three hinds or* (Burke & Burke, *Encyclopaedia of Heraldry or General Armory of England Scotland and Ireland,* 1844). These Hynde armorials were granted in 1583 to Roland Hynd of Hedsor, Buckinghamshire (Foster & Rylands, *Grantees of Arms named in Docquets and Patents to the end of the Seventeenth Century*, 2016) although Emma Hynde's connection to this family is unclear; her father is thought to be Morgan Hynde, a wealthy London brewer (Sedgwick, 'Gough, Henry 1681–1751, of Endfield, Middlesex,' *History of Parliament: The House of Commons 1715–1754*, 1921).

The three Chinese armorial commissions of Johnson, Pitt and Gough represent the genesis of Chinese armorial porcelain on decorative wares, a dinner service and a coffee service. Johnson and Pitt moved in the same business circles: their tenure as Members of Parliament overlapped, and they had worked together during Parliament's attempts to curtail the East India Company's monopoly in the early 1690s. Indeed, Henry Johnson and Thomas Pitt were friends; in 1700 Pitt wrote to Johnson thanking him for the gift of a butt of sherry (Dalton, *The Life of Thomas Pitt*, 1915). It is unlikely that Pitt had seen Johnson's armorial Chinese porcelain jardinières as he was abroad from 1693 onwards, but it highly probable he would have known about the commission.

Henry Johnson, Thomas Pitt and Harry Gough had much in common. Firstly, all three profited greatly from their involvement in the East India Company. Both Henry Johnson and Harry Gough can be considered 'second generation' East India Company officials, as Henry Johnson's father and Harry Gough's uncle had had previous involvement with the company and had introduced them into the Company as an act of patronage. The lives of all three patrons illustrate the opportunities to create great wealth that the East India Company could offer. Having created their fortunes, both Henry Johnson and Thomas Pitt used their new wealth to acquire landed wealth. Henry Johnson's second marriage to Martha Lovelace in 1694 brought him landed property, and her armorials were included on the Chinese porcelain jardinières. Thomas Pitt used his wealth to purchase properties, including the manor house of the town he was born to St Marys Blandford. Only Harry Gough appears to have lived mainly in London, where he died, at St Andrews Holborn, in 1751. The descendants of Johnson and Pitt married into the titled nobility: Henry Johnson became the father-in-law of Sir Thomas Wentworth, 1st Earl of Strafford, whilst Thomas Pitt's descendants included the 1st Earl of Londonderry and 1st Earl of Chatham. Gough's son Richard Gough (1735–1809) became a noted antiquarian, who amassed a considerable collection of books, manuscripts, prints, maps and drawings (*Dictionary of National Biography*, vol 8, 1922b).

Having gone to so much effort, not to mention expense, in order to acquire armorial Chinese porcelain, the commissioning patrons intended their items to be seen. Visitors and guests being entertained in the homes of Johnson, Pitt and Gough would have been expected to notice and admire the porcelains.

Thomas Pitt is the first known patron to have commissioned an armorial dinner service, and he started the trend for what would be over 3,000 commissions of English armorials on Chinese porcelains. The majority of these are documented in the reference works by David Sanctuary *Howard Chinese Armorial Porcelain* Vol 1 and Vol 2 published in 1974 and 2003, although additional commissions do occasionally emerge. The sheer volume and size of these commissions demonstrates the appetite for such personalised luxury items throughout the century, as well as the considerable expenditure made on armorial porcelain.

7.3 Armorial Chinese Porcelain Commissions: 1715–1730

In the period 1715–1730, the ability to acquire a personalised armorial Chinese porcelain required a connection to the East India Company, either directly as a Company official or indirectly via a personal contact as family or close friend. The official trade of the East India Company was tea, silks, and general porcelain. Armorial commissions were managed as a private trade or side line business by the supercargoes and captains of the East India Company ships. The commission process itself was slow.

Orders had to be carried by ships sailing East, the porcelain had to be manufactured and decorated by the Chinese factories in Jingdezhen, then the order would be transported to the docks to wait for a ship travelling westwards. Porcelains were typically shipped into London then transport had to be arranged from the dockside to the customer. It was usual for commissions to take at least two to four years to be fulfilled, but the inconvenience of placing the order and waiting for its arrival did not deter patrons, however. If anything, this added to the lustre and exclusivity of the armorial porcelains, and the practical difficulties of obtaining armorial porcelain seem to have simply added to its desirability.

7.3.1 The Armorial Chinese Porcelain Commission of Robert Child (1674–1721)

The Child armorial dinner service, commissioned around 1720, is thought to have been commissioned by Sir Robert Child (1674–1721), Chairman of the East India Company.

The service plates depict the Child family arms *gu a chevron ermine betw three eagles close, arg* in the centre, with the family crest on the rim (Finn & Smith, *The East India Company At Home 1757–1857*, 2018).

The Child family had long been settled at Northwick, Worcestershire; Richard Le Child was lord of the manor of Northwick in 1320 (Child, *Genealogy of the Child Family*, 2018). Ten generations of the Child family were documented in the heraldic Visitation of Worcester in 1569, where the arms of William Childe of Northwick are given as as *gules a chevron ermine between three eagles close argent* and the crest was given *as an eagle with wings expanded argent, enveloped round the neck with a snake vert* (Phillimore & Phillimore (eds.) *The Visitation of the County of Worcester Made in the Year 1569*, 1888). William Child was lord of the manor of Northwick in 1634, and it is believed to be his sons, Thomas of Northwick, William, and John, who moved to London (Child, *Genealogy of the Child Family*, 2018).

The armorials decorating the service show that Robert Child he was a descendant of this family, however he appears to belong to another branch of the family that had settled at Heddington, Worcestershire before 1642, when his grandfather Francis Child (1642–1713) was baptised there. Francis was the son of Robert Child (b. c. 1602) and Jane Titcombe (d. 1690). Robert and Jane had at least 11 children, and Francis was their eighth child. He had moved to London by 1656 when he was apprenticed to William Hall, goldsmith. He married Elizabeth Wheeler in 1671, who was from a family of goldsmiths dating back to John Wheeler in 1575. Her father William had died in 1663, and her mother had married Robert Blanchard who took over the business. Robert Blanchard and Francis Child were in partnership before 1677 and operated as goldsmith-bankers. Francis Child ultimately inherited the business, and by 1690 he had dropped the goldsmith trade to focus solely on banking, and thus established the first London banking house (Van den Berg, *Cantillon's Essay on the Nature of Trade in General*, 2015). Francis Child served first as Alderman of London, then Sherriff then Lord Mayor in 1698. He was also Member of Parliament for constituencies in London and Devizes. He became very wealthy and bought Osterley Park in 1711, two years before his death. He is not thought to have lived at Osterley Park, but his son Robert Child, who inherited the banking business Child & Co lived there until his death in 1721. Robert Child also served as Member of Parliament for Devizes, and he was Chairman of the East India Company in 1715. When he died unmarried in 1721, Robert Child was succeeded by his brother Francis (1684–1740) who inherited Osterley Park as well as becoming the head partner in Child & Co. He also served as Lord Mayor of London in 1731 and was a Director in the East India Company.

It should be noted that another branch of the Northwicke Child family had settled in London a generation or two earlier, which was also deeply involved with the East India Company. Sir Joshia Child 1st Baronet (1630–1699) was the second son of Richard Child, a merchant of Fleet Street, and he was baptised at St Bartholomew Exchange, in London. Joshia Child used the same Child armorial that Robert Child would have decorated on the service commissioned for Osterley Park. He is listed as Sir Joshia Child of Wanstead House, *gules, a chevron ermine between three eagles close, arg* (Guillim & Mackenzie, *A Display of Heraldry*, 1724). The precise relationship between Sir Joshia Child and Sir Francis Child is unclear, however their armorials display their familial connection. By 1655 Joshia Child was Deputy Treasurer of the Fleet, responsible for naval stores. He was Mayor of Portsmouth before becoming MP for Petersfield, Dartmouth, and Ludlow. He was a Director of the East India Company by 1677 and received his baronet in 1678. He went on to become Governor of the East India Company 1681–1683 then again 1686–1688. He is said to have been the richest member of the East India Company, holding one third of its stock (Child, *Genealogy of the Child Family*, 2018). He had bought Wanstead House by 1675, where he died in 1699. Sir Joshia's brother John Child (d. 1690) was also involved in the East India Company. He was Governor of Bombay and was knighted as Sir John Child of Surat in 1685 (*Dictionary of National Biography vol 4*, 1922).

7.3.2 *The Armorial Chinese Porcelain Commission of Anthony Duncombe (1695–1763): The* **Duncombe Cornwallis and Verney Service** *c. 1724*

An armorial Chinese porcelain service was commissioned by Anthony Duncombe (1695–1763) around 1724 (Howard, *Chinese Armorial Porcelain*, vol 1, 1974). The Duncombe family originated in Barley End, Buckinghamshire, being settled in the area by the time of Henry III. Anthony's direct ancestor, William Duncombe of Irvinghao had 4 sons, the second son being Charles Duncombe (1648–1711).

Charles transformed the family fortunes. He was a goldsmith and a banker. He held numerous civic roles, including serving as a Clerk in Chancery 1682–1683,

Alderman of London 1683–1686, then Member of Parliament for Hedon, Yorkshire from 1685–1687. He subsequently served as Member of Parliament for Downton, Wiltshire, was appointed Knight in 1699 when he served as Sheriff of London then Alderman for Bridge Ward in 1700. He went onto become Lord Mayor of London in 1708. He bought the Helmsley estate in 1695 from George, 2nd Duke of Buckingham (Anon, *A Description of Duncombe Park, Rivalx Abbey and Helmsley Castle*, 1829). Charles died without direct heirs, so his fortune passed to his nieces and nephews on his death in 1711.

Charles' sister Ursula had married Thomas Brown before 1680. Their daughter Mary Brown (1680–1716) married John Campbell, 2nd Duke of Argyll (1678–1743) in 1701, and she inherited a significant portion of her uncle's wealth on his death in 1711. Ursula inherited Duncombe Park which in time passed to her son, also Thomas Browne, who took the name Duncombe. Thomas married Sarah Slingsby of Scivers in 1714, who was maid of honour to Queen Anne and daughter of Sir Thomas Slingsby 4th Bart (Foster, *Pedigrees of the County Families of Yorkshire*, vol 2, 1874).

Charles younger brother Anthony Duncombe had predeceased him in 1708, so the inheritance passed to Anthony's son, also Anthony, on Charles' death (Burke, *A Genealogical and Heraldic History of the Commoners of Great Britain and Ireland, Vol. I*, 1834). He served as Member of Parliament for New Sarum and for Downton, and he was created Lord Feversham Baron of Downton in 1747.

Anthony Duncombe married Margaret Verney (d. 1755) in 1714 at St Georges Chapel, New Windsor Berkshire, daughter of George Verney 12th Lord Willoughby de Broke (c. 1660–1728) and Margaret Heath (b. 1665). Their armorial Chinese porcelain service was commissioned around 10 years later, in 1724. It is decorated with the Duncombe arms *per chev engrailed gu and ar 3 talbots heads erased counter-charged* quartering those of Cornwallis *sa guttee d'eau on a fess arg three Cornish choughs ppr* impaling Verney *gu. three crosses recercelée or, a chief vair erm. and ermines* (Burke & Burke, *Encyclopædia of Heraldry Or General Armory of England, Scotland, and Ireland*, 1844).

7.3.3 The Armorial Chinese Porcelain Commission of Robert Blatchford (1699–1729): A Gift for James Brydges, 1st Duke of Chandos (1673–1744)

James Brydges (1673–1744) was a member of the old East India Company by 1700 and served as Member of Parliament for Hereford from 1698 to 1714. He married three times, firstly Mary Lake (1666–1712) followed by Cassandra Willoughby (1670–1735) and then Lydia Catherine Hattem (1693–1750).

The Chandos armorial Chinese porcelain service has been dated to around 1729 (Howard, *Chinese Armorial Porcelain*, vol 1, 1974). James Brydges father, also James Brydges (1642–1714), was the 8th Baron Chandos of Sudeley. The Chandos family goes back to Robert de Chaundos (d. 1302) who lived in Herefordshire. His son Roger de Chaundos (d. 1353) was 1st Lord Chaundos. John Brydges (1492–1557) became 1st Baron Chandos of Sudeley, and John's grandson Sir Giles Brydges (1573–1637) was created 1st Baronet Brydges of Wilton in 1627.

The armorial Chinese porcelain service contains the armorials for Brydges impaling Willoughby and Middleton. These are for the Duchess of Chandos, Cassandra Willoughby (1670–1735). Cassandra and James were first cousins, their mothers having been sisters. Elizabeth Bernard (1642–1719) had married James Brydges, 8th Baron Chandos, and Emma Barnard (1644–1725) had married Francis Willoughby (1635–1672).

The Willoughby family can be traced back to Richard Willoughby (b. c. 1290) and Cassandra Willoughby's ancestors include Anne Grey (1520–1548), daughter of Thomas Grey 2nd Marquis of Dorset (1477–1530) and descendant of Elizabeth Woodville (1437–1492) who married Sir Henry Willoughby (1517–1549). The Middleton armorial recognised descent from Margaret Freville (b. c. 1397) who conveyed the Middleton estates to the Willoughby family by her marriage to Hugh Willoughby (1380–1448).

The Willoughby and Brydges families were close-knit, as can be demonstrated by the letters of Cassandra, who was a frequent correspondent with her brother, nieces, nephews, and cousins (O'Day, *Cassandra Brydges, Duchess of Chandos, 1670–1735: Life and Letters*, 2007). These close kinship ties may well explain the existence of this armorial Chinese porcelain service, as evidence shows that the commission was actually placed on behalf of the Duke of Chandos and indeed, that he was unaware of its existence for quite some time.

The commission for the service appears to have been placed by Robert Blatchford (1699–1729). Blatchford was a distant relation of the Duke; his mother Anne Brydges of Tiberton (b. c. 1670) had married Robert Blatchford in 1695. The Duke of Chandos had used his influence to secure Blatchford a role within the East India Company. Blatchford had served on the *Cardigan*'s voyage in 1717 and he was second mate on the Caernarvon's voyage to China in 1720. On the second voyage, he commissioned several items adorned with the Duke's arms including sufficient satin for a bed, two Japan screens and a china service, by way of showing his appreciation to his patron. The Duke of Chandos describes the commission in a letter to his steward dated 1730 ('Huntington Library, Stowe Papers, STB Box 13, Letters to J. Farquharson, Steward, 55,' cited by Ferguson, P. F., *Chinese Armorial Wares for James Brydges, 1st Duke of Chandos: Self-fashioning or artifice?* 2017).

This service appears to be one of the earliest armorial commissions that was ordered as a gift; however it should be noted Blatchford did expect the Duke to pay for it! The gift appears to have caused James Brydges some surprise and consternation. He was a noted connoisseur, however, and doubtless he would have valued the service. He used his great wealth to rebuild Cannons, in Little Stanmore, which he had acquired through his first marriage to Mary Lake. Her grandfather, Sir Thomas Lake (1567–1630) who served as Secretary of State to James I had bought the property in 1604. Brydges demolished the existing house and replaced it with a magnificent mansion that was much admired (Lysons, *The Environs of London: Middlesex*, 1811).

He hired the best architects of the day to design the house and its gardens, including James Gibbs (1682–1754) whose work for Canon House is considered one of his highest achievements. He was a noted art collector (Jenkins, *Portrait of a Patron: The Patronage and Collecting of James Brydges 1st Duke of Chandos (1674–1744)*, 2016). The splendour of the interior design was noted by Daniel Defoe who described the mansion as 'royally furnished'. Brydges style of living was flamboyant. When dining at Canon House, he reportedly had flourishes of trumpets announcing each change of dishes (Pope, *The Poetical Works*, vol 4, 1854). The build included a new chapel, and Brydges engaged the composer George Frederick Handel (1685–1759) as director of music at Canons, and it was during his time at Cannons that Handel composed the Chandos Anthems.

7.3.4 The Chinese Porcelain Armorial Service of Humphrey Parsons (1676–1741)

Humphrey Parsons ordered an armorial Chinese porcelain service around 1720 (Howard, *Chinese Armorial Porcelain*, vol 1, 1974). His grandfather John Parsons had been a brewer in East Smithfield, who established the Red Lion Brewery. The business was inherited by his son also John Parsons (1639–1717) who became a victualler to the Navy in 1677. He was wealthy enough to purchase the Priory estate in Reigate and became the Member of Parliament for Reigate in 1685. He was knighted by James II in 1687 and served as Lord Mayor of London in 1703.

John Parsons married Elizabeth Beane (1645–1677) and their son Humphrey was born in 1676. Humphrey inherited the brewery on his father's death in 1717, along with the Reigate Priory estate. He developed a drink known as Parsons Black Champagne, which he supplied to the French court (Hutchings, *The Red Lion Brewery*, 2013). The Red Lion brewery and its drink was sufficiently well-known to be referenced in the works of Oliver Goldsmith in 1762 (Goldsmith, *The Vicar of Wakefield: Together with the Poems of Oliver* Goldsmith, 1854).

Humphrey Parsons also served as a Member of Parliament from 1722 until 1741. He was a member of the Wax Chandlers Company before transferring to the Grocers Company when he became Lord Mayor, a position he held twice, serving in 1730 and again in 1740. He was in his second term of office as Lord Mayor at the time of his death.

Humphrey Parsons married Sarah Crowley (d. 1759) in 1719. She was the daughter of Sir Ambrose Crowley (1658–1741). Her grandfather, also Ambrose Crowley (b. c. 1635) had been a blacksmith, and her father had founded the Crowley Ironworks in Winlaton, County Durham c. 1682. This venture had been sufficiently successful for Ambrose Crowley to have become Sheriff of London in 1706, the same year he received a knighthood from Queen Anne. He also served as a Member of Parliament, being elected for Harwich in 1722.

A surviving plate from the serviceshows that on the dexter side, the armorial has the arms for Parsons *gu 2 chevrons ermine between 3 eagles displayed or* (Burke & Burke, *Encyclopaedia of Heraldry, or General Armory of England, Scotland and Ireland*, 1844). This indicates that Humphrey Parsons was related to the Parsons family of Great Milton, who were first recorded in the Visitations in 1574 (Fetherston, *Visitations of the County of Oxford*, 1871). The Parsons armorial is quartered with another armorial *az 2 chevrons or between 3 goats heads erased arg.* Marshall identified these arms as being for the Yarker family (Christie, Manson & Woods, *The G. W. Marshall Collection of Chinese Armorial Porcelain and other Fine Chinese Ceramics, Enamels and Works of Art, Auction Catalogue*, 1978) however a familial connection between Humphrey Parsons and a Yarker ancestor has not been traced. Furthermore, the arms quartered with Parsons are not listed in Berry's *Encyclopaedia Heraldica or Complete Dictionary of Heraldry* published in 1828. The Yarker arms are described in Robson as *sa a chev between 3 goats heads erased, arg* (Robson, *The British Herald, or Cabinet of Armorial Bearings of the Nobility and Gentry of Great Britain and Ireland*, 1830), therefore if the armorial is indeed for Yarker, there are two discrepancies with its depiction on the service: namely, the black background of the armorial rather than blue, and two chevrons instead of one.

On the sinister side of the armorial, the arms for Crowley are shown. The Crowley armorials, granted in June 1707, were *vert, on a chev or, an estoile of sixteen points, betw. two roses gu* (Burke & Burke, *Encyclopaedia of Heraldry or General Armory of England Scotland and Ireland*, 1844).

This commission, therefore, represents a couple whose wealth came from a combination of victualling connections to the Navy and industrialisation. Whilst the Crowley armorials were newly acquired following a family fortune made via industry, the Parsons armorials date back to an Elizabethan county gentry family, demonstrating the social mobility opportunities that trade and industrialisation could create. Humphrey Parsons and Sarah Crowley had two daughters and a son; their eldest daughter Ann (d. 1769) married Sir John Hinde Cotton, 4th baronet of Landwade (1717–1795). Sarah Crowley's brother John (d. 1727) married Theodosia Gascoyne, daughter of the Rev Joseph Gascoyne and Anne Theobald, and by this marriage acquired the estate of Barking, Essex. Their daughter Elizabeth Crowley (1727–1781) married John Ashburnham (1724–1812) and thus became the 2nd Countess of Ashburnham.

Humphrey Parsons may not have been directly involved with the East India Company, but he was well acquainted with senior East India Company officials. Both Humphrey Parsons and Francis Child of Child & Co served as Sherriff of London in 1722. His time as a Member of Parliament also overlapped with that of Francis Child, who was elected to represent London in 1722 whilst Parsons was elected the same year to represent Harwich. Parsons then represented London from 1727 onwards and remained as a Member of Parliament until 1740, whilst Child represented Middlesex for the same period.

Parsons served as Lord Mayor of London, and it is noticeable that Lord Mayors of London formed a significant group of people who commissioned armorial services. Indeed, Howard estimates that some 48 services were commissioned by London's

Lord Mayors (Howard, 'From Canton To the City: Export Porcelain for London', *Art Antiques*, 2013).

The office of Lord Mayor, established in 1189, was an elected role. The City of London was split into 26 wards, and every ward elected an Alderman. The Aldermen appointed 2 Sheriffs each year and elected a Lord Mayor. In order to become Lord Mayor, candidates needed to be an Alderman and to have already served as Sherriff. The role of the Lord Mayor was to promote and represent the interests of trade and commerce as organised through the livery companies. There were some 70 livery companies active in 1700, the most senior being the Worshipful Company of Mercers (general merchants). By tradition the Lord Mayor was associated with one of the 12 senior livery companies. The nature of the Lord Mayor's role ensured that the office holder interacted with the monarch and the aristocratic families at court. Humphry Parsons was a popular Lord Mayor as demonstrated by his being appointed twice, which was rare.

The patrons who were ordering armorial services were a close-knit group. Kinship networks often overlapped with these business networks. Ambrose Crowley (1658–1713) had 2 daughters: Letitia, and Sarah. Letitia married John Hynde Cotton 3rd Baronet, whilst Sarah married Humphrey Parsons. These family links were further strengthened in the next generation: Sarah's daughter, Ann Parsons, married Mary's son John Hynde Cotton, 4th Baronet (Wotton et al., *The Baronetage of England: Containing a Genealogical and Historical Account of all the English Baronets Now*, 1771).

7.4 Armorial Service Commissions Within Families

7.4.1 The Armorial Porcelain Commissions of the Pitt Family

Following Thomas Pitt's 1702 commission, two more services were commissioned by his immediate descendants. His son Thomas Pitt (1688–1729) 1st Earl of Londonderry commissioned an armorial Chinese porcelain service around 1720. The style of the decoration is similar to that of Pulteney's service, in that the armorial is large and its depiction takes up most of the body of the plate. The crest again appears in the border at the four compass points, with floral decorations in between. The armorial depicted in this service is that of Pitt impaled with Ridgeway—*sa a pair of wings conjoined and elevated arg* (Burke & Burke, *A Genealogical and Heraldic History of the Extinct and Dormant Baronetcies of England*, 1838). The Ridgeway armorial is for Frances Ridgeway (b. c. 1693) who married Thomas Pitt in 1717. She was the daughter and heiress of Robert Ridgeway 4th Earl of Londonderry.

The second commission was placed by William Pitt, 1st Earl of Chatham (1708–1778), who served as Prime Minister from 1766 until 1768. Often known as Pitt the Elder, to distinguish him from his son also William Pitt and also Prime Minister, this service displays the Pitt arms impaling Grenville. William Pitt had married Hester

Grenville (1720–1803) in 1754 (Howard, 'The Pitt Services: Porcelain for a Political Dynasty,' *American Ceramic Circle Journal*, 2015). These services are discussed in more detail in Chapter 1.

7.4.2 The Armorial Commissions of the Gough/Hynde Families

Harry Gough is known to have placed multiple armorial commissions. In addition to the early coffee service c. 1710, he commissioned another four armorial Chinese porcelain services around 1720 to mark his marriage to Elizabeth Hynde (1698–1774) all decorated in blue and white (Howard, *Chinese Armorial Porcelain*, vol 1, 1974).

His sister Isabella Gough (1650–1734) had married Sir Eldred Lancelot Lee (1560–1734) in 1713. A service was commissioned for them c. 1735 (Howard, *Chinese Armorial Porcelain*, vol 1, 1974) which was the same time as a portrait of Isabella and Eldred with their children was commissioned. '*The Family of E Lancelot Lee*' by Joseph Highmore (1692–1780), now in Wolverhampton Art Gallery, was completed after Eldred's death in 1734. The armorials on this service are Lee of Coton, Shropshire, quartered with Astley of Staffordshire in polychrome enamels and gold.

The Lee armorials are *gu a fesse componée or and az between eight billets ar* and the Astley of Staffordshire armorials are *az a cinquefoil erm within a bordure engrailed of the second.* Above the armorial is the Lee crest: *On a staff raguly, a squirrel cracking a nut, from the dexter end of the staff an oak branch fructed, all ppr.* The family motto is also displayed: *virtus vera est nobilitas* (Burke, & Burke, *Encyclopaedia of Heraldry or General Armory of England Scotland and Ireland*, 1844).

The armorial decoration completely fills the base of the plate, leaving no room for decoration. The border of the plate is depicted with scenes of London on one side, and Canton on the other, separated by 2 small panels of floral decoration, which depicts chrysanthemums.

The armigerous Lee family is complex, with multiple branches; Burke's *Armory* lists 54 different coats-of-arms for the family, many very similar with perhaps a different tincture or crest (Burke, & Burke, *Encyclopaedia of Heraldry or General Armory of England Scotland and Ireland*, 1844).

Eldred Lancelot Lee's maternal grandfather, John Eldred (1552–1632), had made a five-year voyage on the *Tiger* which took in Bassorah, Baghdad, Aleppo, Antioch, and Bethlehem, before returning to London in 1583 with a vast cargo of goods which was the foundation of his fortune. John Eldred subsequently became an Alderman of the City of London, and he purchased Great Saxham Suffolk in 1597. He was also an early subscriber to the East India Company and a member of its first court of directors (Eldred, *Eldred Family; particularly the descendants of William Eldred of*

Yarmouth, 1940). The Lee armorial dates back 9 generations, having been granted to Richard Lee, High Sheriff of Shropshire in 1479. Richard Lee had married Margaret Astley, daughter of Thomas Astley around 1475.

7.4.3 Walpole Family Commissions: Robert Walpole (1676–1745) and Horatio Walpole (1723–1809)

Multiple armorial commissions were placed by the Walpole family in the eighteenth century.

Prime Minister Robert Walpole (1676–1745) was in Parliament from 1702 onwards and was serving as Secretary of War by 1708. He made his fortune in the 1710s through the South Seas Company and obtained sufficient wealth to build Houghton Hall, Norfolk (Coxe, *Memoirs of the Life and Administration of Sir Robert Walpole Earl of Orford*, 1800). He had commissioned an Imari patterned service around 1715. Another service was ordered by Horatio Walpole 2nd Baron Walpole Earl of Orford (1723–1809) to mark his marriage to Rachel Cavendish (1727–1805) in 1742 (Howard, *Chinese Armorial Porcelain*, vol 1, 1974). This service was a blue and white commission, with a Japanese style decoration. The central pattern depicts a long-tailed tiger and bamboo, and the border is divided into eight panels decorated with Japanese imagery. The arms are on the underside of the dishes.

The armorial on this service shows Walpole impaling Cavendish. The Walpole armorials are quartered on the dexter side with those for Robsart. The Walpole armorials are *or on a fess between 2 chevrons sable 3 crosses crosslet of the first*, and Robsart are *vert, a lion ramp, or, vulned gu.* The Cavendish armorial depicted on plate is *sa 3 stags heads cabossed ar. attired or.* The crest is for Walpole—*the bust of a man in profile couped, ppr, ducally crowned or; and from the coronet flowing a long cap, turned forwards, gu, tasselled gold, charged with a catherine wheel of the last* (Burke & Burke, *Encyclopaedia of Heraldry or General Armory of England Scotland and Ireland*, 1844).

The Robsart armorial goes back 7 generations for Horatio Walpole, to the marriage of Edward Walpole (1483–1558) and Lucy Robsart (1490–1560) in 1508. She was the daughter of Sir Terry Robsart of Syderstone, Norfolk (1465–1496) and the aunt of Amy Robsart (1532–1560) the first wife of Robert Dudley, Earl of Leicester (1533–1588). The property Syderstone manor was conveyed to the Earl at the time of the marriage, and he retained a life interest in it. Following his death, the manor reverted to John Walpole the heir of Edward Walpole and Lucy Robsart (Palmer, *The History of Great Yarmouth*, vol 2, 1856) and continued in the Walpole family. George Walpole, 3rd Earl of Orford, inherited the manor, and it duly passed to the 4th Earl, Horace Walpole (1717–1797) and his wife Rachel, who had commissioned the armorial service. The fact that the armorial service included the Robsart armorial demonstrates that the Walpole family valued the connection to the Robsart family and held it in some regard. It is unusual for the armorial to be painted in a non-prominent position, however, where they are not immediately visible. The armorials on the underside of the dishes would only be seen when they were carried into the dining room or being removed. By the 1750s there were usually two courses served, with multiple dishes for diners to select from, which would be removed after each course (Brooks, *The Importance of British Material Culture to Historical Archaeologies of the Nineteenth Century*, 2016). This fashion had supplanted the earlier style of dining where all the food was already in dishes on the table or sideboard at the start of the meal.

7.5 Early Armorial Chinese Porcelain Commissions Followed by Later Armorial English Commissions

It was not always the case that a commission of an armorial Chinese porcelain items was followed by a second armorial commission on Chinese porcelain. There are several instances where an early armorial Chinese porcelain commission was followed by one on English porcelain.

7.5.1 Armorial Chinese Porcelain Service for Thomas Bootle (1685–1753) & Later Armorial English Commission by Mary Bootle Wilbraham (1734–1813)

Thomas Bootle of Melling (1685–1753) was a lawyer and a Member of Parliament. He held office as Chancellor to Frederick Prince of Wales, and was wealthy enough to purchase Lathom House, Lancashire, in 1724, which had previously belonged to the Stanley Earls of Derby. The 9th Earl of Derby had begun rebuilding the house; however, it was Thomas Bootle who completed the work. His brother Robert Bootle had become a Captain with the East India Company and would become a Director in the Company in 1741. Around 1730, Robert Bootle commissioned an armorial Chinese porcelain service.

Robert's daughter Mary was the heiress of both Thomas and Robert Bootle and she duly inherited Lathom House. Mary married Richard Wilbraham of Rode, Member of Parliament for Chester, who took the additional surname Bootle in accordance with Thomas Bootle's will (Draper, *The House of Stanley: Including the Sieges of Lathom House, with Notices of Relative and Co-temporary Incidents*, 1864). Mary was an avid porcelain collector, who acquired many pieces of English porcelain (McKeown, *English Ceramics: Two Hundred and Fifty Years of Collecting at Rode*, 2006) She also inherited the family armorial Chinese porcelain service, which had suffered some losses by 1780. She sought replacements for the Chinese service from Duesbury's Derby factory commissioning a replacement tea service with the family armorial *gules on a chevron engrailed between three combs argent with a cross patée-fitchée in the field* (Edwards, *Welsh Armorial Porcelain*, 2022).

7.5.2 The Chinese Armorial Service of William Talbot (1685–1737) and the Later Talbot Services

Commissioned in 1705, the Talbot service is amongst the earliest known armorial Chinese porcelain commissions. It is a dinner service, believed to have been commissioned by the Rt Rev William Talbot (1685–1737) and was previously thought to have been the first such service (Fuchs, *Made in China: Export Porcelain from the Leo and Doris Hodroff Collection at Winterthur*, 2005).

William Talbot was a direct descendant of John Talbot 2nd Earl of Shrewsbury (1413–1460) and Elizabeth Butler (1420–1473) on a cadet branch: his ancestor Gilbert Talbot (1452–1518) was a younger brother of John Talbot, 3rd Earl of Shrewsbury (b. 1448). The earldom passed down John's descendants until Edward Talbot, 8th Earl of Shrewsbury (1561–1617) died without a direct heir. The earldom then moved to Gilbert's descendants, and his great-great grandson George Talbot (1566–1630 became the 9th Earl). George Talbot also died without a direct heir, so the earldom went to his nephew John who became the 10th Earl. John's son became the 11th Earl, and it was his son Charles, the future Duke of Shrewsbury, who bestowed the Rectorship of Burghfield on William Talbot in 1682. He also owed his appointment as Dean of Worcester to his Shrewsbury connection. In 1699, Talbot became Bishop of Oxford, and as such he had a seat in the House of Lords. He held both roles in 1702 when his second wife Catherine King died; her memorial tablet is in Worcester Cathedral. He went on to marry again in 1703, in Worcester Cathedral, his third wife being Agnes Hartopp (d. 1730) the daughter of Sir William Hartopp (1626–1692) of Rotherby Leicestershire, and Member of Parliament for Leicester. Sir William was knighted by Charles II in 1660, and he was descended from the Hartopp family of Little Dalby, Leicester, an armigerous family whose coat-of-arms was *sa a chev. erm. betw. three otters pass. of the second* (Burke & Burke, *Encyclopaedia of Heraldry or General Armory of England Scotland and Ireland*, 1844).

The commissioned service was decorated in an under-glazed blue, and the armorial depicted is solely for William Talbot. The *Encyclopaedia of Heraldry* lists 32 armorials for Talbot. William Talbot appears as Talbot of Stourton Castle co Stafford, Bishop of Durham and his armorials as *gu a lion ramp. within a bordure engr. or, with due diff* (Burke & Burke, *Encyclopaedia of Heraldry or General Armory of England Scotland and Ireland*, 1844). A crest for William Talbot is not listed, however the crest of the earldom of Shrewsbury is *on a chapeau gu. turned up erm a lion statant or, the tail extended* (Burke & Burke, *Encyclopaedia of Heraldry or General Armory of England Scotland and Ireland*, 1844) which matches the crest displayed on the service; one of William's descendants did ultimately become Earl of Shrewsbury;

his great-great-grandson, Henry John Chetwynd-Talbot (1803–1868) became 18th Earl of Shrewsbury after other lines died out.

William Talbot was aged around 46 in 1705, when he commissioned the armorial service. The reason for the commission is unclear; it may have been to mark his marriage, or perhaps it was associated with his career progression within the Church. He was already a prominent clergyman, and he would go on to hold multiple senior appointments. As Bishop of Oxford in 1707, he gave the sermon at St Pauls Cathedral before Queen Anne on 1st May 1707, to mark the Act of Union coming into force. He preached at the Coronation of George I in 1714 and became Dean of The Chapel Royal that same year. He became Bishop of Salisbury in 1715 and then Bishop of Durham in 1721 (Stephen, *Dictionary of National Biography*, vol 19, 1922).

William Talbot and Catherine King had a daughter, Catherine born c. 1700. In 1723 she married Exton Sayer (1692–1731). He studied at Cambridge University before being admitted to Lincolns Inn, and he was an official of the Archdeaconry of Wiltshire. Following the marriage, he was Chancellor of Durham 1724–1731. He also served as Member of Parliament for Helston and for Totness. Other appointments included the commissary of the London diocese, and Surveyor General of Land Revenue. At least some of these can be attributed to the influence and patronage of his father-in-law, William Talbot who was Bishop of Durham by this time. Exton Sayer and Catherine Talbot commissioned an armorial Chinese porcelain service around 1725 in a blue underglaze style very similar to the service of William Talbot (Howard, *Chinese Armorial Porcelain*, 1974).

Exton Sayer was the son of George Sayer (b. c. 1760) and Mary Exton (b. c. 1765) who married in London in 1687. The armorial depicted on the plate is that of Sayer of Worsall, Yorkshire. The connection of George Sayer to the Sayers of Worsall is unclear; the Sayer family had held the manor of Worsall since 1468 but had lost the property during the Civil War. The Sayer armorials for the Worsall family are the same as for the Sayers of Pulham, Norfolk: *gu a chev betw three seapies arg* but with the addition of *a chief erm* (Burke & Burke, *Encyclopaedia of Heraldry or General Armory of England Scotland and Ireland*, 1844).

Charles Talbot, 1st Baron Talbot of Hensol (1685–1737), was the son of William Talbot, Bishop of Durham. He went to Oxford University before being admitted to the Inner Temple and subsequently called to the bar. In 1708, he had married Cecil Matthew (d. 1720) granddaughter and heiress of David Jenkins of Hensol. At Hensol, he built a palatial mansion known as The Castle. He became a member of Lincoln's Inn and was appointed as Solicitor General to the Prince of Wales in 1717. He was elected as Member of Parliament for Tregony, Cornwall in 1720. In 1724 he ordered a Chinese armorial porcelain service, which shows the arms of Talbot (Howard, *Chinese Armorial Porcelain*, 1974). In 1733 he became Lord Chancellor and was raised to the peerage as Baron Talbot of Hensol, Glamorganshire.

7.5.3 *William Pulteney Earl of Pulteney (1684–1764) and an Armorial Chinese Porcelain Service Commission C. 1722*

William Pulteney was one of the first to commission a Chinese armorial service who did not have a direct link to the East India Company, although William was a prominent Whig politician. After Oxford University, he did a Grand Tour before taking a seat in Parliament as representative for Hedon, Yorkshire and he continued as Member of Parliament from 1705 until 1734. He was appointed as Secretary of War in 1715 by George I (Stephen, *Dictionary of National Biography*, vol 11, 1922). He was a close political friend to Robert Walpole until 1726, when a personal quarrel split their friendship.

Pulteney's service was commissioned around 1722 (Howard, *Chinese Armorial Porcelain*, 1974). He was descended from the Pulteney family of Misterton, Leicester. The Visitation of 1619 describes the Pulteney family arms as *argent, a fess dancettée gules, in chief three leopards' faces sable* (Camden, *The Visitation of the County of Leicester in the Year 1619*, 1870).

One of the earliest recorded members of this family was John de Pulteney (d. 1349) who was a member of the Drapers Company and served as Lord Mayor of London. He owned Penshurst Place which would later be owned by the Stafford Dukes of Buckingham, and, from 1552 onwards, the Sidney family. The direct line of Pulteneys of Misterton continued until Sir John Pulteney (1585–1617) who married Margaret Fortescue (d. 1613). John and Margaret had a son, John, who died in 1637 and the Misterton estate was then divided amongst his 3 sisters. John used the same armorial as the Pulteneys of Misterton, however his crest differs. The 1619 Visitation describes the Pulteney crest as *a lions head erased guardant sable,* but William Pulteney's crest contains a crown as well.

William Pulteney's grandfather, also William (1624–1691), was Member of Parliament for Westminster. This William was the second son of Michael Pulteney (b. c. 1600) of Bray, Berkshire, which indicates that William Pulteney's connection to the Misterton line was a few generations back.

His armorial service depicted the Pulteney arms *argent, a fess dancettée gules, in chief three leopards' faces sable* with his crest above. Unlike other services, Pulteney did not include the arms of his wife's family. William Pulteney had married Anna Maria Gumley (1694–1758) in 1715. She was the daughter of John Gumley of Isleworth (1670–1728) who had built Gumley House in 1700 which he left to Anna Maria on his death in 1727, however the Gumley of Isleworth coat-of-arms was not included on the armorial service.

The armorial and crest are decorated across the entire diameter of the plate, rather than being confined to the centre. His crest is also depicted at the four compass points within the border of the plate. The plate's decoration is a floral arrangement of chrysanthemums with other florals and foliage in gold. The armorial design dominates the decoration, unlike the Pitt service where the crest was taken into the decorative pattern. In this service, the Chinese elements of the pattern are subordinate to the armorial.

William Pulteney is a representative example of several patrons who held high office and commissioned armorial Chinese porcelain services in this period. He was the first and last Duke of Pulteney. At the time of his death in 1764, he had no surviving children; his son William Pulteney had died the previous year aged 32, and his daughter Anna Maria had died aged 15 in 1741. His heir was his brother, General Harry Pulteney, but he died only 3 years later in 1767. The Pulteney fortune accordingly passed to the descendants of William and Harry's uncle, John Pulteney (1661–1726) who had married Lucy Colville. John and his eldest son Daniel Pulteney (1684–1731) were deceased, but Daniel had married Margaret Deering Tichborne (1699–1763) and it was their daughter, Frances Pulteney, who inherited. She had married Sir William Johnstone, who took on the name Pulteney under Royal Licence, as did their daughter Henrietta Laura Pulteney (1766–1808) who duly became Countess of Bath Her husband Sir John Murray of Dunerne 7th Baronet (1751–1811) also took the name Pulteney under Royal Licence. On Henrietta's death, he retained a life interest in her estates, which passed on his death to Henrietta's cousin, Elizabeth Evelyn Sutton in 1811.

Elizabeth Evelyn Sutton was Henrietta's cousin on her mother's side of her family; Margaret Deering Tichborne had a sister Judith (1685–1749) who had married Robert Sutton 1st Baronet (1671–1746). Their son Robert Sutton, 2nd Baronet, had died in 1802, and Elizabeth was his daughter.

In less than 50 years since the death of William Pulteney in 1764, the armorial commissioning patron, the Pulteney estate passed to the cadet branch of the Pulteney family and then to the Sutton family: Elizabeth was the granddaughter of William Pulteney's first cousin's sister-in-law. Those who inherited the Pulteney estate assumed both the armorials and the surname of Pulteney, and Elizabeth Sutton was no exception. She had married John Fawcett (1766–1849) in 1803, and the couple took the name Pulteney by Royal Licence in 1813. Although the armorial of William Pulteney, 1st Earl of Bath, was *ar a fess indented gu, in chief three leopards faces sa.* Crest: *a leopard's head affrontée and erased sa, ducally gorged or,* the armorial of Elizabeth and John Fawsett Pulteney was *ar a fess indented gu, in chief three leopards faces sa, on a canton of the second an estoile of the first.* Crest: *a leopard's head erased sa gorged with a ducal coronet and charged with a crosse crosslet* (Burke, *The General Armory of England, Scotland, Ireland and Wales* 1844). Having assumed the Pulteney name and armorials, the couple commissioned an armorial English porcelain service from Flight, Barr and Barr c. 1810–20 decorated with the Pulteney crest (Woolley & Wallis, *Fine Porcelain and Pottery Auction*, 2016).

7.6 Conclusion

By the seventeenth century, trade and commerce made it possible for an individual to acquire great wealth without following the traditional routes of significant service to the monarch, or inter-marriage with a great family and an inheritance. Both these routes had, de facto, resulted in social mobility being generational rather than achieved in one's own lifetime, particularly in times when life expectancy was uncertain. Wars, the dangers of childbirth, high infant mortality rates, accidents and untreatable illnesses all meant that neither longevity nor producing heirs who would thrive to adulthood, were a given: aristocratic titles would often pass from brother to brother, to nephew, to cousin or become extinct within a couple of generations. Wealth would often be tied to the title, but not always; there are multiple instances of a title passing to a distant kinsman whilst a daughter inherited the family fortune.

Armorial usage ensured that familial links were remembered and passed down the generations, and kinships associations were retained and passed on for centuries. Kinship networks still played a key role in society; the ability to call upon a distant cousin for their influence or patronage to secure a position was all-important.

From 1695 onwards, armorials were appearing on Chinese porcelain products such as vases, breakfast services, tea and coffee services, dinner services and dessert services. The very earliest commissioning patrons such as Johnson, Pitt and Gough were from armigerous families. The East India Company offered significant opportunity for individuals to acquire significant wealth, and a large number of those who were able to take advantage of this came from the armigerous families. That is not to say that there were no instances of self-made individuals who received armorials, however they were within the minority of the commissioning patrons of armorial Chinese porcelain, when these are considered as a group.

The most expensive Chinese armorial porcelain commission was that of the dinner service, particularly one decorated with a full armorial rather than just the crest. These services could include hundreds of individual pieces of porcelain, and armorials were applied to lids, stands and trays as well as the main service dishes. Armorial Chinese porcelain dinner services would have been used at fashionable dinner parties and balls: the incorporation of domed covers into dining services shows that porcelain services had become the fashionable way to serve guests according to the newest dining styles prevalent in the 1720s, and owning an armorial Chinese porcelain dinner service was a sign of great wealth. The effort and cost, not to mention the long wait to receive the end product, indicates how prized these armorial Chinese porcelain items were. The purpose of a service such as this, was to impress. Owning Chinese porcelain was an indication of wealth, but commissioning armorial Chinese porcelain products was a display of social standing and stability, as it linked the commissioning patron back to prominent ancestors from the aristocratic nobility and landed gentry.

The desire for armorial Chinese porcelain, and the prices that commissioning patrons would pay for it, would doubtless have been apparent to the owners of the earliest English porcelain manufactories. Once the English porcelain manufactories were established in the 1740s, armorial commissions followed. The commissioning patrons for these products are discussed in Chapter 8.

References

Anon. (1829). *A Description of Duncombe Park, Rivalx Abbey and Helmsley Castle*. British Library.

Avery, L. (1944, May). Chinese Porcelain in English Mounts. *The Metropolitan Museum of Art Bulletin New Series, 2*(9), 266–272. The Metropolitan Museum of Art.

Bray, W. (Ed.). (1827). *Memoirs of John Evelyn, Comprising His Diary from 1641 to 1705–1706*. Colburn.

Brooks, A. M. (2016). *The Importance of British Material Culture to Historical Archaeologies of the Nineteenth Century*. University of Nebraska Press.

Buckler, B. (1764). *Stemmata Chicheleana: Or A Genealogical Account of Some of the Families Derived from Thomas Chichele of Higham Ferrers*. Oxford Clarendon Press.

Burke, J. (1834). *A Genealogical and Heraldic History of the Commoners of Great Britain and Ireland* (Vol. 1). H. Colburn.

Burke, J., & Burke, J. B. (1838). *A Genealogical and Heraldic History of the Extinct and Dormant Baronetcies of England*. Scott, Webster & Geary.

Burke, J., & Burke, B. (1844). *Encyclopædia of Heraldry or General Armory of England, Scotland, and Ireland*. H. G. Bohn.

Burke, J., & Burke, B. (1850). *A Genealogical and Heraldic Dictionary of the Landed Gentry* (Vol. 2). H. Colburn.

Camden, W. (1870). *The Visitation of the County of Leicester in the Year 1619*. Harleian Society.

Child, E. (2018). *Genealogy of the Child Family*. Franklin Classics.

Christie Manson & Woods Limited. (1978). *The G.W. Marshall Collection of Chinese Armorial Porcelain and Other Fine Chinese Ceramics, Enamels and Works of Art*. Auction Catalogue, Christie Manson & Woods.

Claxton, J. (2010). The Countess of Arundel's Dutch Pranketing Room: 'An Inventory of all the Parcells or Purselin, Glasses and Other Goods Now Remayning in the Pranketing Roome at Tart Hall, 8th Sept 1641'. *Journal of the History of Collections, 22*(2), 187–196.

Coxe, W. (1800). *Memoirs of the Life and Administration of Sir Robert Walpole Earl of Orford*. Caddell & Davies.

Dalton, C. N. (1915). *The Life of Thomas Pitt*. Cambridge University Press.

Draper, P. (1864). *The House of Stanley: Including the Sieges of Lathom House, with Notices of Relative and Co-temporary Incidents*. T. Hutton.

Edwards, H.G.M. (2022). *Welsh Armorial Porcelain: Nantgarw and Swansea Crested China*. Springer Nature.

Eldred, O. W. (1940). *Eldred Family; Particularly the Descendants of William Eldred of Yarmouth*. Privately Printed.

Evelyn, J., Browne, R. Clarendon, E. H., Forster, J., & Nicholas, E. (1859). *Diary and Correspondence of John Evelyn*. H.G. Bohn.

Fetherston, J. (Ed.). (1871). *Visitations of the County of Oxford*. Taylor and Co.

Finn, M., & Smith, K. (Eds.). (2018). *The East India Company At Home 1757–1857*. UCL Press.

Foster, J. (1874). *Pedigrees of the County Families of Yorkshire* (Vol. 2). J. Forster.

Foster, J., & Rylands, W. H. (2016). *Grantees of Arms named in Docquets and Patents to the End of the Seventeenth Century*. The Armorial Register Limited.

Fuchs, R. W. (2005). *Made In China: Export Porcelain from the Leo and Doris Hodroff Collection at Winterthur.* Henry Francis du Pont Winterthur Museum.
Goldsmith, O. (1854). *The Vicar of Wakefield: Together with the Poems of Oliver Goldsmith.* Routledge.
Guillim, J., & Mackenzie, G. (1724). *A Display of Heraldry.* T.W.
Howard, A. (2013). From Canton to the City: Export Porcelain for London. In *Art Antiques London.* Available online https://shorturl.at/inuHK. Accessed January 20, 2024.
Howard, A. (2015, December 11). The Pitt Services: Porcelain for a Political Dynasty. *American Ceramic Circle Journal, XVIII.* ArtHist.net. Accessed January 29, 2024, https://arthist.net/archive/11716
Howard, D. S. (1974). *Chinese Armorial Porcelain* (Vol. 1). Faber & Faber.
Huntington Library, Stowe Papers, STB Box 13, Letters to J. Farquharson, Steward, 55, cited by Ferguson, P. F. (2017). *Chinese Armorial Wares for James Brydges, 1st Duke of Chandos: Self-fashioning or Artifice?* Transactions of the Oriental Ceramic Society.
Hutchings, V. (2013). *The Red Lion Brewery.* Umbria Press.
Jenkins, S. (2016). *Portrait of a Patron: The Patronage and Collecting of James Brydges 1st Duke of Chandos (1674–1744).* Routledge.
Le Neve, P. (1873). *Le Neve's Pedigrees of the Knights Made by King Charles II, King James II, King William III and Queen Mary, King William Alone, and Queen Anne.* Harleian Society.
Lubbock, B. (2015). *The Blackwall Frigates.* Europaicher Hochschulverlag.
Lysons, D. (1811). *The Environs of London: Middlesex.* T. Cadell & W. Davies.
McKeown, J. (2006). *English Ceramics: Two Hundred and Fifty Years of Collecting at Rode.* Philip Wilson Publishers.
O'Day, R. (2007). *Cassandra Brydges, Duchess of Chandos, 1670–1735: Life and Letters.* Boydell & Brewer.
Lach, D. F. (2010). *Asia in the Making of Europe, Volume II A Century of Wonder. Book 1: The Visual Arts.* University of Chicago Press.
Palmer, C. J. (1856). *The History of Great Yarmouth* (Vol. 2). L.A. Meall.
Peck, L. L. (2005). *Consumer Splendour: Society and Culture in Seventeenth Century England.* Cambridge University Press.
Phillimore, W., & Phillimore, W. (Eds.). (1888). *The Visitation of the County of Worcester Made in the Year 1569.* Harleian Society.
Pierson, S. J. (2007). *Collectors, Collections and Museums—the Field of Chinese Ceramics in Britain 1560–1969.* Verlag Peter Lang.
Pope, A. (1854). *The Poetical Works* (Vol. 4). Nathaniel Cooke.
Reeves, D. D. (1956). Sir Richard Gough and His Broadside Collection. *Harvard Library Bulletin, X*(1) (Winter), 94–118.
Robson, T. (1830). *The British Herald; Or, Cabinet of Armorial Bearings of the Nobility & Gentry of Great Britain & Ireland, from the Earliest to the Present Time.* Sunderland.
Rutton, W. L. (1891). *Three Branches of the Family of Wentworth. I Wentworth of Nettlestead, Suffolk. II Wentworth of Gosfield, Essex. III Wentworth of Lillingstone Lovell.* Mitchell and Hughes.
Sedgwick, R. (Ed.). (1921). Gough, Henry 1681–1751, of Endfield, Middlesex. In *History of Parliament: The House of Commons 1715–1754.* Boydell and Brewer.
St. George, H., & St. George, Sir R. (1880). *The Visitation of London Anno Domini 1633, 1634 and 1635.* Harleian Society.
Stephen, L. (Ed.). (1921–1922). *Dictionary of National Biography* (Vol. 4). Oxford University Press.
Stephen, L. (Ed.). (1921–1922). *Dictionary of National Biography* (Vol. 8). Oxford University Press.
Stephen, L. (Ed.). (1921–1922). *Dictionary of National Biography* (Vol. 11). Oxford University Press.

Stephen, L. (Ed.). (1921–1922). *Dictionary of National Biography* (Vol. 19). Oxford University Press.
Stone, L. (1958, November). The Inflation of Honours 1558–1641. *Past & Present, 14*(1), 45–70.
Thursh, A., & Ferris, J. P. (Eds.). (2010). *The History of Parliament: The House of Commons 1604–1629*. Cambridge University Press.
Van den Berg, R. (Ed.). (2015). *Cantillon's Essay on the Nature of Trade in General* (Variorum ed.). Taylor & Francis.
Wotton, T., Kimber, E., & Johnson, R. (1771). *The Baronetage of England: Containing a Genealogical and Historical Account of all the English Baronets Now.*

Online Sources

'*Coffee Cup with the arms of Gough*,' Museums at Washington and Lee University. Online Exhibits. Available online https://exhibitsmuseums.omeka.wlu.edu/items/show/299. Accessed January 24, 2024.
Joseph Highmore. (1692–1780). '*The Family of E. Lancelot Lee*,' Wolverhampton Art Gallery. Art UK. Available online https://artuk.org/discover/artworks/the-family-of-e-lancelot-lee-19451. Accessed January 30, 2024.
Woolley & Wallis. (2016). 'An Extensive Flight Barr and Barr Armorial Dinner Service c. 1810–20,' Lot 141, *Fine Porcelain and Pottery Auction*. Available online https://www.woolleyandwallis.co.uk/departments/british-and-continental-ceramics-and-glass/pg230216/view-lot/141/. Accessed January 30, 2024.

Chapter 8
The Commissioning Patrons for Armorial British Porcelain

Abstract This chapter considers the commissioning patrons for British armorial porcelain. It discusses the social changes in England during the seventeenth century, and the role of the livery companies. The earliest known armorial English porcelain commissions are discussed: many of which carried livery company armorials. The earliest family armorial commissions are reviewed, and the commissioning patron is identified where possible. The first armorial English hard paste porcelain armorials of William Cookworthy and Richard Champion are reviewed, along with a representative selection of armorial porcelain commissions from different porcelain manufactories.

Keywords Fire of London · House-building · Social change · Skilled labour · Livery companies · East India company · Decorative items · Shopping as leisure · Chinoiserie · Bow · Merchant Taylors Company · Liverpool · Holbwrch Llowarch & Corbet · Vauxhall · Axtell · Marke · Worcester · David Gavin & Christina Hearsey · Viscount Howe of Langar · George Gordon · William Calmady · Bristol · Edmund Burke & Jane Nugent · Joseph Smith & Jane Pope · Harford & Lloyd · Duke of Clarence · Mary Bootle · Duke of Hamilton · John Egerton · William Howley · Derby · Georgiana Duchess of Devonshire · Horatio Nelson 1st Duke of Bronte · Nantgarw · Windham Henry Quinn & Wyndham · Swansea · John Graham Clarke and Mary Parkinson · Spode · William Amscot Ingilby · Charles Kemeys Tynte & Elizabeth Swinnerton.

8.1 Introduction

By the time of the establishment of the early porcelain manufactories in England in the 1740s, there was already a long history of commissioning armorial porcelain from China, which illustrates the demand for the product. Once the manufacture of porcelain began in eighteenth century England, armorial commissions followed.

The Great Fire of London in 1666 had led to a significant rebuilt of London's streets, and continuing expansion of London saw continual building as the city

R. L. Denyer et al., *Armorial Porcelain*, https://doi.org/10.1007/978-3-031-63745-2_8

expanded well beyond the original City Mile. Architecture had developed to create new buildings that inspired and incorporated the latest fashions of the Renaissance. Over the course of the eighteenth century, there were a number of general changes in how people lived. More private rooms were created within homes, and there were dedicated spaces for dining. Improved transportation, particularly the development of the canal network, enabled bulky goods to be moved more readily around the UK. 1700s onwards saw an expansion of country estates and great houses being built either as new buildings, or remodelled as earlier properties were enlarged or pulled down completely.

At the same time, skilled labour settled in London; French Protestants from the French Wars of Religion and Flemings from the Dutch war with the Spanish. Huguenots with skills in goldsmithing and bookmaking came to London, and the second wave of settlers after the Revocation of the Edict of Nantes in 1685 led to settlement in Spitalfields, the centre for the silk textile trades (Peck, *Consumer Splendour: Society and Culture in Seventeenth Century England,* 2005).

London trade at this time was dominated by the livery companies, which managed all the skilled trades. Every livery company had its armorial banners and hall, and the livery companies provided the city officials; Aldermen, Sheriffs, and Lord Mayors of London. Very early commissions of armorial ceramics by the livery companies include pieces by the Pickleherring Pottery, which was established around 1618.

The ability to commission armorial Chinese porcelains, particularly in the early years to 1725, depended very much on patrons having both the wealth to afford the items plus the connections with the East India Company to place the order.

From the 1720s onwards, standards of living began to rise, especially in towns and cities where there were more opportunities. As new housing developed, and the overall number of homes began to increase, there was an increase in the number of material goods and personal possessions that people owned, particularly amongst the middling sort which included professionals such as solicitors, medics, and the clergy as well as administrators, clerks, small business owners, merchants and those engaged in commerce and transportation. Decorative items as well as the functional became more common in these new homes, and fashions in home decor began. Shopping became a fashionable leisure activity in Georgian towns such as Bath, where the wealthy congregated to enjoy social activities as well as the spa. Shops began to specialise in luxury and high-end goods and services and vied with each other to offer the very best to their clientele. In this climate, consumerism flourished, and more shops appeared. Shops themselves evolved; bow fronts began appearing on shop windows, which enabled better displays, and window-dressing became an art to entice wealthy customers in. Shop interiors were often sumptuously laid out with displays of stock available for order and dispatch.

The English porcelain manufactories tended to be sited in the best place to produce the porcelain, rather than near their customer base. Chelsea was situated near to the fashionable Ranelagh Gardens, which were opened in 1742, but other manufacturers took shops in London or Bath. Larger premises, known as warehouses, carried significant volumes of stock, which provided customers with an extensive range of choices and options to personalise and bespoke their orders. Josiah Wedgwood, for example, initially had a showroom in Charles Street, Grosvenor Square, London, but had already outgrown this by 1767 when the company took a warehouse on St Martins Lane. Wedgwood also had a showroom in Westgate, Bath (Meteyard, *The Life of Josiah Wedgwood: from his private correspondence and family papers*, 1866).

Chinese porcelains continued to be imported into London throughout the eighteenth century, however it is no coincidence that most of the early English porcelain manufactories produced products that were decorated in imitation of the Chinese styles; the decoration being an acknowledgement of the appetite of the consumer market for Chinese porcelain by the mid-century. Indeed, this was the peak of the Chinoiserie style, with Chinese patterns appearing on wallpaper, fabrics, and all manner of decorative items. Some of these English manufactories also took armorial porcelain commissions, and the role of the commissioning patron for English armorial products is now considered.

8.2 Worshipful Company of Merchant Taylors: Commission of an Armorial Mug Bow c.1750–1755

The Merchant Taylors livery company had been founded by 1300 and had a Royal Charter by 1327. Its origins lay in the lining and quilting of armour; the original name being Taylors and Linen Armourers (Hazlitt, *The Livery Companies of the City of London*, 1892) and it was one of the 'Great 12' livery companies, from which the Lord Mayor of London was traditionally selected.

Around 1750–1755, an armorial mug was commissioned from the Bow manufactory (Woolley & Wallis, *English & European Ceramics & Glass*, 2020). It is decorated with the armorial for the Merchant Taylors Guild *argent, a tent royal between two parliament robes gules, lined ermine. On a chief azure, a lion of England*. This is one of the earliest known armorial English porcelain commissions, and whilst the identity of the commissioning patron is unknown, it is likely it was commissioned for a senior official of the guild, or presented as a gift. Key London offices were filled by the livery companies, such as Alderman, Sheriff of London and Lord Mayor, and the mug may have commissioned for such a post-holder. The Merchant Taylor company did not put forward a Lord Mayor between Sir Robert Westley in 1743 and John Ansley in 1807, however Sir Nathaniel Nash of the Merchant Taylors guild was elected as Alderman in 1758 and served as Sheriff in 1761 (Noorthouck, *A New History of London Including Westminster and Southwark*, 1773). It is possible the commission was to mark one of these appointments.

8.3 Holbwrch Llowarch–Corbet Armorial English Porcelain Punch Bowl, Liverpool c.1756–1758

One of the earliest known armorial English porcelain items is a punch bowl, believed to be by William Reid of Liverpool. The bowl dates to c.1756–1758 (Hillis, *Liverpool Porcelain 1756–1804*, 2011) which makes it one of the earliest known armorial English porcelain commissions. The armorial consists of two coats of arms—*vert a stag trippant argent attired or* on the dexter side and *or a raven sable* on the sinister side.

The dexter arms appear to be those of Holbwrch Llowarch *vert a stag trippant argent attired or.* This is an ancient armorial; he was the Treasurer of Griffith ap Llewellyn, the last Prince of Wales (d. 1282) (Burke, *The General Armory of England, Scotland, Ireland, and Wales*, 1884).

The sinister arms are those of the Corbet family, however there were three Corbet baronetcies associated with the armorial *or a raven sable:* Corbet of Stoke, Corbet of Sprowston and Corbet of Moreton Corbett, Shropshire. Two of these baronetcies were already extinct by the time of the commission: the Corbet of Sprowston baronetcy had ended with the death of Sir Thomas Corbet in 1661, and the Moreton Corbet baronetcy had ended in 1688 with the death of the 3rd baronet Vincent Corbet (1670–1688). The Corbet of Stoke baronetcy would pass from Sir Henry Corbet, who died in 1750, to his nephew Corbet Avenant, who was a minor in the 1750s (Burke & Burke, *A Genealogical and Heraldic history of the Extinct and Dormant Baronetcies of England, Ireland And Scotland*, 1844b).

It is unclear who placed this particular commission. Punch bowls were used for the popular punch drink, which was made from imported spices, rum and brandy. It was often drunk by groups, clubs, and societies. There are several examples of punch bowls decorated to mark elections. Pubs and taverns were often named after the local landowning families and displayed their armorials on signage. It is possible it was commissioned by a group or society who met at a particular pub that used these armorials. It is also possible that this armorial combination was used by a descendant who was several generations removed from the original marriage; perhaps a direct descendant of a cadet branch of one of these families. The armorial punch bowl could have been commissioned to mark a descendant's coming of age or even to mark the inheritance of property that had once been owned by the original holders of these armorials. Whatever the meaning behind the commission, it was significant enough for the purchaser to invest a significant sum of money in ordering this customised punch bowl, and it would have been a prized possession.

8.4 Axtell Armorial English Porcelain Ale Jug, Vauxhall c.1755

Ale jugs came in many sizes in the Georgian period. There are many instances of ale jugs in pewter, glass, or ceramic, and they were often personalised with initials or armorials.

An armorial English porcelain ale jug was commissioned around 1755 with the Axtell armorial *az three axes arg handles or* (Papworth, *An Alphabetical Dictionary of Coats of Arms belonging to Families in great Britain and Ireland* vol 1, 1874). It was produced by the Vauxhall porcelain manufactory in London and is one of the earliest known armorial English porcelain items (see Fig. 8.1).

The Axtell armorials are not associated with an aristocratic title: the armorial is believed to have been granted to Colonel Daniel Axtell (c.1648–1650) (Axtell, *Axtell Genealogy*, 1945) who played a prominent role in the trial of Charles I and was one of the Regicides executed for treason in 1660, following the restoration of the monarchy. He is believed to have received the armorials around 1648.

Daniel Axtell is thought to have been the son of William Axtell (1587–1638) and Thomasine Cutler (1591–1675) from Berkhamsted, Hertfordshire. He was originally

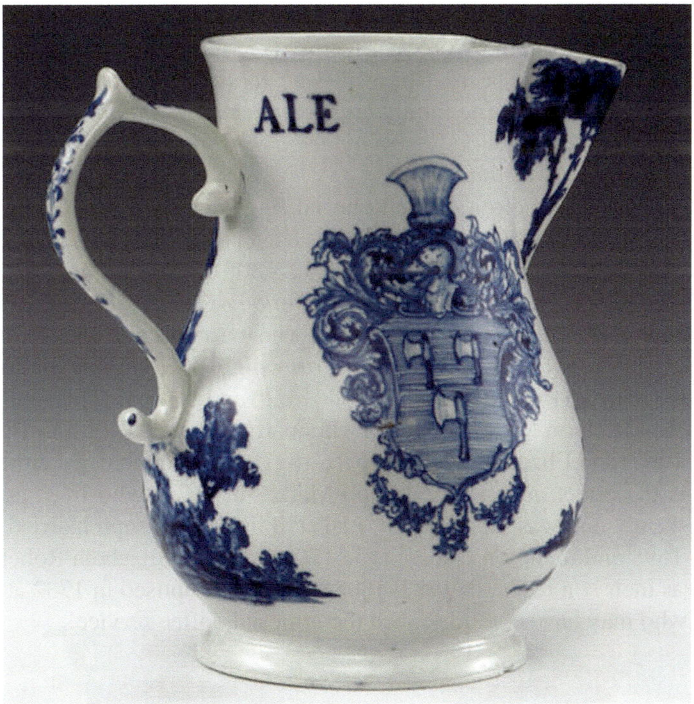

Fig. 8.1 Vauxhall Ale Jug with the armorials for Axtell. Reproduced with the permission of Fergus Gambon, Bonhams Antiques and Valuers, Bond Street London

in the service of Sir Henry Vane before supporting the Parliamentarian cause. He married Mary Marsham (b. 1622) in 1639 and the couple had at least two sons who survived into adulthood; Daniel (1640–1683) and William (b. 1643).

Daniel Axtell (1640–1683) married Rebecca Holland (1625–1720) who was the daughter of Cornelius Holland (1599–1671) another of the Regicides of Charles I. Daniel and Rebecca emigrated to South Carolina, USA (Reed, 'A Tale of Two Regicides: Daniel Axtell and Cornelius Holland [and Their Son and Daughter, Who Helped Save the Carolinas],' in *The American Genealogist*, 81, 2004). William Axtell was born in 1639, and he married twice: firstly, Elizabeth Adkins (b. c.1662) and secondly Sarah Maletrat (b. c.1664). He had at least one son, also named William Axtell, who was baptised 21 November 1686.

The jug may have been commissioned as a stand-alone item or as part of a service. It is probable that the jug was commissioned by a descendant of Colonel Daniel Axtell, but the commissioning patron has not been identified.

8.5 Marke Armorial English Porcelain Coffee Service, Worcester, c.1758

One of the earliest known armorial English porcelain coffee services was commissioned around 1758 from the Worcester factory (see Fig. 8.2). It bears the armorials for the Marke family of Liskeard, Cornwall—*gu, a lion ramp. within an orle of eight fleur-de-lis, a canton erm.* The family crest is *a demi lion holding a fleur-de-lis in his dexter paw* (Burke, & Burke, *Encyclopaedia of Heraldry or General Armory of England, Scotland, and Ireland,* 1844a).

The commissioning patron of the armorial English coffee service has not been identified with certainty, but it seems to have been around the time of the marriage of Richard Marke of Liskeard, Cornwall (1732–1808).

The Marke family appears in the 1620 *Visitation of Cornwall*, where the earliest noted ancestor is William Marke, born c.1525 who married Jane Andrew of Liskeard (St George, *The Visitation of the County of Cornwall in 1620*). Their son John Marke (1550–1614) married Emeline Vivian (1554–1629) and the couple had multiple children including Richard Marke (d. 1674) who married Francis North in 1618, James Marke (b. 1605) and Elizabeth Marke (1575–1621). It is probable that either Richard or James is the direct ancestor of Joseph Marke (b. 1672) who lived at Woodhill manor, Liskeard, and was the Mayor of Liskeard in 1691. Joseph married Elizabeth Vesper in 1698 and their son George (b. c.1700) married Elizabeth Roberts around 1728. It was their son Richard (1732–1808), who was baptised in 1732 at Liskeard, Cornwall who may have commissioned the armorial coffee service.

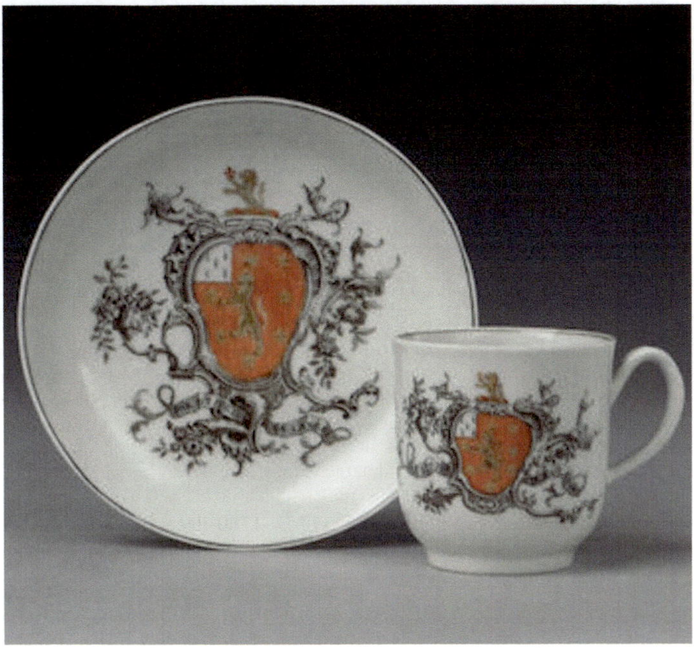

Fig. 8.2 Worcester armorial cup and saucer c.1758–1760 decorated with the armorials of Marke of Liskard. Reproduced with the permission of Fergus Gambon, Bonhams Antiques and Valuers, Bond Street London

8.6 Viscount Howe of Langar Armorial English Porcelain Tea Service, Worcester c.1758–1760

An armorial English porcelain tea service was ordered by Viscount Howe of Langar (1726–1799). The service is decorated with a coat-of-arms and a crest (see Fig. 8.3). The coat-of-arms is for the Scrope family *az a bend or*. This is a very simple, early armorial, believed to date back to a grant from Edward the Confessor (1003–1066). The last of the direct line was Lord Scrope of Bolton, Emanuel 1st Earl of Sunderland (1584–1630). He had married Elizabeth Manners (1588–1653) daughter of John, 4th Earl of Rutland. They had several children, but all died young. On his death, the barony and entailed estates passed to distant relatives but the Earl's personal wealth and estates went to his illegitimate children. His third daughter Annabella Scrope (1629–1704) received the Langar estate, the Scrope family having held the manor of Langar since 1373.

In 1647, Annabella married John Grubham Howe (1624–1728) and Langar House became the seat of the Howe family. Their son Scrope Howe (1648–1713) became 1st Viscount Howe and Baron Glenawley in the Irish peerage in 1701. His son Emanuel Scrope Howe (1700–1735), 2nd Viscount Howe, was the father of Richard Howe (1726–1799).

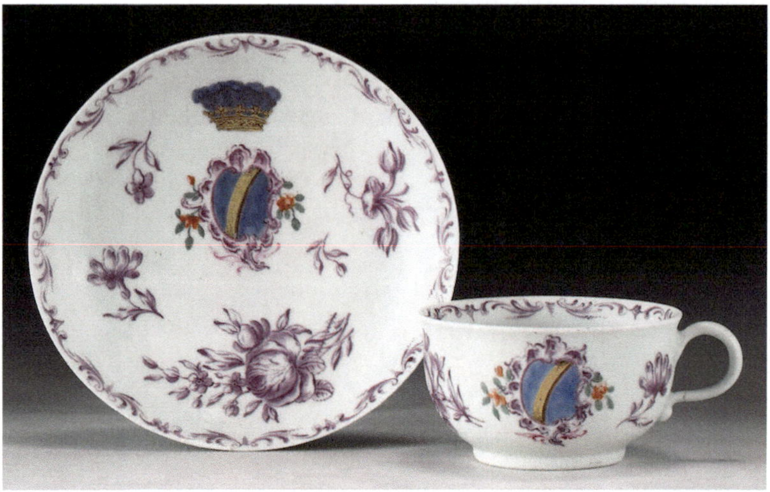

Fig. 8.3 Worcester armorial tea cup and saucer c.1757–1760 decorated with the armorials for Viscount Howe of Langar. Reproduced with the permission of Fergus Gambon, Bonhams Antiques and Valuers, Bond Street London

Richard Howe joined the navy in 1739 and had a long naval career, ultimately becoming Treasurer of the Navy and Admiral of the Fleet in 1796. In March 1758 he married Mary Hartopp (1732–1800), and in July that year he succeeded his brother to become 4th Viscount Howe. He also became a Member of Parliament that year, representing Dartmouth. He is thought to have commissioned the tea service around this time.

Mary Hartopp was the daughter and heiress of Colonel Chiverton Hartopp of Welby Leicester (1655–1759). Her family had been settled in Leicestershire for generations and appear in the *Visitation of Leicester* in 1619 as Hartopp of Burton Lazars, with line of descent from Raphe Hartopp. *temp R.2* (Camden, *The Visitation of the County of Leicester in the year* 1619). Although there are multiple armorials for various branches of the Hartopp family in Leicester, her family line does not seem to have been one of those. Her great-grandfather Thomas Hartopp (1594–1654) and her grandfather Sir William Hartopp (1616–1692) were both knighted, in 1624 and 1660 respectively, but do not seem to have had armorials. Acquiring armorials was a separate process which would usually follow receiving a knighthood, if the person were not already armigerous; it entailed making an application to the College of Arms and the payment of a fee.

The armorial decoration on this tea service is of particular interest as the Viscount Howes had their own armorial—*or, a fesse betw three wolves heads, couped, sa.* His crest was—*a gauntlet, lying fesseways ppr. lined gu holding a falchion ar hilt or the middle part of the blade debruised by a wolf's head, couped, sa paw* (Burke & Burke, *Encyclopaedia of Heraldry or General Armory of England, Scotland and Ireland,* 1844a) yet this was not used on the service. Instead, Richard Howe chose to have the service decorated with armorial of his grandmother, Arabella Scrope. It is possible that there was armorial decoration already on the building at Langar House, therefore the use of the same armorial on the tea service underlined the connection to the Scrope family and was intended to highlight the current owners relationship to the family; the Scrope armorial pre-dated the Norman knights, and over the centuries the Scrope family had produced 1 Archbishop, 2 Bishops, 5 Knights of the Garter, 1 Chancellor, 4 Treasurers, 2 Chief Justices, 2 Earls, 20 Barons and numerous baronets (Burke & Burke, *Encyclopaedia of Heraldry or General Armory of England, Scotland and Ireland,* 1844a). The coat-of-arms on the service displays the Scrope armorial with the *band or* is painted as *bar sinistre* to denote an illegitimate descent. Above the coat-of-arms is a crest—*out of a ducal coronet, a plume of ostrich feathers.* This is the crest assumed by the sons of Sir Richard Le Scrope, 1st Lord Scrope of Bolton (1327–1403), the original crest having been *issuing out of a ducal coronet, a crab* (Burke & Burke, *Encyclopaedia of Heraldry or General Armory of England, Scotland and Ireland,* 1844a).

The tea service is decorated with a floral pattern in a Meissen style. German porcelain imports were restricted in the 1750s therefore floral designs on English porcelain in this style were very popular.

Another point of interest is that Richard Howe chose to commission his tea service from an English manufactory, rather than having a Chinese armorial service. With his naval connections, he would have had the ability to order such a service, plus he had the wealth. Chinese patterns were available from Worcester during the 1750s, so it would also have been an option to have a Chinese-style pattern from Worcester had he wanted one. This makes the selection of a floral Worcester tea service a personal choice, and it indicates that the quality and reputation of Worcester by this time put their armorial porcelain services on a par with the imported armorial Chinese porcelain services. Another consideration may have been the reduced time to receive the commission; a service manufactured in Worcester would have been decorated and delivered in a shorter timescale.

8.7 David Gavin (b. c.1720) and Christina Hearsey (1730–1767) Armorial English Porcelain Dinner Service, Worcester, 1765

An armorial porcelain service was commissioned c.1765 carrying the arms of Gavin impaling Hersey c.1765 (Payne & Marks, *British Heraldry from its Origins to c.1800,* 1978). The armorials on the plate are those of David Gavin (b. c.1720) and Christina Hearsey (1730–1767) who married around 1765. This commission is amongst the earliest known armorial English dishware or flatware services with surviving pieces. It is thought to have been a dinner service rather than a dessert service as surviving plates range from 19.5 cm to 22 cm in diameter.

The Gavin arms are for Gavin of Langton. The *General Armory* describes the armorial as *ar a sword in pale, az ensigned with a mullet gu surmounted by a saltire couped sa* and the crest *in a sea a 2 masted ship, in full sail ppr*. The family motto was *by industry we prosper.* The *Armory* describes the Hearsey family arms as *gu, a chev ar. Crest a stag sejant ppr* (Burke & Burke, *Encyclopaedia of Heraldry or General Armory of England, Scotland and Ireland,* 1844a).

David Gavin's family was from Lunan, Montrose. His earliest identified direct ancestor is his grandfather James Gavin (b. c.1660) who held the position of parish beadle. James Gavin had at least one daughter, Catherine (b. c.1688) and a son, Alexander (b. c.1686). Catherine Gavin married a Dutch skipper who had been shipwrecked in Lunan Bay, and she had then moved to Holland with him (Taylor, *The Great Historic Families of Scotland* vol 1, 1887).

Alexander Gavin followed in James' stead and served as sexton and beadle of Lunan parish (Warden, *Angus or Forfarshire, the Land and People* vol 4, 1884). Alexander was also a merchant and retailer of provisions; the family moved from Peat Loch in Lunan to Denhead around 1723 and opened a small shop (Bushnell, *From Bricks to Books*, 1949).

Alexander's son David spent time in Holland with his aunt and uncle. By that point, his uncle had ceased skippering and set up in commerce. David Gavin became a partner in this business and married his cousin, Christina Hearsey (c.1730–1767). In time, he duly became a merchant in his own right. By 1755 David Gavin was trading from Glasgow as David Gavin & Co, dealing in tea, alcohol and cloth. The family ventures proved profitable. Alexander Gavin bought Easter Braikie in the 1750s, and David Gavin was wealthy enough to purchase the Langton estate in Berwickshire from the Cockburn family in 1758. The Langton property had been held by the Cockburn family since 1314, and William Cockburn had been granted a baronetcy by Charles I in 1627 (Brown, *The Parish of Langton*, 1845). Following his purchase, David Gavin demolished the existing manorial house and village. He rebuilt the mansion and relocated the village to a new site named Gavinstown (modern-day Gavinton).

It is likely that the armorial dinner service was commissioned for use at Langton, their primary residence. The Gavin-Hearsey service was commissioned, therefore, by a wealthy couple whose wealth had been derived from commerce and trade, and who were socially mobile. This mobility took the form of investing in land, housebuilding, and property development, and then securing a marriage for their eventual heiress into an aristocratic titled family. David and Christina are not thought to have direct descendants, however after Christina's death in 1767, David remarried. His second wife was Elizabeth Maitland (1751–1826) daughter of James Maitland, 7th Earl of Lauderdale (1718–1789), and Mary Turner Lombe (1732–1789). David and Elizabeth Maitland had at least 4 daughters; their eldest daughter Mary Turner Gavin (1771–1845) married Lt General John Campbell, 4th Earl of Breadalbane, 1st Marquess of Breadalbane (1762–1834), and duly inherited the Langton estate. David Gavin thereby became the grandfather of John Campbell, 2nd Marquess of Breadalbane (1796–1862).

8.8 Commission by George Gordon of Gight 12th Lord of Gight (1741–1779) of an Armorial Porcelain Mug from West Pans, c.1764–1770

George Gordon was the 12th Lord of Gight. His parents were Margaret Duff (1720–1801) and Alexander (1716–1761). His grandmother Mary Gordon (1680–1740) had married Alexander Davidson (1680–1740) in 1701 and he had taken the name of Gordon.

Following his father's death in 1761, George Gordon had inherited the family estates, and in 1763 he married Catherine Innes (1740–1805). The armorial mug from West Pans was commissioned around 1764–1770; it may have been a marriage gift for the couple or bought by them, making this one of the earliest known Scottish armorial porcelain commissions (see Fig. 8.4). The mug carries the Gordon crest—*stag's head erased ppr*—with the motto above the crest: *by dand*. Their daughter Catherine (1764–1811) married Captain John Byron (1756–1791) and was the mother of the poet, George Gordon Lord Byron (1788–1824).

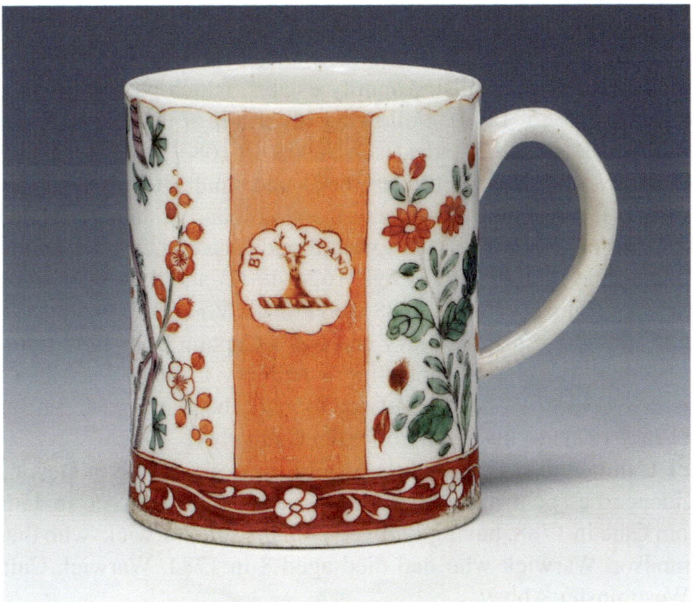

Fig. 8.4 West Pans armorial mug c.1764–1770 decorated with the armorial crest and motto for Gordon of Gight. Reproduced with the permission of Fergus Gambon, Bonhams Antiques and Valuers, Bond Street London

8.9 William Calmady (1751–1779) and the Commission of an Armorial English Porcelain Tea Service and Dessert Service, Worcester c.1772

A tea service and dessert service from Worcester, decorated with the Calmady armorials, was commissioned around 1772 (Spero & Sandon, *Worcester Porcelain, 1751–1790: the Zorensky Collection,* 1996). This is one of the earliest known armorial English porcelain dessert service commissions, and in conjunction with a matching armorial tea service, it is one of the largest armorial English porcelain commissions placed before 1775.

William Calmady (1751–1779) is believed to be the commissioning patron for the services. The Calmady arms consist of *azure a chevron between 3 pears or* (Burke & Burke, *Encyclopaedia of Heraldry or General Armory of England, Scotland and Ireland*, 1844a). In addition to the family armorial, service pieces carry the initials *W C.*

The Calmady family are believed to have originated from Calmady in Poundstock, Cornwall. In 1564, Vincent and Richard Calmady acquired the manor of Langdon, Devon. Vincent Calmady died in 1579 and his son Josias (1565–1611) inherited the family estate. He married Katherine Courtney (d. 1605) who was the granddaughter of Philip Courteney of Ugbrooke (d. 1552) and Elizabeth Shilston (d.1605) and thus the heiress to the Shilston family estates, which passed to her eldest son Shilston Calmady (1585–1646). Following his death in the Civil War, the family estates initially passed to his eldest son Josias (d.1683), then to his second son Shilston Calmady of Leawood (1621–1688) and finally to his sixth son Francis Calmady (1627–1704). Francis Calmady married Joan Fursland (b. 1638) and they had at least 7 children who survived into adulthood. One of their sons, another Francis (1664–1704) married Cecilia Pollexfen (1674–1747) and they were the parents of Captain Warwick Calmady (1711–1788) and Francis Vincent Calmady (1705–1765).

Warwick Calmady joined the Navy and was a Captain from 1743 until 1757. He went on to become Assistant Comptroller in the Navy. Captain Warwick Calmady is believed to have commissioned the service to mark the marriage of his son, another Warwick (1751–1779) to his niece Pollexfen Calmady (b. 1755).

Warwick Calmady had inherited the manor of Mothecombe in Devon, but there is little evidence he spent much time there. He died at his home in James Street, Buckingham Gate in 1788, having outlived both his son Warwick, who died in 1779, and his grandson Warwick who had died aged 8 in 1784. Warwick Calmady was buried in Westminster Abbey.

Connections between those who ordered armorial porcelain are frequently found, so it is worth noting that a later armorial English dinner service commissioned from Wedgwood in 1779 was ordered by a distant cousin of Warwick Calmady. In 1779, Frances Courteney (b. 1763) married Sir John Honywood, 4th Baron Evington (1756–1806). To mark the marriage, a dinner service was commissioned, with the armorials for Honywood and Courtenay. Frances Courtenay and Warwick Calmady were both descendants of William 'the Great' Courtenay of Powderham Castle (1485–1535).

William had married twice; Frances was a descendant of William and his first wife Mary Gainsford, whilst Warwick Calmady was descended from William's second marriage to Margaret Edgcombe.

8.10 Edmund Burke (1729–1797) and Jane Nugent (1738–1812): The *Burke Service* & The *Smith Service*: Two Armorial English Hard Paste Tea Services, 1774, Bristol

In 1774, two armorial services were made by the Bristol porcelain manufactory owned by Richard Champion: the *Burke Service* and the *Smith Service*. These services represent the first known armorial hard paste English porcelain commissions.

William Cookworthy (1705–1780) had discovered the kaolin and china stone deposits at St Austell, Cornwall, and patented the process for the manufacture of hard paste porcelain in 1768. The patent was licensed to Richard Champion (1743–1791) in 1772.

Edmund Burke was the son of Richard Burke (1705–1761) an attorney in Dublin, and Mary Nagle (1705–1770). He studied at Trinity London before moving to London to study law. He did not pursue law as a profession, and instead turned to writing. He became a secretary to William Gerard Hamilton, who was in turn Chief Secretary to the Lord Lieutenant of Ireland. In 1765 he became Member of Parliament for Wendover, a position he held until 1774. In the elections of that year, he stood for Bristol, then a city with an independent electorate of 5,000 voters which made it the third largest constituency behind London and Westminster. The size of the city meant that electioneering was very expensive, particularly if multiple candidates stood. A local arrangement between the Union Club (Whig) and the Steadfast Society (Tory) meant that each party selected a single candidate to support at the hustings. Robert Nugent, Viscount Clare (1709–1788), had been the Whig representative but there was growing dissatisfaction with him, and by 1774 the party was looking for another representative. Richard Champion was a Quaker and an urban merchant. He saw Burke as a strong candidate for Bristol and used his influence to get him accepted as the candidate (SkjöNsberg, 'Richard Champion and the Rockingham Whigs: The Aristocratic Politics of a Bristolian Quaker Merchant In the Age of the American Revolution' *The English Historical Review,* Vol 138, 2023).

Richard Champion had the *Burke Service* made for Jane Burke, and it was presented to her as a joint gift from Richard Champion and his wife Judith Lloyd (1741–1790) (Owen, *Two Centuries of ceramic art in Bristol, being a history of the manufacture of "the true porcelain" by Richard Champion,* 1873).

The service is enamelled with the armorials for Burke *or a cross gu in the dexter canton a lion ramp sa* impaling Nugent *erm, two bars gu* (Burke & Burke, *Encyclopaedia of Heraldry or General Armory of England, Scotland, and Ireland,* 1844a). Jane Nugent (1738–1812) was the daughter of Dr Christopher Nugent (1698–1775) and Jane Leake (Burke, *The Public and Domestic Life of the Right Hon Edmund* Burke, 1854). The Nugent armorial indicates a familial connection to Robert Nugent 1st Earl Nugent (1709–1788).

The service itself is decorated with mythical figures. This was a bespoke and expensive gift from Champion, given in the early years of his porcelain manufacturing business, and it illustrates the value that he placed on his relationship with the Burkes.

During the election campaign for Bristol, Edmund Burke was invited to stay with Joseph Smith (1742–1825) who was a friend of Richard Champion. The *Smith* tea service was commissioned by Edmund Burke from Champion as a gift for Mrs Smith (Sarah Pope). (Owen, *Two Centuries of ceramic art in Bristol, being a history of the manufacture of "the true porcelain" by Richard Champion,* 1873). The service is enamelled with the armorial for the Pope family from Marnhull, Dorset—*arg two chevs gu on a chief of the second an escallop or within a bordure gu* (Burke & Burke, *Encyclopaedia of Heraldry or General Armory of England, Scotland, and Ireland,* 1844a). Sarah Pope's ancestors can be traced back to Thomas Pope of the Privy Chamber of Henry VI and Joan Weston of Buckstead, Sussex, and she was related to the Pope family of Hendall, Sussex who had the same armorial (Berry, *County Genealogies, Pedigrees of the Families in the County of Sussex,* 1830).

The tea service was decorated with a floral design, and it was further personalised with Mrs Smith's initials enamelled in flowers. The quality of the workmanship and decoration of both services is significant, and the association with the election allows precise dating of their production. Both were executed 'without reference to either time, labour or expense' and demonstrate that Champion's Bristol porcelain manufactory had the potential to rival Sevres and Dresden (Owen, *Two Centuries of ceramic art in Bristol, being a history of the manufacture of "the true porcelain" by Richard Champion,* 1873).

Richard Champion also made an armorial plaque for Joseph Smith and Sarah Pope, with the Smith armorials impaled with those of Pope. The plaque is likely to have been a gift; Champion's manufactory produced a small number of decorative floral plaques and armorial plaques that are thought to have been made as gifts for friends (L.W., *English Pottery and Porcelain; Being a Concise Account of the Development of the Potter's Art in England,* 1871).

Another example of a decorative armorial plaque was made for Mark Harford (1738–1798) & Sarah Lloyd (1733–1798) around 1775 (Owen, *Two Centuries of ceramic art in Bristol, being a history of the manufacture of "the true porcelain" by Richard Champion,* 1873). The Harfords were originally from Marshfield, Gloucestershire. Charles Harford (1631–1709) had moved to Bristol where he married Mary Bushe. Their son Edward (1658–1706) married Elizabeth Jones (d. 1758) and they had 3 sons: the eldest son Edward Harford (1691–1779) married Elizabeth Lloyd (d.1729), and Mark Harford (1695–1788), married Love Andrew (b. c.1707) and it was their son also Mark (1738–1798) who married Sarah Lloyd (1733–1798) daughter of Samuel Lloyd. The third son Charles (1704–1747) was the father of Joseph Harford (1741–1802) who was a close friend of Edmund Burke. Joseph Harford served as Justice of the Peace for Gloucestershire and was Sheriff of Bristol in 1779 and 1786. Joseph Harford was in partnership with Richard Champion, providing financial backing to him for the porcelain business, as did Mark Harford (Harford, *Annals of the Harford Family*, 1909). It is possible that there was also a kinship connection between Champion and the Harfords; Richard Champion's wife Judith Lloyd was from Bristol and thus likely to have been related to Sarah Lloyd. It is highly likely, therefore that this armorial plaque was also made for Mark Harford and Sarah Lloyd as a personal gift from Champion.

8.11 Royal Influence and Armorial Porcelain Commissions: Duke of Clarence (1765–1837) Commission of an Armorial Porcelain Dining Service, Flight & Barr Worcester 1789

Royal residences began to contain English porcelain from Chelsea from c.1750. These pieces were soft paste porcelain. Prince William the Duke of Cumberland (1721–1765) is not known to have commissioned porcelain from any UK manufactories, although he is thought to have owned some pieces from the Chelsea factory (Edwards, *Porcelain Analysis and its role in the Forensic Attribution of Ceramic Specimens,* 2022). The earliest Royal commission for a porcelain service was that placed by George III and Queen Charlotte to mark the occasion of the marriage of the Queen's brother, the Duke of Mecklenburg-Strelitz in 1763. This service, however, was not an armorial service. The first royal armorial commissioned service was not until 1789, when Prince William later King William IV (1765–1737) commissioned Flight Worcester to produce the *Duke of Clarence* service, to mark his new honours (Denyer, Denyer & Edwards, *The Pendock Barry Porcelain Service: A Forensic Evaluation*, 2023). The service plates were enamelled with his armorial and the motto of the Order of the Thistle in the centre of plate, with the heraldic badges of the Order of the Thistle and the Order of the Garter forming part of the decorative border pattern. Other items within the service, such as the wine cooler, carried his armorial and motto. The Duke of Clarence commissioned a second service called *Hope* from Flight Worcester in 1790, which was non-armorial.

The Royal family, therefore, did not drive demand for armorial porcelain by leading the fashion for such items. By the time the Duke of Clarence placed his armorial commission in 1789, some 30 years had elapsed since the first armorial English porcelain had been commissioned, and there are numerous examples of bespoke orders for porcelain services from the armigerous nobility without armorial decoration being specified. The Duke of Northumberland, for example, commissioned a non-armorial porcelain dinner service from Derby in 1779. This service evidently saw good usage, as he ordered replacement pieces from Derby in 1791 (see Fig. 8.5).

By the 1780s, English manufactories were sufficiently adept at producing armorial porcelain that Mary Bootle (1734–1813) commissioned some replacement pieces from Derby for her inherited Chinese armorial porcelain service (Edwards, *Welsh*

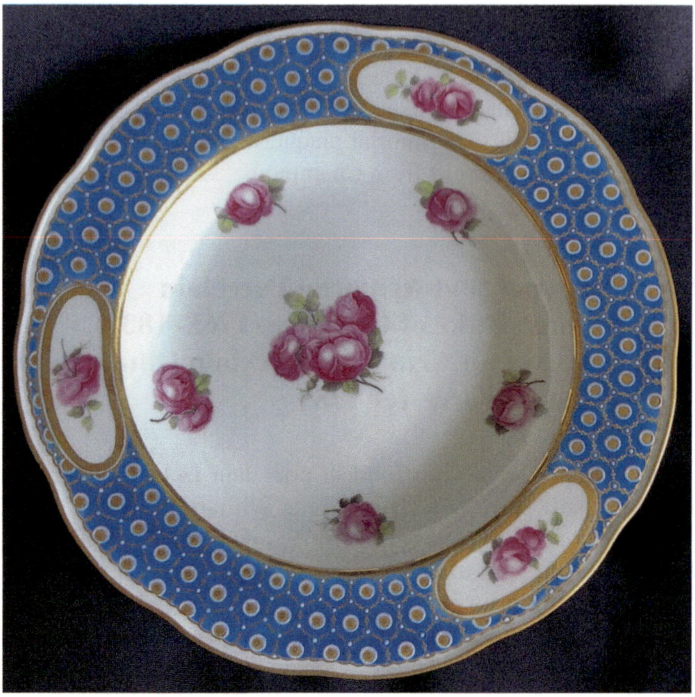

Fig. 8.5 Duke of Northumberland Service dinner plate, commissioned 1791 from Derby, an example of a non-armorial English porcelain dinner service commission. Decorated with roses by William Billingsley. Private Collection

Armorial Porcelain, 2022). This demonstrates that home market armorial porcelain wares were seen as of sufficient quality to be a like-for-like replacement for Chinese armorial porcelains, and that commissioning patrons did not distinguish between Chinese and English armorial porcelain. Nor was Mary Bootle the only person to do this; the 8th Duke of Hamilton also ordered replacement pieces from Derby for an earlier Chinese armorial porcelain service that had been commissioned by the 5th Duke of Hamilton (Edwards, *Derby Porcelain: the Golden Years 1789–1830,* 2017).

8.12 Clerical Commissions: John Egerton (1721–1787) & An Armorial English Porcelain Tea Service Commission, Chelsea-Derby, 1782

Senior clergy were typically drawn from the aristocratic and noble family throughout the eighteenth century. The church was seen as a suitable occupation for younger sons, who would attend Oxford or Cambridge before taking up an appointment. Landed families often had rectorships within their gift and would provide for junior branches by conferring such positions on them. John Egerton was no exception. He was born in 1721, the son of Rev Henry Egerton (1681–1746) and Elizabeth Bentick (1703–1765). He was a grandson of John Egerton, 3rd Earl of Bridgewater (1646–1700), and Jane Powlett/Paulet (1656–1716) who was the daughter of Charles, 1st Duke of Bolton (1630–1699). He studied at Oxford then he was appointed to the Rectorship of Ross on Wye, Herefordshire, by his father, who was Bishop of Hereford at the time. He became the King's Chaplain in 1749, then Bishop of Bangor in 1756. He was appointed Bishop of Durham in 1771, a role he held until his death in 1787.

Whilst Bishop of Durham, he commissioned an armorial English porcelain tea service from Derby, around 1784 (Edwards, *Welsh Armorial Porcelain: Nantgarw and Swansea Crested China,* 2022). The armorials decorated on the service are those for the bishopric of Durham—*az a cross between four lions rampant, or. The mitre over the arms is encircled with a ducal coronet.* Bishops and Archbishops are entitled to bear the arms of their see impaled with their own family arms (Burke, *A Genealogical and Heraldic Dictionary of the Peerage and Baronetage of the British Empire,* 1865) yet the Egerton armorial is not included on the service decoration. John Egerton would have been entitled to use the armorials of the Egerton Earls. Indeed, if he had outlived the 6th Earl of Egerton, who died with no direct heirs in 1803, he would have succeeded him as the next Earl. In the event, the earldom passed in turn to his two sons, John William Egerton (1752–1823) and Francis Henry Egerton (1756–1829), who were successively the 7th and 8th Earls of Bridgewater. John Egerton had married Anna Sophia Grey (1729–1780) daughter and co-heiress of Henry Grey, 1st Duke of Kent, so the Egerton armorials could have been impaled with Grey to represent the couple and their descendants.

The Bishop of London, William Howley (1766–1848), commissioned an armorial porcelain service from Chamberlain Worcester (see Fig. 8.6). He was Bishop of London from 1813 until 1828, when he became Archbishop of Canterbury, so the service must have been commissioned prior to that date. Having been appointed Archbishop of Canterbury, he was entitled to use the armorials of the archbishopric—*Az, an episcopal staff, in pale, or, and ensigned with a cross-patée, ar, surmounted of a pall, of the last, charged with four crosses, formée, fitchée, sa, edged and fringed, gold* (Burke & Burke, *A Genealogical and Heraldic Dictionary of the Peerage and Baronetage of the British Empire,* 1845).

William Howley had applied for a grant of arms for Howley in 1813—*azure an eagle displayed erminois charged on the breast with a cross fleury gules* (Howard, *Miscellanea Genealogica et Heraldica* vol 4, 1884). In contrast to John Egerton's service, the armorials on Howley's porcelain service display those for the Bishop of London *gu two swords in saltier ar pommels, or* impaling Howley.

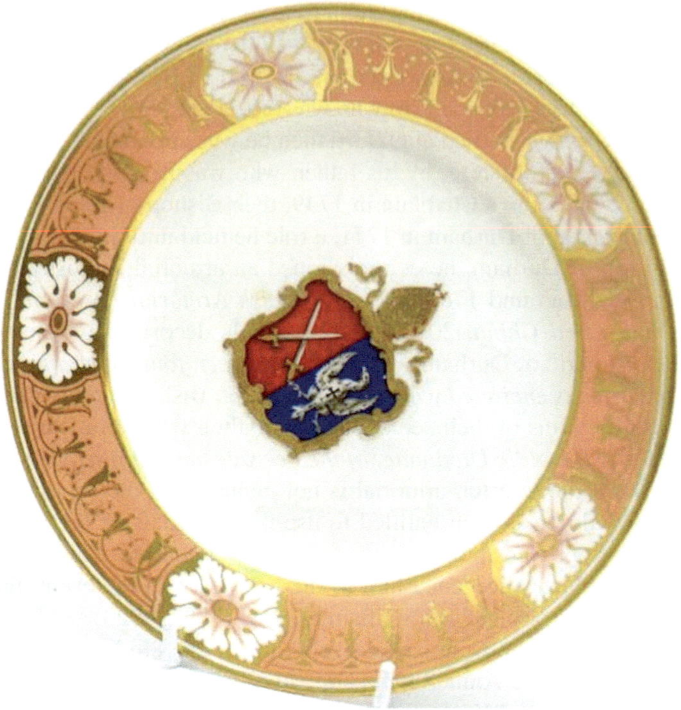

Fig. 8.6 Chamberlain Worcester plate decorated with the armorials for Hawley and the bishopric of London. Reproduced with the permission of Jo Fletcher-Lee, Tennants of Yorkshire Antiques and Fine Art Auctioneers Ltd, Leyburn, North Yorkshire

8.13 Georgiana Duchess of Devonshire (1757–1806) and a Unique Armorial Service: A Derby Armorial Caudle Service

In the Devonshire Collection at Chatsworth House is a pair of lidded early Derby armorial caudle cups and matching tray, on saucers or stands decorated in gilt with a blue coronet and cipher *GD,* which has been dated to 1783 (Fig. 1.32). It was apparently made in celebration of the first-born child of Georgiana, the Duchess of Devonshire, who was also named Georgiana, namely Lady Georgiana Dorothy Cavendish (who later became the Countess of Carlisle), on 12 July 1783. Lady Georgiana Spencer of Althorp House married William Cavendish, the 5th Duke of Devonshire, on her 17th birthday on 7 June 1774, when he was 25 years of age, in Wimbledon Parish Church. She was a noted Georgian beauty, but the marriage was not a happy one with marital transgressions taking place on both sides. Caudle was a potent warm drink comprising sack or wine, mixed with eggs, milk, oatmeal, spices and sugar and was usually drunk on occasions such as weddings, baptisms and births: the lid of the two-handled caudle cup was designed to keep the liquid contents warm. The Chatsworth armorial version is especially rare in having two caudle cups with a matching tray picked out with a seemingly haphazard gilt background design, simulating a crinoid marble which actually pays homage to the Duchess Georgiana's love of minerals and fossils: she built a significant collection of fossils and minerals at Chatsworth during her lifetime there. The simple coronet displays the five visible strawberry leaves of a Duke or Duchess and although Georgiana was a Cavendish by marriage the cipher displays a GD, signifying Georgiana Devonshire. The caudle cups and tray and are in excellent condition and, next to the Bishop of Durham tea service, must represent one of the earliest armorial outputs from the Derby China Works. A rather curious addition to the original caudle service are several more caudle cups, covers and stands which bear the coronet and monogram *ED* for Elizabeth Devonshire, the second wife of the 5th Duke, who bore him another two children. Elizabeth, as Lady Elizabeth Foster was a very close friend of Georgiana, Duchess of Devonshire, and the Duke, whom she married in 1809 after the death of Georgiana in 1806. The Duke died in 1811. It appears, therefore, that the additional caudle cups bearing the monogram *ED* must date from between 1809 and 1811. The authors are appreciative of information about the Chatsworth armorial caudle service from Sash Giles, Curator of the Devonshire Collections at Chatsworth House, Derbyshire (Fig. 8.7).

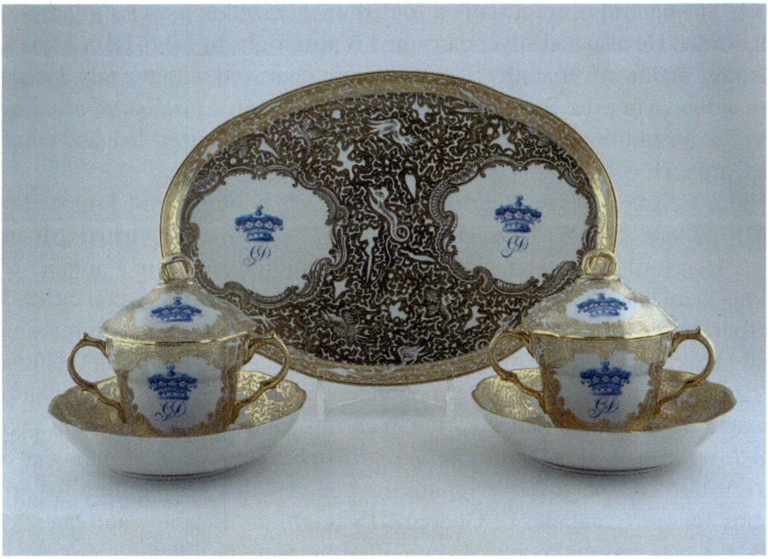

Fig. 8.7 Derby porcelain caudle service, ca. 1783, made in celebration of the first-born child of Georgiana, the Duchess of Devonshire, who was also named Georgiana namely Lady Georgiana Dorothy Cavendish (who later became the Countess of Carlisle), on the 12th July 1783. Courtesy of Sash Giles, Curator of the Devonshire Collections, Chatsworth House, Derbyshire

8.14 Vice-Admiral Horatio Nelson 1st Duke of Bronte, 1st Viscount Nelson (1758–1805) and English Armorial Porcelain Services

The Nelson family can be traced back to Mawdesley in Lancashire, where Richard Nelson lived in 1508, but the Nelsons had moved to Norfolk where they were settled by the time of Thomas Nelson (1580–1654). Nelson's mother was Catherine Suckling (1725–1767). Her mother Ann Mary Turner (1693–1768) was the daughter of Mary Ann Walpole (1673–1730) sister of Robert Walpole, Prime Minister & 1st Earl of Orford (1676–1737). Nelson owed his initial introduction to the Navy in 1771 to the influence and patronage of his uncle, Captain Maurice Suckling (1725–1778). He served on the HMS *Raisonnable* and in 1777 he received his commission as lieutenant. By 1779 he was Post-Captain and rose to be Rear Admiral with the Order of Bath in 1797. Nelson was handsomely rewarded for his military victories, and his armorial bearings changed as he acquired fresh honours and rewards. After the Victory of the Nile, he was created Baron Nelson of the Nile and of Burnham-Thorpe. After Copenhagen in 1801, he received the Viscounty and Barony. His final victory was at Trafalgar in 1805, where he was fatally wounded on board the HMS *Victory* during the battle.

Admiral Nelson had multiple porcelain services. Several of these were gifts, such as the Baltic service: a dessert service in French porcelain and a tea service in Coalport. The central decoration is a fouled anchor within a victor's laurel wreath and gilt border. He also had silver carrying his armorials. In 1796, he ordered a set of silver from London silversmiths Eley and Fearn engraved with the *San Joseph* crest on the handles. After the Battle of the Nile, he augmented his coat of arms with the Chelengk as an additional crest. He also received a silver *entrée* dish and cover from Lloyds Coffee House with his coat-of-arms engraved.

In 1802, Nelson was on a tour of England, in the company of Emma Hamilton (1765–1815) and Sir William Hamilton (1730–1803). He visited Monmouth, Swansea, Hereford, Worcester, Coventry and Birmingham where a number of civic receptions were held, during which he received the freedom of several cities. Whilst on the tour, he visited the Coalport Worcester factory, where he commissioned an armorial breakfast service (Edwards, *Welsh Armorial Porcelain,* 2022). This service was a 12 place-setting, 150-piece service, decorated in the Fine Old Japan pattern with Imari colours and floral decoration. It carries a naval crown, viscount's coronet, the *San Josef* crest and the motto *Faith and Works.* This was a magnificent service and, given the 'Nelson fever' gripping the country, this set the tone for armorial porcelains for the remainder of George III's reign, the Regency and later reign of George IV when armorial porcelain commissions were prolific.

8.15 Windham Henry Quinn (1782–1850): Commission of an Armorial Nantgarw Service c.1817–1822

Windham Henry Quinn is thought to have commissioned a Nantgarw armorial service between 1817 and 1822, making this one of the earliest Welsh armorial services from the Nantgarw manufactory (Edwards, *Welsh Armorial Porcelain,* 2022).

Windham Henry Quinn was the grandson of Windham Quinn (1717–1789), whose maternal grandfather was Henry Windham of Limerick, Ireland (b. 1667). Windham Henry Quinn married Caroline Wyndham (1790–1870) in 1810. She was the daughter of Thomas Wyndham (1764–1814) of Dunraven Castle, Glamorganshire, and Anna Charlotte Ashby (b. 1768) who also held lands in Gloucestershire;

Windham Henry Quinn was granted the additional surname of Wyndham by Royal Licence in 1815, along with the arms for Wyndham of Orchard Wyndham, through Caroline's right of descent after the death of her father Thomas the previous year; Windham and Caroline were distantly related, their common ancestors are believed to be Sir John Wyndham (1568–1645) and Joan Portman of Orchard (1565–1633).

The armorials on the service are for Quinn *az a chev or, three lions heads erased with gilt fetterlock chain.* The Quinn motto *au bon droit* is enamelled. The reason for the commission is unclear; the date falls several years after the marriage of Windham Henry Quinn and Caroline Quinn and the birth of their first son Edward Richard Windham Wyndham-Quinn in 1812. Windham Henry Quinn served as Member of Parliament for Limerick from 1806–1820. He became 2nd Earl of Dunraven and Mount-Earl in 1824, when he also acquired the titles of 2nd Baronet Quinn of Adare, 2nd Viscount Mount-Earl and 2nd Viscount Adare.

8.16 John Graham Clarke (1782–1862) Commission of an Armorial Swansea Service in 1818

In 1818, John Graham Clarke placed a commission for an armorial Swansea service. This service therefore represents one of the first Swansea armorial porcelain commissions (see Chapter 1, Figs. 1.9 and 1.10).

John Graham Clarke was born in Newcastle upon Tyne, and he was an industrialist and shipowner. He married Mary Elizabeth Parkinson, of Kinnersley Castle, Hertfordshire in 1812. Her father Leonard Parkinson had purchased Frocester Manor in Gloucestershire in 1803, which Mary inherited on his death in 1818. The couple subsequently lived in Frocester, where John served as a Magistrate. The inheritance is likely to be the reason for the commission of the armorial service from Swansea in 1818. The service is likely to have been a dinner service, and it is decorated with the armorials of Graham Clarke of Newcastle impaling Parkinson (Edwards, *Welsh Armorial Porcelain,* 2022).

8.17 Sir William Amcotts Ingilby Bart of Ripley Castle (1783–1854): Commission of a Spode Service c.1825

Sir William Ingilby, 2nd Baronet, was a Member of Parliament. He represented East Retford, Lincolnshire and Lincolnshire North. He also served as Sheriff of Yorkshire 1821–1822. His family can be traced to Thomas Ingilby Member of Parliament for York in 1349, and Sir William Ingilby, Knight, who was created a baronet in 1642.

Sir John, the 4th baronet, died in 1772 without a direct heir, however the title was revived, and a baronetcy conferred on his illegitimate son John Ingilby of Ripley on 24 March 1781. John married Elizabeth Amcotts, heiress of Wharton Amcotts of Kettlethorpe in Lincolnshire. Wharton Amcotts had been created a baronet in 1796, with remainder to the eldest son of his daughter, Lady Ingilby. On his death that same year, therefore, William Ingilby inherited his grandfather's lands and title. He then inherited his father's baronetcy on his death in 1815. The Ingilby estates included Ripley Castle. The Ingilby armorial is *sa an estoile ar* (Burke, *Genealogical and Heraldic Dictionary of the Peerage and Baronetage of the British Empire* vol 3, 1830). For Amcotts, the armorial is *argent a tower triple towered between three covered cups, two and one, azure.* William commissioned his service from Spode with the two crests: for Amcotts *a boar's head couped and erect, argent, issuant out of the mouth an estoile or* and for Ingilby *a squirrel sejant gules, cracking a nut and collared, or* (Debrett, *Debrett's Baronetage of England,* 1835).

William Amcotts Ingilby commissioned his service around 1825. It may have been associated with his marriage in 1822 to Louisa Atkinson (Stockdale, *Stockdale's Baronetage of the United Kingdom for the year 1833,* 1833) After his death in 1854 without a direct heir, the Ripley estates went to his cousin, Rev Henry John Ingilby who became a baronet in 1866, and the Amcotts family lands in Lincolnshire passed to his nephew Weston Cracroft Amcotts.

8.18 Charles Kemeys–Tynte (1800–1882) and Elizabeth Swinnerton (1795–1838): Commission of an Armorial Service from Derby, 1820

Charles Kemeys Tynte and Elizabeth Swinnerton married in 1820. A large armorial porcelain service was commissioned from Derby to mark the marriage; either by the couple or as a gift. The service plates are decorated with a large and intricate coat-of-arms for the couple consisting of the Kemeys Tynte armorials with the Swinnerton armorials inescutcheon surrounded by a border consisting of wide band of cobalt blue with gilt decoration (see Fig. 8.8).

The first quarter of the armorial displays the Tynte armorial counter-quartered with those for Kemeys. The Tynte armorials are *gules, a lion couchant between six crosses crosslet arg* and the Kemeys armorials are *vert on a chev ar three pheons sable* (Burke, *General Armory of England, Scotland, Ireland and Wales,* 1884).

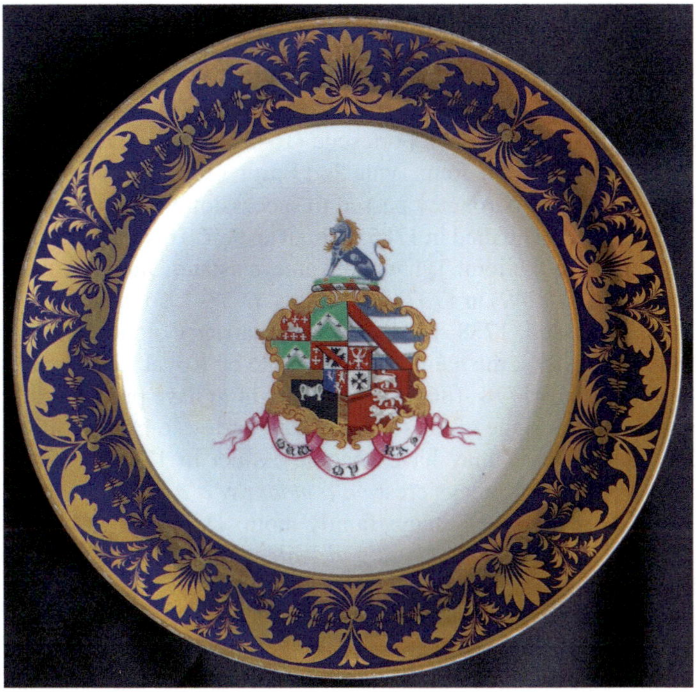

Fig. 8.8 Armorial plate c.1820 from an English armorial porcelain dinner service commissioned in celebration of the marriage of Charles Kemeys Tynte and Elizabeth Swinnerton. Private Collection

The Tynte and Kemeys armorials relate to the marriage of Sir John Tyne, 2nd Baronet (1683–1710) and Jane Kemeys (1679–1747). The Kemeys family had long been settled in Monmouthshire, however it was Jane's great-grandfather Sir Nicholas Kemeys (1593–1648) who became the 1st Baron Kemeys of Kevanmably. He married Jane Williams and their son Sir Charles Kemeys (1614–1658) duly became the 2nd Baron. He married Margaret Whitmore and their son Charles (d.1702) became 3rd Baron Kemeys. Charles married Mary Wharton (1649–1699) daughter of Philip 4th Baron Wharton and Jane Goodwin (1619–1658). This marriage is reflected in the Wharton armorial within the third quarter coat-of-arms: *sable, a maunch argent, on a border or, an orle of lions gambs erased in saltire gules* (Fox-Davies, *Armorial Families: a Complete Peerage, Baronetage and Knightage* part 1, 1895). Charles and Mary's son Sir Charles Kemeys (1710–1785) became the 4th Baron however he died without direct issue, therefore his sister Jane (1685–1747) who had married Sir John Tynte, 2nd Baronet Tynte (1683–1710), was the heiress to the Kemeys lands and estates although the baronetcy became extinct.

The Tynte family was from the Somerset and Gloucestershire area and can be traced back to Edward Tynte (1570–1629) who married Ann Gorge (1590–1660). Their son John (1617–1669) married Jane Halsewell (1630–1649). This marriage is represented by the Halsewell armorial in the second quarter: *azure, three bars wavy argent over all a bend gules.* Jane was the daughter of Hugh Halsewell of Halsewell, where the family were settled by 1285. Her direct ancestor Nicholas Halsewell (born c.1512) served as Member of Parliament for Bridgewater. Jane and John's son, Halsewell Tynte (1649–1702) became the 1st Baron Tynte in 1673.

Halsewell Tynte married Grace Fortescue (1645–1695) and the Tynte baronetcy passed in turn to their son Sir John Tynte 2nd Baron (1683–1710) who had married Jane Kemeys. The Tynte baronetcy passed to succession to their three sons; Henry, John and then Charles, who died in 1785 without an heir. This left their daughter Jane Tynte as the heiress to the joint Tynte and Kemeys estates. Jane had married Major Ruishe Hassell (1707–1749) in 1737 and had died in 1741, so the inheritance went to her daughter Jane Hassell (1738–1824). Jane had married Colonel John Johnstone (d. 1795) and the couple took the name Kemeys Tynte by Royal Licence; it is their son Charles Kemeys Tynte (1778–1860) who is thought to have purchased the armorial service to mark the marriage of his son, another Charles, to Elizabeth Swinnerton.

The Kemeys baronetcy may have become extinct in 1735, but Jane Kemeys' descendants adopted the Kemeys crest *on a mount vert a unicorn segant az armed and crined or* as well as the Kemeys family motto *Ddw-Dy-Ras* (Burke, *General Armory of England, Scotland, Ireland and Wales,* 1884). The crest and motto are displayed on the armorial porcelain service.

The fourth quarter of the armorial displays *gu, three lions passant, arg.* This armorial may be for Gifford, and it potentially represents Gifford of Burstall (Burke & Burke, *A Genealogical and Heraldic History of the Extinct and Dormant Baronetcies of England, Ireland and Scotland,* 1844b), however a connection to that family between the Kemeys Tyntes, or the families that the Kemeys Tyntes inter-married with, has not been identified. As a point of interest, in the coat-of-arms of Charles Kemeys Tynte (1822–1891), the eldest son of Charles Kemeys Tynte and Elizabeth Swinnerton, this quarter displays *az a wolf's head erased arg* which is the armorial for Hugh d'Avranches (d. 1101), known as Lupus, Earl of Chester (Burke, *Genealogical and Heraldic History of the Commoners of Great Britain and Ireland, Enjoying Territorial Possessions or High Official Rank: But Uninvested with Heritable Honours* vol 4, 1838).

The armorial for Elizabeth Swinnerton is displayed inescutcheon, indicating that she was an heiress. Her parents were Thomas Swinnerton (1770–1836) and Mary Milbourne (1772–1795). Her armorials display the Swinnerton armorial in the first quarter: *arg a cross-formée flory sa.* The Swinnerton family was an old Staffordshire family and took its name from the lordship of Swynnerton. Her earliest direct ancestor is John de Swinnerton, Lord of Swinnerton and Sherriff of Staffordshire (Burke, *A genealogical and heraldic dictionary of the landed gentry of Great Britain and Ireland for 1852*). Her paternal grandmother was Mary Abnet (b. c.1708) who married Thomas Swinnerton (b. c.1710) and this marriage is referenced in the coat-of-arms by the Abnet armorial *gu an eagle displayed between 2 pheons arg.*

Elizabeth's mother, Mary Milbourne (1772–1795) was also an heiress; she was the daughter of Charles Milbourne (1736–1775) and Martha Harley (1736–1774). Mary's paternal grandparents were George Milbourne (b. 1698) and Mary Gunter (1711–1803) therefore Mary's armorials were Milbourne quartering Gunter. Mary's maternal grandparents were Edward Harley, 3rd Earl of Oxford (1699–1755), and Martha Morgan (1707–1774) and she was the niece of Edward, 4th Earl of Oxford (1726–1790). George Milbourne is believed to be the son of another George Milbourne who married Christian Herbert (Williams, *The Parliamentary History of the Principality of Wales,* 1895) and through this marriage there was a connection to the Herbert Earls of Pembroke. The third quarter displays the Herbert armorial: *per pale az and gu, three lions rampant ar* (Burke, *A Genealogical and Heraldic Dictionary of the Peerage and Baronetage of the British Empire*, 1852).

The Kemeys Tynte armorial porcelain is a good example of how complex heraldic badges had become by 1820, and it also illustrates how distant the connection could be between the commissioning patron of a service and the original ancestor represented within in the coat-of-arms. The fact that such family pedigrees were maintained and handed down, generation to generation, is an indication of the importance attached to descent by armigerous families. Armorial porcelains such as the Kemeys Tynte commissions also show the actual heraldic badges used by family members at specific points in time, rather than the ascribed heraldic devices recorded for the family, making them a valuable primary historical source.

8.19 Conclusion

The start of the Early Modern period saw significant changes in the opportunities available for the acquisition of wealth beyond the traditional method of service to the crown in one format or another. Trade and commerce could provide lucrative opportunities to establish a fortune, and many family fortunes were established or re-established in this way.

Monies thus acquired were often but not always put into landed estates and property; the wealth established though trade and commerce enabled and facilitated the purchase or re-building of substantial properties. Having acquired or re-developed a property, the internal decoration of the home for entertaining and networking became key. Heraldic display was a way of linking the prosperity of the current owners back to earlier times and thus creating a sense of status, stability and belonging. The old distinction between 'the gentry and the merchants' became very blurred during the eighteenth century, as did the lines between 'the aristocracy and the gentry'. Commerce and trade enabled the re-establishment of 'the family fortune' by cadet branches of old families, as well as creating new opportunities for merchants to secure great wealth. Wealth generation opportunities created a ripple effect as ancillary services such as law, medicine and administration saw an uptake in demand. The middling section of society was able, and willing, to spend on wants rather than simply on needs, and this in turn created the demand for decoration; fashions in decoration such as the Chinoiserie style seen in the eighteenth century, and the demand for porcelain, laid the foundations for the establishment of the English porcelain manufactories in the 1740s, and armorial porcelain as a product.

In an era of armigerous families whose wealth was transmitted by descent and inter-marriage, individual families would have been very aware of their forebearers and any titles or properties they held. It was not unusual for the fortunes of a cadet branch to be substantially transformed by a failure of heirs in the direct line of descendants from a titled ancestor.

It is striking how many of the earliest known armorial English porcelain commissions dating to the 1750s were for smaller drinking vessels, such as mugs and jugs, with armorials for the livery companies. It does appear that the first customers for armorial English porcelain items were not the wealthy armigerous families, who continued to order Chinese armorial porcelains. The earliest known armorial English porcelain dinner service was ordered in the late 1760s, two decades after establishment of the first commercially viable English porcelain manufactories. Consumer demand may also explain why so many of the early English armorial porcelain pieces were made at Worcester; the factory set up by Dr Wall was principally concerned with producing a porcelain that could be used for hot drinks without cracking. Therefore, Worcester porcelain was the best material for services that would be used to hold tea, coffee and hot chocolate, and it does appear to have been the major English manufacturer for armorial porcelains of this type during the early period up to 1770 and the start of hard paste or 'true' porcelain production in Britain.

The availability of armorial English porcelains did not replace the demand for Chinese armorials by any means. Orders continued to be placed for armorial Chinese porcelain dinner services until the 1790s. An order for a single dinner service could be for hundreds of pieces of porcelain, and they were still very expensive to commission due to the costs associated with the armorial decoration, particularly when every piece, including stands and covers, carried the armorial devices.

By 1780, there are instances of Chinese armorial porcelain services being augmented and supplemented with porcelain from British manufactories. This indicates that not only was armorial British porcelain considered to be an equivalent product to armorial Chinese porcelain by this point, but that the social cachet of commissioning and owning an armorial Chinese porcelain service was now giving way to more practical considerations around price, distance, ease of transport, and speed of receiving the service. The commission of the *Duke of Clarence* service in 1789 underlines the change that had already taken place in consumers behaviour, rather than leading or driving that behaviour.

In the 1790s, the East India Company decided to cease imports of Chinese porcelain, thus ending the commission of armorial Chinese porcelains. From this point onwards, consumer demand for armorial porcelain was met by the British porcelain manufactories. 1800–1820 saw significant volumes of armorial porcelains being ordered; by this time, the number of British porcelain manufactories had grown, and armorial porcelains were available from newer manufactories such as Spode, Nantgarw and Swansea. There were established ways of working that supported larger-scale porcelain production, and also the development of the waterways had created a transport network that supported moving fragile products more easily.

The genesis of the armorial English porcelain commissions lies in what was being ordered; the armorial porcelain drinking vessels ordered in the 1750s were smaller, and thus less expensive, than the armorial porcelain dinner services being commissioned by the very wealthy from China. The British manufactories were also more accessible to prospective buyers, particularly once shops selling porcelain were established. By the 1800s, armorial porcelain had become an established product in its own right, and it continued to exert a cachet long after the eighteenth-century fashion for Chinoiserie had disappeared. Highly decorative and very personal to the original owner, armorial porcelains are a product of their commissioning patrons connection to their kinship networks, and a reflection of their wealth and social standing. Families passed down their armorial porcelain wares from one generation to another, and often re-purposed it for decorative display once the functional, practical use of a piece was redundant. The use of functional armorial porcelain pieces as decorative items takes armorial porcelain as a product back to its original purpose of display.

References

Axtell, C.A. (1945). *Axtell Genealogy*. Fairhaven.

Berry, W. (1830). *County Genealogies, Pedigrees of the Families in the County of Sussex*. Sherwood, Gilbert and Piper.

Brown, Rev J. (1845). *The Parish of Langton*. William Blackwood & Sons.

Burke, B. (1852). *A Genealogical and Heraldic Dictionary of the Landed Gentry of Great Britain and Ireland for 1852*. Colburn and Company.

Burke, B. (1865). *A Genealogical and Heraldic Dictionary of the Peerage and Baronetage of the British Empire*.

Burke, B. (1884). *The General Armory of England, Scotland, Ireland, and Wales*.

Burke, J. (1830). *Genealogical and Heraldic Dictionary of the Peerage and Baronetage of the British Empire* (Vol. 3).

Burke, J. (1838). *A Genealogical and Heraldic History of the Commoners of Great Britain and Ireland, Enjoying Territorial Possessions or High Official Rank: But Uninvested with Heritable Honours* (Vol. 4). H Colburn.

Burke, J. B. (1852). *A Genealogical and Heraldic Dictionary of the Peerage and Baronetage of the British Empire*. Colburn.

Burke, J., & Burke, J. B. (1844a). *Encyclopaedia of Heraldry or General Armory of England, Scotland, and Ireland*.

Burke, J., & Burke, B. (1844b). *A Genealogical and Heraldic History of the Extinct and Dormant Baronetcies of England, Ireland and Scotland*. J. R. Smith.

Burke, P. (1854). *The Public and Domestic Life of the Right Hon Edmund Burke*.

Burke, B., & Burke, A. P. (1898). *A Genealogical and Heraldic History of the Landed Gentry of Great Britain and Ireland* (Vol. 2). Harrison and Sons.

Bushnell, G. H. (1949). *From Bricks to Books*. Grafton.

Camden, W. (1870). *The Visitation of the County of Leicester in the Year 1619* (J. Fetherston, Ed.). Harleian Society, Taylor & Co.

Debrett, J. (1835). *Debrett's Baronetage of England*. J.G & F. Rivington.

Denyer, R. L., Denyer, M. C. T., & Edwards, H. G. M. (2023). *The Pendock Barry Porcelain Service: A Forensic Evaluation*. Springer Nature.

Edwards, H. G. M. (2017). *Derby Porcelain: The Golden Years 1789–1830*. Lulu Press.

Edwards, H. G. M. (2022). *Porcelain Analysis and Its Role in the Forensic Attribution of Ceramic Specimens*. Springer Nature.

Edwards, H. G. M. (2022). *Welsh Armorial Porcelain*. Springer Nature.

Fox-Davies, A. C. (1895). *Armorial Families: A Complete Peerage, Baronetage and Knightage Part 1*. Grange Publishing Works.
Harford, A. (1909). *Annals of the Harford Family*. Westminster Press.
Hazlitt, W. C. (1892). *The Livery Companies of the City of London*. Swan Sonnenschein & Co.
Hillis, M. (2011). *Liverpool Porcelain 1756–1804*. Maurice Hillis.
Howard, J. J. (1884). (Ed.). *Miscellanea Genealogica et Heraldica* (Vol. 4). Hamilton, Adams and Company.
L.W. (1871). *English Pottery and Porcelain; Being a Concise Account of the Development of the Potter's Art in England*. Alfred Bradley.
Meteyard, E. (1866). *The Life of Josiah Wedgwood: From His Private Correspondence and Family Papers*. Cornmarket Press.
Noorthouck, J. (1773). *A New History of London Including Westminster and Southwark*. R Baldwin.
Owen, H. (1873). *Two Centuries of Ceramic Art in Bristol, Being a History of the Manufacture of "the True Porcelain" by Richard Champion*. Bell and Daldy.
Papworth, J. W. (1874). *An Alphabetical Dictionary of Coats of Arms Belonging to Families in Great Britain and Ireland* (Vol. 1). T. Richards.
Payne, A., & Marks, R. (Eds.). (1978). *British Heraldry from Its Origins to c.1800*. British Museum Publications Limited.
Peck, L. L. (2005). *Consumer Splendour: Society and Culture in Seventeenth Century England*. Cambridge University Press.
Reed, P. C. (2004, April). A Tale of Two Regicides: Daniel Axtell and Cornelius Holland (and Their Son and Daughter, Who Helped Save the Carolinas). *The American Genealogist, 81*, 81–98, 192–198.
SkjöNsberg, M. (2023, February–April). Richard Champion and the Rockingham Whigs: The Aristocratic Politics of a Bristolian Quaker Merchant in the Age of the American Revolution. *The English Historical Review, 138*(590–591), 157–184.
Spero, S., & Sandon, J. (1996). *Worcester Porcelain, 1751–1790: The Zorensky Collection*. Antique Collectors Club.
St. George, H. (1874). *The Visitation of the County of Cornwall in 1620* (J. L Vivian & H. H Drake, Eds.). Harleian Society, Michell & Hughes.
Stockdale, W. (1833). *Stockdale's Baronetage of the United Kingdom for the year 1833*. William Stockdale.
Taylor, J. (1887). *The Great Historic Families of Scotland* (Vol. 1). J. S. Virtue.
Warden, A. J. (1884). *Angus or Forfarshire, the Land and People* (Vol. 4). Charles Alexander & Co.
Williams, W. R. (1895). *The Parliamentary History of the Principality of Wales*, privately printed for the author by Edwin Davies & Bell, Brecknock.

Online Sources

Woolley & Wallis. (2020, June 17). 'A Large and Rare Armorial Bow Mug,' Lot 407. *English & European Ceramics & Glass*. https://issuu.com/jammdesign/docs/ww_17june2020. Accessed 30 January 2024.

Chapter 9
Conclusion

Abstract Armorials were originally used to identify knights on the battlefield, then they became a sign of ownership, kinship and belonging. Armorials appeared on earthenware in the early 1600s, and on stonewares after 1650. Chinese porcelain was being imported into England by the East India Company from the 1690s, when armigerous patrons began to commission pieces and services adorned with their own armorials. The reasons for armorial porcelain commissions are reviewed. Porcelain factories were established in Germany and France under Royal patronage from 1710 onwards; however, this did not happen in England and there was a hiatus until the 1740s before the earliest porcelain manufactories established. Some reasons for this delay are considered. A further gap of some 20 years before the first known armorial dinner service commission from a British factory was made, in 1767, and the reasons for this are considered. By the 1750s, multiple fledgling porcelain manufactories were operating, with varying degrees of commercial success. Livery company officials were early patrons for armorial porcelains from the British manufactories, with commissions from armigerous families following several years later; some reasons for this are proposed. The East India Company ceased the import of Chinese porcelain to Britain in 1796; the factors leading to this are also discussed.

Keywords Armorials · Heraldic display · Social change · Genesis of porcelain · Porcelain decorative developments · Trade links · Transport development · Chinese armorial porcelain · Commissioning patrons · East India Company · Delftware · Delftware armorials · British porcelain manufactories · English armorial commissions · Reasons for commissions · Discrepancies in armorials · Armorial porcelain as a product

R. L. Denyer et al., *Armorial Porcelain*, https://doi.org/10.1007/978-3-031-63745-2_9

9.1 Origins of Armorials

The use of armorial badges in Western Europe dates to the Late Middle Ages. Armorials were granted by the crown to the knightly families who supported the monarch; the original purpose of an armorial was to differentiate between knights on the battlefield. These armorials were handed down to subsequent generations who may have had less direct contact with the reigning monarch, or who served the crown in its executive functions of government as administrative systems developed. Heraldic badges thus became an image that connected descendants back to their noble forebearers.

This connection was particularly important in an era of a landowning nobility; in Britain, the eldest son would inherit the family patrimony, but all descendants could use the family armorial. Heraldic badges therefore enabled increasingly distant connections to illustrious or noble ancestors to be maintained for many generations, which in turn facilitated the recognition and acknowledgement of kinship connections to very distant cousins as time progressed.

In England, the College of Arms was founded in 1484 to manage and govern the administration of heraldic badges, which had become more frequently used and misused. Heralds undertook Visitations into the counties to verify and update their pedigrees of descendants of armorial families, and these Visitations also gave newly established county families the opportunity to apply for heraldic arms and to register their pedigrees for a fee. Over time, this process of acquiring armorials by application and payment weakened the traditional association of arms belonging to those who fought for the king in battle. It also increased the number of armigerous families; acquiring a family armorial became a desirable status symbol for wealthier individuals who did not already have one. True connections to increasingly distant armigerous ancestors became harder to prove or disprove as the centuries progressed; by the seventeenth century, forgery of illustrious pedigrees and documentation supplied to the Heralds was becoming common.

By the 1600s, English society had already begun to move away from the traditional structure of a land-based nobility and gentry class who served the monarch either in battle or at court. The Civil War saw many landed families lose their lands by forfeiture and attainder, and although some retrieved their former property once the monarchy was restored, others did not; or the heir had been killed in the fighting. A lack of funds had caused James I to create a new rank of baronetcy, which was bestowed on any who would pay the 1,000 guinea fee. Further laws compelled those with incomes over £40 a year to be knighted in 1630. Many who acquired knighthoods also registered an armorial if they were not from an armigerous family, and the Visitations held during the 1680s bear testament to the varied occupations of those who either held armorials or who were desirous of acquiring them—for the payment of the necessary fee.

9.2 Armorial Display

Armorial display moved from knightly surcoats onto buildings and the personal possessions of the armigerous. Armorials appeared in the Great Halls of the nobility and gentry, typically over the fireplace, which incorporated them into the very fabric of buildings and the centre point of the principal public room. The earliest window armorials were commissioned by Henry Ferrers of Baddesley Clinton around 1575, and armorials were frequently placed on gates, in porches and above front doors as well as in the public rooms of homes. Armorials were also included in portraits, which were then hung in the public spaces of homes.

Valuable personal possessions acquired armorial decoration; books carried the armorial bookplates of their first owners. Nicholas Bacon (1509–1579) placed his armorials on books he donated to Cambridge University and Sir Thomas Tresham (1543–1605) used an engraved armorial bookplate on his books. Books were rare, expensive items that were only owned by the very wealthy, and they were expected to be handed down to subsequent generations. Armorial bookplates ensured that the book's association with its original owner would not be lost, no matter how many hands it subsequently passed through. Most books until the eighteenth century were factual tomes, concerned with religion, law, culture and history. Marking a book with an armorial device, as opposed simply writing a name on the flyleaf, linked the family name to the book and created an association between the family and the power and authority of learning and the law. This was a potent association, especially in a time when most people could not read and would never possess a book within their lifetime. The armorial served as a visual indicator that a family had had the wealth to own books, which in turn was a symbol of the family's status as members of the armorial-bearing, land-owning gentry (Denyer, Denyer & Edwards, *The Pendock Barry Porcelain Service: A Forensic Evaluation,* 2022).

9.3 Social Change in the Early Modern Period

Moving the criteria for knighthoods and armorials onto a commercial, transactional basis, rather than being acquired through service to the crown, had the effect of opening up both knighthoods and armorials to more of the population. The overall number of armigerous families remained a small section of society, but it was possible for someone to acquire armorials and a knighthood if they had the wealth. Going to London to seek one's fortune was an aspiration for many in the seventeenth century; it is no coincidence that the earliest versions of the story of Dick Whittington date to c.1605, which ties into the opportunities that were available through trade and mercantile shipping.

Commerce and trade opened up new avenues to make wealth, and professional occupations also expanded considerably during the century. The demand for solicitors, clerks and administrators grew as transport and trade developed. Learning and scientific understanding grew, as did medical knowledge. Society was still organised on kinship, patronage and networks however, and polite society remained a very small world that was very inter-connected. Patronage and letters of introduction were essential for anyone desiring to advance, but advancement was possible.

Following on from the Great Fire of London in 1666, there was significant rebuilding and regeneration of the city. New styles of architecture were incorporated into building designs, and aristocrats such as the Dukes of Bedford built and developed land they owned, resulting in London undergoing significant expansion by the 1700s. In addition to the rebuilding and expansion of London, the late seventeenth / early eighteenth century saw the building of great houses. Many county families either re-built or had commissioned new properties by the early part of the eighteenth century. Architectural designs inspired by the Renaissance were used to create small classical houses in the country; these were smaller than the traditional landed seats, but still properties of considerable size, large enough for gracious living as well as entertaining. These properties were decorated and furnished in the latest styles; decorative fashions began, whereby houses were dressed to impress guests and visitors. The arrangement of space within the property, and its furnishings, reflected a family's position within society as well as stability and social traditions. Improved transport and trade links meant that the newly built or remodelled town and country houses would contain global goods. The variety of goods that the wealthy acquired in their houses was considerable: armorial porcelains, coffee and spices, luxury furniture such as lacquered screens, and exotic plants to name a few. Traditional English oak furniture was augmented with mahogany furniture, the wood for which was imported from the West Indies (Corfield; *The Georgians: The deeds and misdeeds of 18th Century Britain*, 2022).

Increased opportunities for travel led to greater awareness of the decorative arts. The earliest Grand Tours had been in the 1600s, but they came to be seen as part of a gentleman's education in the eighteenth century. The result of these Tours was the establishment of many collections of art and decorative items such as porcelains. These would be carefully displayed, in order to delight, impress and amaze visitors. It became fashionable for the wealthy to collect enough books to form a private library; many country seats or London townhouses had libraries built, which frequently had armorials installed as a decorative feature over the fireplace.

Some of those who made significant wealth though commerce and trade did buy country estates and assumed the traditional lifestyle of the landed elite, but not all. Social mobility still tended to be generational, rather than achieved in one person's lifetime. Many of those who made their fortune in commerce did not set up as country gentlemen in the sense of changing their income source to come from a landed estate. However, their houses and style of living did mark them as having a social standing commensurate with that of the county gentry, and their children would often marry into noble or gentry families regardless of whether their income came from land, early industry, textile manufacturing, mercantile shipping or commerce.

9.4 The Genesis of Porcelain

Porcelain has always been a highly valued product, with a wide range of uses from the utilitarian to the ornamental. All these uses have been dependant on porcelain's unique properties of appearance, translucency, strength, resilience to rapid changes in temperature, and imperviousness to water and corrosive solutions. The development of this material has spanned the millennia.

The first earthenware pottery began to appear in the early Neolithic period c10,000 years ago. This pottery was formed from locally sourced clays and fired at relatively low temperatures of around 700 °C giving rise to unfused granular porous bodies. The first evidence of technological advancements beyond simple earthenware seems to have been in China during the late Neolithic Xia to Shang period c1800–1600 BCE. By this time, the earliest of the Dragon kilns were emerging. Dragon kilns enabled firing at much higher temperatures of up to 1100–1200 °C giving rise to proto-porcelains or stoneware, consisted of a stronger, finer granular body formed from fused kaolinic clay. This new ceramic was significantly more impervious to water. The earliest sherds of proto-porcelain have been found in Northern China at the Erlitou site in Yanshi, Henan Province. It isn't known if these early proto-porcelains were manufactured in Henan Province because no kiln sites have been identified in this region of China. However, the earliest known proto-porcelain kiln site, the Piaoshan Kiln site, is located in Huzhou in the more South-easterly province of Zhejiang. Sherds of proto-porcelain located at this site closely resemble those found at the Northern Erlitou site, potentially indicating that by this time early proto-porcelain was transported and traded within China.

Technological advancements in Dragon kiln design continued at Zhejiang so that during the Eastern Han period (25–220 CE) the kilns had become steadily longer allowing the kiln temperatures to rise to over 1200 °C. The elevation of the temperature allowed potters to fire finer proto-porcelains, giving rise to the Yue wares which had a glaze containing up to 4% iron oxide, which under the reducing conditions of the cooling Dragon kilns would result in the reduction of iron (III) oxide giving rise to the typical greenish, greyish yellowish coloration of the celadon glaze. Further elongation of the Dragon kilns by the Late Eastern Han period enabled kiln temperatures to rise further, enabling the potters of Zhejiang to fire the first true porcelains. The bodies of these first porcelains were greyish white but had a compact glassy structure with porcelanite crystals along with some fine rounded quartz crystals typical of a true porcelain fired at high temperatures. The first porcelains allowed the manufacture of much finer porcelain celadons. The production of this porcelain was very dependent on the use of locally sourced China stone.

The manufacture of celadon products spread to Northern China, particularly the Yiyongjie workshop located close to the imperial court in Henan Province during the Sui and Tang periods (581–907 CE). These Northern Chinese potters used kaolin rich secondary sedimentary clays along with potassium and calcium rich feldspar fired at temperatures of up to 1380 °C allowing them to produce the first truly white porcelain glazed with what is known as the northern white alkali calcium glaze; this was used to produce the fine white porcelain of the Xing, Gong and Ding kilns of Henan province c.608–907 CE (Li *et al.*, *Ceramics in China,* 2008). This glaze was later adopted in the Yongle period of the Ming Dynasty to produce the sweet white wares of Jingdezhen (1403–1424). Technological advances similar to those made in Northern China were steadily incorporated into the manufacture of porcelain in the South-East of the country at Jingdezhen during the late Song and early Mongolian Yuan period (1227–1279) with the incorporation of kaolin clays along with quarts and muscovite. These porcelains, like celadon, were glazed with glazes that developed a bluish green to yellow colouration during firing, which like the colouration of celadon probably arose due to the reducing conditions within the kilns. These southern porcelains were referred to as qingbai wares and it was the celadon wares and the qingbai wares that formed the bulk of the early Mongolian Yuan period Chinese porcelain export ware, which according to Marco Polo was exported via the Chinese port of Zayton (modern day Quanzhou) across East Asia, India and Persia.

9.4.1 Porcelain Decorative Developments

Chinese celadon and qingbai wares steadily gave way to the manufacture of the Chinese blue and white porcelains of the Yuan period in around 1330. The inspiration for the blue and white ware was a direct result of the expansion of the Mongolian empire following the silk routes into Persia. Persia had its own ceramics industry based around tinglazed earthenware. By the ninth century CE, the Persian artisans of Samarrain developed a technique of decorating their earthenware ceramics with painted underglaze blue designs using cobalt sourced from Al Amarah and Al Hayy.

The expansion of the Mongol empire allowed the Chinese potters to gain access to these ceramics and to the cobalt used in their decoration. One of the problems that the Persians encountered with cobalt was that it tended to diffuse into their glazes during firing, which reduced the fineness of their decoration. The Chinese found that their glazes being thicker prevented this diffusion allowing them to use the Persian technique to decorate their porcelain with finer blue and white Persian inspired designs. The Chinese blue and white wares very soon superseded the celadon and qingbai wares in the Chinese export market. The cobalt used in this early Chinese blue and white porcelain had a high iron content and traces of arsenic and sulfur.

Following the ousting of the last Mongol emperor in 1368 by the founding Emperor of the Ming Dynasty, Hongwo, the Chinese adopted policies of isolationism. This significantly reduced the foreign trade in Chinese porcelain, giving rise to what is referred to as the Ming gap, a period of just over 200 years (1350–1567) when very little Chinese porcelain was exported from China and a period when the majority of blue and white Chinese porcelain was exclusively produced for use within the Imperial court. As a result of this embargo on foreign trade, cobalt was sourced from neighbouring vassal states such as Sumatra and then later acquired from Jiangxi Provence within China itself. Both these later sources yielded cobalt that was high in manganese but low in iron. Manganese rich cobalt tended to diffuse into the glazes during firing, which reduced the finesse of the decoration during the Ming period. This difficulty was only overcome in the Quing Dynasty, under the emperor Kangxi about the turn of the eighteenth century, when the Imperial kilns began to adopt a new superior deep underglaze blue cobalt pigment which consisted of an iron rich cobalt with traces of arsenic, probably arising from mines in Saxony, along with the use of overglaze coloured enamels. The secrets of these new enamels were believed to have been acquired by the Chinese as a result of a Jesuit mission to Beijing, sponsored by Louis XIV of France, during which information about enamels was thought to have been exchanged for information on the manufacture of porcelain. This enabled the Chinese to start producing polychrome decorated porcelains, including polychrome armorials; the earliest known example of which is the armorial Chinese Imari pattern dinner service made for Thomas Pitt, Governor of Madras, in 1702.

9.4.2 Trade Links and Transport Developments

The earliest evidence of a Chinese ceramic trade dates to around 5000 BCE when it seems that the white pottery of the Dawenkuo culture was traded between economic centres within China mediated by an established trade network (Wu *et al.*, 'Geochemical evidence reveals a long-distance trade of white pottery in Neolithic China 5,000 years ago', *Journal of Archaeological Science*, 2022). Evidence of foreign trade is a little harder to find, but it would have occurred between neighbouring tribes, and this trade initially led to the establishment of the earliest silk routes during the Han period (202 BCE–220 CE).

The silk routes functioned as links between trading posts that eventually allowed a steady trickle of goods from the East Asian continent towards India, Persia, North Africa and Turkey. There is also some evidence of a maritime trade between China and Persia during the Tang period (618–907 CE) mediated by Islamic traders who settled on the south coast of China. This provided one of the first routes by which Chinese ceramics found their way into Europe via North Africa. Chinese porcelain sherds dating to around the tenth century CE have been found at Alcazaba, the Islamic fortress of Almería in Spain. Almería was the major Spanish Islamic port up until its sacking in 1147 CE and it is likely that porcelain would have been imported as diplomatic gifts from Syria and North Africa. Porcelain sherds have also been found in Lucera castle in Italy.

The silk routes became more formalised following the defeat of the Northern Jin Dynasty of China by the Mongols under Genghis Khan. He used the silk routes to move from population centre to population centre, conquering all in his path then converting the routes he had used into true roads that could then be used for the transport of military supplies and goods. By 1279 under the emperor Khublai Khan these extended right down to ports in India, Persia, Syria and Turkey. The Mongols used these ports to further accelerate trade, supported by a huge navy which not only transported goods but policed the seas supressing pirates.

The opening of sea silk routes enabled the transport of previously unimagined quantities of Chinese good with a proportion reaching the Levant. By this time Venice had developed into a Mediterranean naval power. Venetians stormed the walls of Constantinople at its fall and were the first to enter the city by force. They inherited colonies from the Byzantines in the Levant and used this to expand their trading influence. By 1221, they established a trade agreement with Genghis Khan which allowed them to ship Chinese goods that had reached the Levant via the maritime and physical silk roads directly to Venice where they could be traded. These goods would have included silks, spice and the occasional piece of highly prized Chinese porcelain. As a result of this, Venice became the Western European centre for the trade in these valuable commodities until the Portuguese colonized Madeira in 1420, the Azores in 1439 and the Cape Verde Islands in 1460.

9.4.3 Trade with Portugal

In 1497, Vasco da Gama set sail on behalf of King Manuel I Portugal in a fleet of 4 ships with the aim of establishing trading links with India. He reached Calicut in May 1498 and eventually arrived back in Lisbon in August 1499, having shown that it was possible to trade directly with the peoples of the East Indies. By 1511, the Portuguese had established control of the Malaysian coastal state of Malacca which became a key Portuguese trading post until 1541.

In 1517 the Portuguese sent their first envoy, Tomé Pires, from Malacca to Guangzhou. Tomé Pires and his party was well received; by 1520 they had arrived in Nanjing with the aim of meeting the Emperor Zhengde in Beijing the following February. The Emperor Zhehgde died, however, and all emissaries were ordered to return to Guangzhou. By the time Tomé Pires arrived in Guangzhou, newly arrived Portuguese had caused relations to sour. They were forcibly evicted in a bloody sea battle and Tomé Pires himself was imprisoned where he later died. Nevertheless, the Portuguese persisted, and built an illicit trade with Chinese merchants.

In 1542 three Portuguese sailors inadvertently discovered the Japanese Island of Tanegashima, just off the coast of Kyushu. Their ships were loaded with Chinese silks, porcelain and other luxury goods, so they exchanged their goods for Japanese goods. This visit was soon followed by other traders and missionaries including Saint Francis Xavier in 1449, which led to the establishment of Christian communities in Kagoshima and Hirado and initiated the Portuguese nanban trade. The nanban trade represented a triangular system of trade in which the Portuguese functioned as middlemen between Japan and China; silks, porcelain and other Chinese luxuries were exchanged for copper, silver and other Japanese goods. For the Portuguese, the nanban trade led to the establishment of trading bases in Macau near the mouth of the Pearl River and the Japanese port of Hirado in 1557. This triangular trade became highly profitable and by 1595 Macau had become Spanish, due to Philip II of Spain inheriting the throne of Portugal. He made Macau a city, and one of the most important ports in the far East.

9.4.4 The Genesis of Chinese Porcelain Decorated with European Armorials

An armorial porcelain can be defined as a porcelain item decorated with one or more heraldic emblems. The heraldic decoration can be simply the crest, or it can be the full coat-of-arms; the escutcheon, badge, coronet, helm mantlings, torse, crest and motto (Edwards, *Welsh Armorial Porcelain,* 2022).

Manuel I of Portugal (1459–1521) is thought to have been the first to commission a Chinese armorial with a European armorial on it. He had received a gift of porcelain in 1499 on the return of Vasco da Gama from his voyage to Calicut. He was so delighted with the porcelain that he ordered an ewer bearing his heraldic coat-of-arms from a Portuguese expedition going to China in 1517 (Edwards, *Welsh Armorial Porcelain,* 2022). It is unlikely that this commission was executed, however; Manuel I died in 1721, and most of the expedition members did not return to Portugal.

9.4.5 The Dutch East India Company: Growing Availability and Popularity of Chinese Porcelain in Europe

In 1601, the first Dutch ships reached Macau. The following year, 1602, the Dutch East India Company (the VOC) was formed by uniting 6 competing Dutch East India Companies. This new company had both a commercial and military remit; the latter focused on harrying the Spanish/Portuguese navies. In 1603 the Dutch captured the huge Portuguese carrack, the *Santa Catarina,* as it sailed from Macau to Malacca. The ship was redirected to Amsterdam, where its highly valuable cargo of 60 tonnes of porcelain, 70 tonnes of gold and 1,200 bales of raw silk was sold. The funds raised were apparently equivalent to half the initial funding of the VOC.

The Dutch trade in porcelain was facilitated by the establishment a trading post near Guangzhou (Canton). The Dutch initially attempted to establish this by taking Macau from the Portuguese. This attempt failed, so they instead established a trading post at Formosa (modern day Taiwan) in 1624, from which the Dutch continued to disrupt the nanban trade. Unlike the Portuguese, the Dutch weren't interested in converting the Japanese to Christianity, so when the Japanese Shogunate under Tokugawa Lemitsu issued a ban on the Japanese leaving Japan around 1635, the Dutch found themselves included as foreign trade partners with Japan whilst the Portuguese were excluded. This meant that the Dutch gained unopposed rights to the nanban trade, and as a result of this, their trade flourished. The sale of porcelain from the *Santa Catarina* represented the opening up of the European market for Chinese porcelain. *Kraak* porcelain, so named because it initially arose from captured carracks, became very fashionable in the Netherlands. The Princes and Princesses of Orange filled their houses with decorative *kraak* porcelain and began combining *kraak* porcelain with purpose-built architectural features.

At about the same time, the Dutch introduced tea to the Netherlands and this further increased demand for Chinese porcelain, particularly the tea wares. By 1637 VOC directors were specifically requesting more porcelain tea bowls to be shipped from China to the Netherlands, and some 25,000 porcelain tea bowls were ordered that year. The Dutch also exported porcelain and tea to France and England. By the mid-seventeenth century, coffee, tea and cacao (chocolate), had arrived in Europe, which further increased the demand for porcelain vessels.

However, China was in revolt and the Manchurians led by Hong Taiji were closing in on Beijing. In 1644 the last Ming emperor Chongzen hanged himself as the Manchu army neared the city.

With the end of the Ming Dynasty, the Manchu Qing Dynasty was established. The war in China continued, after the fall of Beijing with the Manchu armies steadily pushing the Ming armies back into the South of China until in 1662 the Ming forces, supported by the powerful Merchant Zheng family, overthrew the Dutch in Formosa massacring all the Dutch who failed to escape. The Ming forces retained Formosa until it was eventually captured by Qing forces in 1683. During this war the kilns at Jingdezhen were destroyed, however, Qing Emperor Kangxi had them rebuilt and by 1685 had opened up the port of Canton (Guangzhou) to foreign trade. The Dutch trade with China never really fully recovered after the loss of Formosa, partially because the productivity of the Japanese silver mines had diminished sufficiently by this time to end the nanban trade that the Dutch had historically depended on.

The decreased trading activity of the Dutch in China left a trade vacuum, which the English East India Company (EIC) found themselves in the position to fill. Up until around 1700 the EIC's activities in China had been negligible. However, the EIC established a new triangular trade in which they acquired goods such as cotton and opium in India and then exchanged them for silks, tea, and porcelain in China. Tea by this time had become a very profitable commodity in Europe as had the porcelain to drink it from. Thus, huge volumes of tea needed to be transported to England from the Chinese port of Canton. This was achieved by English East India Company's huge cargo ships, the East Indiamen. These ships had holds large enough to load up to 1200 tonnes of cargo in their holds. By 1800 80% of the tea exported from China arrived in Britain via these massive ships, although there were some challenges associated with shipping voluminous, light, perishable commodities such as tea. Firstly, the valuable tea cargo needed to be kept away from the seawater to avoid spoiling; this was a particular hazard in the leaky cargo ships of the day. Secondly, the East Indiamen were a little top heavy and needed a counterweight in the hold; as tea was a lightweight cargo, the ships needed to carry additional ballast on their return trip from China. The answer lay in porcelain. When porcelain was packed into specially constructed cases that fitted snugly into the hold, the cases created a very valuable salt water resistant ballast that could function as a raised floor onto which the more easily damaged goods such as tea and silks could be packed. This provided an incredibly profitable way of shipping porcelain from China back to England; not only did the ships carry porcelain vessels for tea-drinking, but also the specialist porcelains used for drinking coffee, chocolate and punch, as well as ornamental porcelains and dinner wares.

Punch became popular in England from the 1650s onwards. Punch was made from imported ingredients: brandy or rum, citrus fruits, nutmeg and spices. It was typically drunk by men at social gatherings held in taverns. The punch would be made in a large open communal bowl then served in small glasses or cups. Early punchbowls were made from delftware or earthenware and were often inscribed for particular occasions or events, such as elections.

Cocoa was first discovered by Columbus in 1502, but he had no idea what the cocoa beans were; he thought they were used as a currency. Chocolate was shipped back to Spain and Portugal, but cocoa didn't really get to the rest of Europe until the Dutch introduced it in the 1630s. Chocolate was perhaps the least popular of the new drinks to reach England, largely because it was difficult to make. It was often made with milk and needed sweetening with sugar, which meant that drinking chocolate was very expensive; a dish of chocolate could cost three times more than a coffee. The first chocolate house opened in London in 1657 (Strevens, *The Birth of the Chocolate city. Life In Georgian York,* 2014), but chocolate houses were frequented only by the very wealthy.

Coffee and coffee houses were adopted from the coffee houses of Turkey and Cairo where the drink was consumed exclusively by men. The first English coffee house was opened in Oxford by Levantine merchants. By the late seventeenth century coffee houses had become commonplace. Coffee houses were very business-like, and they attracted sober professionals who drank coffee whilst discussing the politics of the day along with important matters; Lloyds of London was established in the Lloyds Coffee House, and the London Stock Market was established in Jonathans Coffee House. The tea houses and tea gardens, such as the Golden Lion established by Twinings of London in 1706, were quite different; they were much more genteel, civilised and feminine. Tea was also drunk in the home and of the new hot drinks, tea was by far the most popular. All these new drinks played a considerable part in driving demand for porcelain from China, the development of the British porcelain manufactories as well as the proliferation of different forms of ceramics for drinking and dining.

9.5 The Genesis of English Armorials on Chinese Porcelain

Vast fortunes were made by EIC employees who could use the triangular trade to invest in Indian goods so that they could be exchanged for Chinese goods which were sent back to England and sold at considerable profit. It was one of these individuals, Sir Henry Johnson (1659–1719), a shareholder in 33 East India Company ships, who commissioned what is thought to be the first Chinese porcelain to be decorated with an English armorial. Around 1695, he commissioned 6 large jardinières decorated in underglaze blue and white and customised them by having 8 heraldic badges enamelled onto each; 4 were his own antecedents, and 4 were for Martha Lovelace (1667–1745) his second wife. The commission was in the tradition of decorative porcelain, but having armorials enamelled onto Chinese blue and white porcelain, an expensive and luxurious product, was no small undertaking. Having 8 heraldic badges enamelled onto the jardinières was a very clear signal as to the couple's social standing; their vast fortune may have been obtained from mercantile trade, but this commission flaunted their familial connections back to the earlier times of a landowning aristocracy.

An armorial service is defined as a number of pieces commissioned for the table, which carry the heraldic enamelling of the commissioning patron or first owner of the service. Dinner services, dessert services, breakfast services and tea or coffee services were all used in the 1700s, with the dinner service being the largest and thus the most expensive to order and have decorated. Dinner services often contained 400 individual pieces of porcelain including lids, stands and trays; dessert services typically contained sufficient pieces for a dozen people, as did tea and coffee services.

The next known armorial commission on Chinese porcelain was a dinner service for Sir Thomas Pitt, Governor of Madras, in 1702. As a senior East India Company official, he would have entertained many dignitaries and officials during his time in office. The service was decorated in an Imari pattern, and it would have created a stunning table display. This service represents the genesis of polychrome armorial Chinese commissions, and it would have sent a clear signal to visiting dignitaries as to their host's wealth and social standing. Bearing in mind that polychrome porcelain was only adopted for use in the decoration of Chinese porcelain around 1700, this service may have been amongst the first Chinese polychrome decorated porcelains to reach Madras.

Other early armorial commissions were made by those who had connections to the East India Company such as Henry Gough, who ordered the earliest known armorial Chinese porcelain coffee service. This was enamelled in polychromes; however, the armorial decoration was limited to Gough's crest rather than his heraldic badge.

Johnson, Pitt and Gough were all members of well-established armigerous families. The Johnson armorials are for Johnson of Friston Hall and Aldburgh, Surrey, whilst Martha Lovelace's armorials show that she was a descendant of the Wentworth family, being descended from Sir Philip Wentworth of Nettlestead, Suffolk (1423–1464). The Pitt family armorial depicted on Thomas Pitt's service had been granted to his great grandfather John Pitt (d.1602) a mercer who served Elizabeth I as Clerk to the Exchequer. Thomas was descended from a cadet branch; the direct heirs of John Pitt were the descendants of his eldest son Sir William Pitt (1559–1636) but Thomas Pitt's grandfather was a younger son. Henry Gough's armorials are for the Gough family of Staffordshire. The line between merchant and gentry families became increasingly blurred during the seventeenth century and beyond. As the opportunities offered by trade and commerce to create wealth grew, individuals were able to establish or re-establish their family fortunes and thus their place in society changed. In this changing society, armorial display was clearly an important social marker, else very few would have gone to the time, expense and trouble of commissioning armorial porcelains; it would have been sufficient to merely mark their porcelains with initials or inscriptions.

A fashion for Chinoiserie began in the early 1700s, and the EIC shipped vast amounts of Chinese porcelain from Canton to London via huge ships known as the

Indiamen in order to meet the demand. The popularity of Chinese porcelain and its prevalence amongst the possessions of the wealthy grew significantly in the first quarter of the eighteenth century. By 1725, it is calculated that 11% of personal inventories included Chinese porcelain (Peck, *Consumer Splendour: Society and Culture in Seventeenth Century England,* 2005). Chinese inspired designs were on textiles, wallpapers, furniture and silver as well as on ceramics. Chinese and English designs were blended; Chinese scenes were on English ceramics, and European items appeared on Chinese wallpaper. English manufactured goods such as Chippendale chairs had ball and claw feet, which were inspired by Chinese designs. William Chambers (1723–1796) visited China and, following that visit, he designed the Great Pagoda at Kew (Peck, *Consumer Splendour: Society and Culture in Seventeenth Century England,* 2005). The fashion continued as the century progressed, reaching its peak in the 1750s. Prominent Georgian towns such as Brighton, Bath and Harrogate developed as places for the fashionable nobility and gentry to mingle, far away from London and the court which had traditionally been the focus point of society life. These towns developed fashionable shopping enclaves, to tempt the wealthy visitors to purchase items during their stay. Shopping for decorative items became a fashionable leisure activity.

The EIC dominated the import market and their shipments accounted for around 75% of the cargo shipped from Canton. However, the time for these journeys should not be overlooked. In addition to the sea voyage, there was also the transportation within China to be factored in. From Canton, manufacturing and decoration instructions for commissioned pieces such as armorial porcelains would have to be sent over 900 kms to Jingdezhen, where the commissions would be manufactured and decorated before being transported to Canton (Finlay, *The Pilgrim Art,* 2010). The transport of porcelain goods from Jingdezhen to Canton was a laborious business and would have involved the porcelain being wrapped in paper and packed into large barrels protected by rice and straw. These barrels weighed over 200 kg when packed, and they had to be loaded into boats so that they could be transported 50 miles along the River Gang to Raozhu. From there, the barrels of porcelain were transferred to junks for shipping across the largest freshwater lake in China, Lake Boyang, to Nanchang from where they would be transported along the River Gan to the Meiguan Pass (otherwise known as the Plum Mountain pass). The 20 km (12 mile) long Meiguan pass provided the only route over the Nanling mountains, so the barrels were ported in barrows to Nanxiong, where they were transported by boat along the Zheng River, south towards the Bei River, along which they were then transported to Canton (Guangzhou). This 900-km journey could be made quite quickly in good weather; the Jesuit Joachim Bouvet reported in 1693 that he made the journey from Nanchang to Canton in only 20 days (Finlay, *The Pilgrim Art* 2010). However, in bad weather it could take considerably longer. As a result, leaving London and sailing to Canton, and then making the return journey with porcelain goods once they had been manufactured, decorated and transported back to Canton, could take 2 or more years.

Chinese porcelain began to face increased competition from the domestic market as the eighteenth century progressed. British potters had developed new forms of stoneware that could rival porcelain, and soft paste British porcelain manufactories were being set up from the 1740s onwards.

9.6 Establishment of the British Porcelain Manufactories

Porcelain is a type of ceramic that is fired at a high temperature. There was a long history of earthenware and stoneware manufacture in Britain, prior to the establishment of porcelain factories, and armorial display on ceramics pre-dates the arrival of armorial Chinese porcelains in England.

Delftware manufactories had been established in London by 1570; Southwark and Lambeth were the main centres of production, and these had produced a range of vessels including plates and mugs which, although functional in purpose, were also decorative pieces. Many pieces of delftware have survived that are marked with dates, initials and inscriptions, and there are also delftware pieces that carry armorials. It is noticeable that the majority of these delftware armorials are for the livery companies; it is likely that these were commissioned by livery company officials such as the aldermen, sheriffs and Lord Mayors. The titled nobility at this time are likely to have commissioned armorial silver or pewter pieces; silver tankards with initials and armorial inscriptions were very popular in Stuart England.

By the 1720s, the Staffordshire potteries were producing a white salt-glaze stoneware product, which was thin and durable. This began to supersede delftware, which was both thick and brittle. White salt-glaze products reached their peak popularity around the 1740s. By this time, new experiments with other stonewares had led to redware, and then in the 1750s creamware was produced by Josiah Wedgwood. Creamware was a cheaper alternative to porcelain. It shared many of porcelain's characteristics in that it was fine, and it could be moulded and glazed with colour through the glaze or applied under the glaze. Another stoneware product was pearlware. Formed by adding flint and Cornish china stone to the clay, pearlware was a whitened form of creamware finished with a tiny amount of cobalt. It proved very popular because the decoration was done under the glaze, which meant it was not subject to scratching or flaking and thus it could withstand hard use. Pearlware remained popular until the 1830s. Armorials were decorated on all these types of earthenware and stoneware.

The earliest British porcelain manufactories were established in the 1740s: Ponoma in Newcastle Under Lyme, Chelsea, Bow and Limehouse were all established by 1745. Of these, Ponoma, Chelsea and Limehouse did not produce armorial porcelains, but Bow did. The earliest Bow armorial decorated pieces are mugs. These were typically produced as one-off items, rather than purchased as part of a set. The earliest Bow armorial commissions carry the armorial bearings of livery companies, making it probable that they were ordered by senior livery officials.

Vauxhall was set up in 1751 and it did produce armorials, however it was a short-lived venture and ceased trading in 1763. Worcester was founded the same year and fared rather better. Worcester was founded by Dr John Wall, who wanted to make a porcelain able to withstand changes in temperature, so that it could be used for hot drinks. Not only did Worcester thrive as a business, but it was also the most prolific domestic factory producing armorial porcelains. This was achieved by outsourcing the armorial decoration to the London enamelling ateliers, specifically that of James Giles. Worcester whiteware was sold to the ateliers in volume, and the ateliers then decorated the pieces in line with orders received from retail shops. The arrangement continued until the 1770s when Giles lost Worcester as a client and went bankrupt.

The importance of producing a porcelain able to withstand changes in temperature and thus be used for hot drinks is underlined by the fate of the Longton Hall porcelain factory. This was set up around 1749 by William Littler and produced a porcelain that was frit-based, which meant it could not withstand temperature changes, and Littler was declared bankrupt in 1760. He then moved to West Pans in Scotland, where his bankruptcy did not stop him operating another business. He ran the West Pans business until around 1777, again producing porcelain containing frits, and he also manufactured functional items such as inkstands.

Lowestoft was set up in 1757, and it was able to thrive until 1803, when a warehouse full of porcelain wares in Rotterdam was destroyed by Napoleonic forces. The loss of this porcelain resulted in Lowestoft closing. Until then, Lowestoft had thrived despite having to ship kaolin to the factory site from Cornwall, which added to its overall manufacturing costs. Lowestoft did produce armorial porcelains, an example being the armorial tea and coffee service made for the Rev Potter, vicar of Lowestoft parish in 1790.

The imported Chinese porcelains were still the favoured option for wealthy armigerous families in the 1750s; Chinoiserie was at its peak when the earliest British porcelain manufactories were producing their first armorial products. It had become quicker to obtain armorial Chinese porcelain by this point, as whitewares produced in the kilns of Jingdezhen were being transported to Canton for decoration there. By relocating the decorators workshops at Canton, the overall time taken to commission Chinese porcelains were reduced from 3 years or more to around two years (Le Corbeiller & Frelinghuysen, "Chinese Export Porcelain' *The Metropolitan Museum of Art Bulletin,* 2003). It must be remembered, however, that the Chinese porcelain was a hard paste porcelain, whereas the early British manufactories were producing a soft paste porcelain. Although there had been experimentation with producing hard paste porcelains, English porcelains sold before 1768 were soft paste porcelains as it was not until the discovery of kaolin and the patent for the manufacture of hard paste porcelain by William Cookworthy (1705–1780), that English hard paste porcelain was available commercially. Therefore, Chinese and English porcelain were two different products until that date.

In Europe, monarchs supported the establishment of porcelain manufactories, and either funded them or commissioned pieces from them. King Augusta, for example, supported the establishment of the Meissen porcelain works. That did not happen in England. From the beginning, porcelain manufactories were seen as commercial ventures, and they had to be commercially viable to succeed.

The first royal commission for English porcelain was made in 1763 by George III and Queen Charlotte from the Chelsea porcelain manufactory, which was given to her brother Adolphus Frederick IV (1738–1794) as a wedding gift. The service took 2 years to make and was completed in 1765 (Edwards, *Welsh Armorial Porcelain,* 2022).

The first royal commission for armorial porcelain was placed by the Duke of Clarence in 1789; he commissioned a Worcester service to mark his new honours. Whilst this was no doubt of significant influence from 1780 onwards, the commission came some 20 years after the first English factories had been set up, and more than 90 yearsafter Johnson's armorial Chinese porcelain commission. Royal influence and patronage, therefore, did not drive either the establishment of porcelain factories in England or the fashion for armorial porcelain.

9.7 The Commissioning Patrons for Armorial Porcelains

For every product, there needs to be a market. Armorial porcelains were bespoke decorated products, and as such these items were made to order. A commissioning patron of armorial Chinese porcelain needed to be wealthy, and also to have the connections to arrange it. This was by no means a straight-forward purchase. Instructions had to be sent out via the East India Company's super-cargoers. In an era when armorial commissions could take months if not years to be delivered, this was no small undertaking.

The key reason for commissioning armorial Chinese porcelains appears to be that ownership conferred status. Armorial display was a sign of social standing in the 1700s. It was indicative of descent from an aristocratic or noble family, and it illustrated a familial link back to the earlier times of an aristocratic patrimony. At a time when society was changing, and there was greater social movement, social status signals were all-important. Henry Johnson, Thomas Pitt and Henry Gough all had direct connections with the East India Company, and they all derived significant personal wealth from mercantile trade, but they were all from armigerous families and armorial display on their luxury Chinese porcelains was a sign of this.

Other early commissions were placed by prominent officers of state: a number of politicians commissioned armorial Chinese dinner services, such as Robert Walpole, Robert Walker and Henry Soames, who all placed orders around 1720. All three were from armigerous families but had made their own fortunes through the East India Company or the South China Sea Company. Other early commissions were placed by banking families, such as the Childs of Osterley Park.

Lord Mayors and Members of Parliament also feature amongst the early armorial Chinese porcelain patrons; there are examples of them ordering armorial porcelain to mark their year of office or election year. It is also noticeable that on the earliest Chinese armorial commissions, the English armorial sat within the Chinese decoration; these patrons did not seek to change the traditional decorated style, rather to add their own personal stamp onto the product.

Later commissions were placed to mark marriages; this was the case for several services Gough is known to have ordered, including one in 1720 with the armorials of Gough and Hynde, which marked his marriage that year. Marriage remained a consistent reason to order both Chinese and English armorial porcelains throughout the century; the Kemeys Tynte service being commissioned from Derby around 1820 to mark the marriage of Charles Kemeys Tynte and Elizabeth Swinnerton.

Another reason that patrons commissioned armorial porcelain was when they were taking a family surname and/or title, as part of an inheritance. When the direct male line failed, a family's wealth would be conveyed to another family through marriage, and it often happened that the inheriting couple would apply to change their name by Royal Licence and acquire usage of the first family's armorials. The Pulteney estate, for example, descended from William 1st Earl of Pulteney to the family of his first cousin's wife; Daniel Pulteney's grand-daughter Henrietta left the estate to a second cousin on her mother's side of the family, Elizabeth Evelyn Sutton. She was married to John Fawcett; when she inherited the Pulteney estate in 1811, the couple changed their surname by Royal Licence to Pulteney, assumed the Pulteney armorials and commissioned an armorial porcelain service from Flight, Barr and Barr in 1813.

The Long Century saw many military conflicts. Military service was another way for the younger sons of gentry families to secure a living. Officer commissions were purchased from the Crown, and fresh honours and awards could be obtained by service in battle. The Duke of Clarence, for example, commissioned an armorial porcelain service to mark the acquisition of fresh honours in 1789, and Nelson had armorial porcelain. Another group of patrons were clergymen: William Talbot, later Bishop of Durham, commissioned an armorial Chinese porcelain service in 1705, whilst William Howley, later Archbishop of Canterbury, commissioned an armorial English porcelain service from Chamberlain Worcester to mark his tenure as Bishop of London from 1813–1828.

The first known English armorial dinner service, the Gavin-Hearsey service, was commissioned as late as 1767, more than 20 years after the establishment of the early British porcelain factories. It is striking that this was associated with the rebuilding of the manor at Gaviston; having built or re-built a large property created the need for a large dining service, with which to entertain guests. The costs of commissioning a porcelain armorial dining service, therefore, may not have appeared astronomical in this context of building or re-building an entire property.

Prior to the Gavin-Hearsey commission in 1767, armorial English porcelain commissions appear to have been predominantly on smaller vessels that would have been used in the home for entertaining, such as breakfast services, tea and coffee services, or on stand-alone items that were used outside the home such as punch bowls, or beer mugs. From around 1770s onwards, when there were more porcelain factories established in Britain and hard paste porcelain was available, armorial English commissions for dinner services grew. By the 1780s, Mary Bootle was happy to order replacement pieces from Derby for an earlier armorial Chinese porcelain service. The 8th Duke of Hamilton did the same; he commissioned replacement pieces for the Chinese armorial service that had originally belonged to the 5th Duke of Hamilton.

By the 1770s, the imported Chinese porcelains faced increased competition from the growing manufacturing capability in Britain, Chinese porcelain was subject to increasing import taxes from the 1770s onwards; and by 1799 these taxes had risen to 109% of the import value. The market for Chinese porcelain in Britain shattered; the crash was so metaphorically loud that by 1798, the East India Company had ceased their trade in Chinese porcelain.

9.8 Discrepancies in Armorial Decoration of Porcelain

Given what an expensive, luxury product porcelain was throughout the Georgian Long Century, a very high degree of accuracy in the armorials depicted on the porcelain would be expected, however this is not the case. When examining the armorials depicted, it is striking how often there is an inconsistency or an error between the armorial decorated on the porcelain, and the official armorial record for a particular armigerous family.

Perhaps the most obvious reason for the inconsistencies within the armorials depicted is that it may not have been technically possible to re-create them. This is particularly likely with the earliest of the Chinese porcelain armorials. The armorial palette is restricted to a small group of colours and typically, a heraldic badge consists of a metal, either silver or gold, and 2 colours. The 1702 *Pitt* service should have had a black background for the heraldic badge, but it is omitted on the service, probably due to a technical difficulty.

There were also issues with depicting the armorial correctly. Henry Johnson's armorial jardinières display 8 heraldic badges within the coat-of-arms, but these are incorrectly sized. The Gough coffee service of 1711 depicts the crest from his coat-of-arms on the cups, but it is incorrect; the crest depicted is an arrow through the boars head when it is supposed to be in the jaws.

There were also issues with the manufacturing process itself. Robert Walker's crest appears on the border of his plate. Somehow this got copied into the design for Soames and also for Robert Walpole, as both their armorial services contain Walker's crest. This indicates that an instruction to 'copy the border pattern' from the plates in one commissioned service to two others, was taken very literally. Doubtless, a lack of knowledge of the images being copied was a challenge, and thus the practice developed of sending bookplates depicting the patron's armorials, along with the commission instructions, in order to reduce the risk of issues and errors being made during armorial decoration.

It should be noted that similar issues could also occur with armorial commissions from the British porcelain manufactories. A Worcester tea service was ordered in the 1760s decorated with Wilson impaling Langton. Burke describes the armorial for the Wilsons of Broomhead as *sa, a wolf rampant and in chief 3 estoiles or (Burke, A genealogical and heraldic history of the landed gentry of Great Britain and Ireland,* 1882). There is an error in the heraldic badge depicted on the service, as the stars are black. H Rissik Marshall has attributed this as being most likely due to the enamelling being outsourced to workshops that were not familiar with the armorial (Rissik Marshall, 'Armorial Worcester Porcelain' *English Ceramic Circle,* 1946).

When errors did occur, the commissioning patrons appear to have simply accepted them; there is little evidence of services being rejected or even returned to have the decorative errors corrected. Quite the contrary; the errors appear to have been accepted quite readily. Indeed, the 5th Duke of Hamilton commissioned an armorial Chinese porcelain service in 1738, and in the 1780s, the 8th Duke of Hamilton commissioned Derby to make some replacement pieces for this service. The coat-of-arms, crest and motto were duly copied from the original, including an error in the wording of the motto whereby 'though' is misspelled as 'thhough' in the original Chinese version, which the Derby decorators copied over onto their pieces so that the replacement pieces would 'match' (Edwards, *Derby Porcelain: the Golden Years* 1780–1830, 2017).

Another aspect of armorial commissioning is that there was no policing of the armorials being depicted; the enamellers would follow the instructions of the commissioning patron without querying their accuracy or correctness. As a result, commissioning patrons were free to place whatever armorial commissions they wanted. Pendock Barry, for example, commissioned a Derby service around 1805 displaying the Barry arms, some 6 years before he was formally granted the right to use the Barry surname and armorials by Royal Licence (Denyer, Denyer & Edwards, *The Pendock Barry Porcelain Service: A Forensic Evaluation,* 2022).

Indeed, there are instances where accurate depiction of armorials appears to have come second to decorative appearance; on the 1702 *Thomas Pitt* service, the crest is below the heraldic badge rather than above it. The crest, a stork, forms part of the decoration by appearing to wade.

All of these errors and inconsistencies in armorial decoration raises the question around the role of the College of Arms and the Heralds in enforcement of armorial usage. The last Visitations by the Heralds from the College of Arms occurred in the 1680s. After this date the heralds do not seem to have had an active role in policing and censoring those who used armorials incorrectly. Doubtless there were many instances of people ordering armorial commissions which included heraldic badges they were not entitled to use. The standard use of ancient armorial badges in coats-of-arms would in any case have exacerbated the challenges around proving descent claims to be incorrect.

9.9 Armorial Porcelain as a Product

Armorials were originally used as an identification on the battlefield for knights, and once this purpose became obsolete, armorial usage could have simply ceased. This did not happen, partially because armorials had become associated with surnames and particular families as they were passed down the generations. In a society with a land-owning nobility, heraldic badges became associated with land, property, and personal possessions. Heraldic badges were combined into new coats-of-arms, reflecting transfers of land and wealth as families intermarried. Armorials came to indicate kinship, allegiance and belonging. Servants wore the livery of their masters; Warwick the Kingmaker decked out all of his retinue in his own livery in the 1460s, and it was said that any servant in his livery could walk into the Royal kitchen and carry away a hunk of meat. The armorial, therefore, served as a very practical sign to indicate who was in the retinue of a great magnate at a time when illiteracy was still widespread.

When the London livery companies were founded, they too acquired coats-of-arms, and the use of armorials spread throughout society, taking on a far wider usage than that of knights on the battlefield. The management of heraldic badges was formalised in England in 1484 by the establishment of the College of Arms. This laid out the rules for acquiring and using coat-of-arms, as well as instigating formal record-keeping.

By the 1600s, armorial usage was still prevalent in society, yet society was changing under the Stuarts. Trade and commerce were beginning to open up new opportunities for wealth generation, and ripples of social change were taking place. Those who were best placed to take advantage of the new opportunities for wealth generation tended to have two things in common: firstly, they were likely to be literate, and secondly, they were likely to have access to an influential patron who could secure them an initial position. Therefore, they were often from a cadet branch of a noble family or from a minor county gentry family; in other words, they were armigerous.

Those who acquired wealth through the East India Company were also those best placed to commission Chinese porcelain for themselves, their families and wider network, particularly in the early years before significant imports of Chinese porcelain were coming into London. The directors and officials of the East India Company were, therefore, amongst the earliest acquirers of armorial Chinese porcelain and to some extent they set the fashion for it.

Armorials had, however, already appeared on English made ceramics; delftware potteries had been established in London by 1570, and a cluster of such manufactories were set up in the 1600s in Southwark and Lambeth. One of the earliest known armorially-decorated delftware products is a rectangular salt, dated to 1635 and attributed to the Pickleherring or Montague Close works, which carries the armorial for the City of London (Christie's, *Important American Furniture, Folk Art and Silver,* 2020). An armorial delftware candlestick was made for the Fishmongers Company in 1648, probably for a senior official.

Family armorials were also appearing on delftware in this period: a delftware charger dated for 1649 carries the armorials for Markham impaling Faringe. Although a charger is a practical piece, its use would have been decorative, due to the fragility of delftware; any attempt to use the piece would result in scratches and irrevocable damage to the decoration.

The earliest blue and white porcelains shipped from China were principally decorative and used as part of decorative arrangements. It was in this tradition that Henry Johnson commissioned his armorial jardinières, c.1695, whereas Thomas

Pitt's dinner service was a novel commission and set the trend for some 3,500 armorial Chinese porcelain dinner services to be imported into London in the eighteenth century.

Chinese porcelain itself was a very expensive product, so armorial Chinese porcelain commissions were only possible for the very wealthy; and armorial Chinese dinner services, which could be up to 400 pieces of porcelain, could only be afforded by those with vast wealth. Others commissioned dessert services, breakfast services, tea and coffee services, caudle cups, punch bowls or hot chocolate services.

English delftware gave way to stoneware in the first part of the eighteenth century, and those who could not afford imported Chinese porcelain bought similar services made from white salt-glazed stoneware. This new product was an improvement on delftware in that it was fired at a higher temperature which meant it was a thinner and finer product, but it wasn't until 1750 when creamware was developed, that stoneware was able to challenge porcelain as a true alternative. Queen Charlotte's patronage of Wedgwood and the subsequent renaming of creamware to Queensware led to it becoming widely popular as an alternative to porcelain; so much so, Wedgwood was unable to keep up with demand and had to outsource the production to other manufacturers who thus learned the recipe for the product.

The development of pearlware in the 1770s created the closest stoneware product to porcelain; it had a cobalt-containing glaze that gave an under-glaze blue decorated ceramic a finish that was very similar to a Chinese blue and white porcelain. By this time, English hard paste porcelain products were also available following Cookworthy's patent of 1768.

The availability of English hard paste porcelains, plus creamware and pearlware products, meant that there were good quality, cheaper alternatives to Chinese porcelain by the 1770s. There was also a change in taste around this time, away from the Chinoiserie styles to the sparse geometric Neo-classical styles (Peck, *Consumer Splendour: Society and Culture in Seventeenth Century England*, 2005). These factors went a considerable way to reduce the demand for imported Chinese porcelains, and this combined with an increase in taxes on imported Chinese goods, eventually killed the import trade; the East India Company had stopped its direct trade in Chinese porcelain by 1796. Without the Chinese imports, commissioning patrons for armorial porcelains turned to the home markets. From the 1790s through the Regency and beyond, there were many armorial commissions made by the various Worcester factories, Spode, and Derby as well as the newly established factories of Swansea and Nantgarw.

Armorials were a signifier of social status throughout the Early Modern period and into the Modern period. The genesis of armorial porcelain needs to be seen within the context of historical armorial display. Armorial porcelains provided a visual connection to the earlier era of a land-based wealth system dominated by aristocratic noble families, and they were a symbol of familial connection to the families of the past. The heraldic badges depicted on armorial porcelain may not always have been correctly depicted, or factually accurate in every instance, but they represent the commissioning buyer's view of their antecedents, and they were an expensive, customised product that were, and still are, highly prized.

References

Corfield, P. J. (2022). *The Georgians: The Deeds and Misdeeds of 18th Century Britain*. Yale University Press.

Denyer, R. L., Denyer, M. C. T., & Edwards, H. G. M. (2022). *The Pendock Barry Porcelain Service: A Forensic Evaluation*. Springer.

Edwards, H. G. M. (2017). *Derby Porcelain: the Golden Years 1780–1830*. Penrose Antiques Ltd.

Edwards, H. G. M. (2022). *Welsh Armorial Porcelain,* Springer Nature.

Finlay, R. (2010). *The Pilgrim Art*. University of California Press.

George, A. (2015). Direct Sea Trade Between Early Islamic Iraq and Tang China: From the Exchange of Goods to the Transmission of Ideas. *Journal of the Royal Asiatic Society, 25*(4), 579–624.

Li, J., Leung, P. L., Yu, K. N., Wu, R. (2008). Ceramics in China. In H. Selin (Eds.), *Encyclopaedia of the History of Science, Technology, and Medicine in Non-Western Cultures*. Springer.

Le Corbeiller, Clare, and Alice Cooney Frelinghuysen. (2003). Chinese Export Porcelain. *The Metropolitan Museum of Art Bulletin 60*(3): 1–60.

Peck, C. S. (2005). *Society and Culture in Seventeenth Century England*. Cambridge University Press.

Rissik Marshall, H. (1946). Armorial Worcester Porcelain. *English Ceramic Circle Transactions, 2*(9), 189.

Strevens, S. (2014). *The Birth of the Chocolate city*. Life In Georgian York Amberley Publishing.

Wu., Xiaotong, Ji, F., Zhang, X., Wang, F., Feng, F., Lu., Qingyu, Zhao, S., Zhang, Y., Wang, C., Huang, F., & Zhang, X. (2022). Geochemical Evidence Reveals a Long-distance Trade of White Pottery in Neolithic China 5,000 years ago. *Journal of Archaeological Science: Reports, 44*, 103533.

Online Sources

Christie's, 'A London Delft Armorial Blue and White Rectangular Salt' Lot 278, Auction 17810 *Important American Furniture, Folk Art and Silver,* January 2020. Available online at https://www.christies.com/en/lot/lot-6250610. Accessed January 30, 2024.

Glossary

Armorial china: High-fired ceramics decorated with the heraldic achievements of an arms-bearing family which could constitute the full coat-of-arms comprising the escutcheon, motto, supporters, helm, lambrequins, crest and torse or any number of these or more usually perhaps just consisting of a crest and torse. In some cases, the crest and torse, with or without a motto, is encompassed in other decoration and may be accompanied by additional gilding and other geometric motifs.

Biscuit porcelain: The unglazed product of the high-temperature firing process in the manufacture of porcelain using a "biscuit kiln" which operated typically at a temperature in the region of 1100–1400°C. The resultant porcelain is of a pure creamy-white or ivory colour and possessed a texture which, if blemish-free and perfectly shaped, was very highly prized by ceramic artists and modellers particularly for the construction of ornamental figurines in simulation of marble statuary as first perfected by the Meissen manufactory in the mid-18th Century. In the mid- to late -18th Century, the finest biscuit porcelain artefacts were commonly placed on dining tables for admiration by the guests and to stimulate conversation during dinner. Generally, the biscuit porcelain, after cooling from the kiln firing process, was painted, glazed and gilded—during which process any small defects or blemishes in the body or the glaze could often be masked by strategically placed enamelling: hence, painted and enamelled figurines are found much more commonly than their biscuit analogues, which if found to be imperfect after their firing, were usually destroyed or sent to be decorated and glazed. The figures were designed by a "modeller" or sculptor and assembled in the workshop by a "repairer".

Blemish: A blemish is a fault in the fired porcelain, whether this is in biscuit or glazed form, arising from several problems which could arise during the firing process such as inadequate kiln temperature control leading to firing cracks, the introduction of contaminants such as carbonaceous matter which then burned to leave soot deposits and the emission of gases during the firing process which created bubbles in the vitrified paste or the glaze. Serious blemishes resulted in porcelain items being unsaleable, whereas it was a common practice to mask

R. L. Denyer et al., *Armorial Porcelain*, https://doi.org/10.1007/978-3-031-63745-2

minor blemishes, such as small spots of soot or little bubbles in an otherwise perfectly shaped piece, by the shrewd and strategic placement the in enamelling of small devices or motifs such as insects, butterflies or moths prior to a final glazing process taking place.

Body paste: The wet mixture of raw material components which comprise the formulaic recipes for porcelain bodies. Texturally moist to the touch, the paste could be moulded or formed into shapes preparatory to the firing process taking place in the biscuit kiln. This term is also applied to the fired porcelain body prior to its chemical analysis, when of course it is then not a wet "paste" at all but rather a hard ceramic material.

Burnishing: The gentle polishing of gilt decoration on a glazed ceramic surface to a highly polished reflective coating. Early gilding was accomplished using "honey gilding" whereby 24-carat gold leaf was applied to the surface with an adhesive medium of honey, or with the use of applied resins such as gum arabic, which were both replaced in the late 18th Century by mercury gilding using an amalgam of mercury which had been triturated with the gold leaf. During the final firing of the ceramic piece in the "glost" kiln at low temperatures the organic adhesive carrier component or the mercury in the amalgam was volatilised off into the atmosphere, leaving a dull golden finish which was then hand-polished by skilled gilders using a finely powdered jeweller's rouge, iron (III) oxide Fe_2O_3, to give a brilliant finish. The skill of the gilder was much admired in porcelain workshops as it offset and complemented the applied enamelling decoration but it was always an expensive part of the finishing process. Some service commissions received only a gilded decoration without accompanying enamelling decoration whereas others were not gilded at all. In some manufactories, such as Derby, the gilder was a recognised artistic master and was allowed to place an assigned numeral indicative of their work on the underside of the piece. To economise upon the decorating costs for clients, a simple edge gilding was usually employed or perhaps comprised enamelled colours

Chemical descriptors: Care should be taken to differentiate between the descriptors of chemical compounds and their component elements. For example, silica is SiO_2 and contains silicon and oxygen: likewise, calcia is calcium oxide CaO (lime), alumina is aluminium oxide Al_2O_3, magnesia is magnesium oxide MgO, potash is potassium oxide K_2O and soda is sodium oxide Na_2O. Other chemicals are described precisely as they occur, such as phosphorus pentoxide P_2O_5—also known as phosphorus (V) oxide or phosphorus pentoxide, relative to phosphorus (III) oxide P_2O_3—and sulfur dioxide SO_2. Iron oxide is troublesome as this can mean both ferrous oxide FeO and ferric oxide Fe_2O_3, where the iron is in the oxidation state (II) and (III), respectively, and the former is a green colour and the latter orange-red.

China: A generic term for high-fired ceramics, which may be glazed or unglazed, and used for decorative or utilitarian purposes around the home. Believed to originate from descriptors of ceramics that were made in simulation of imported Chinese porcelain, this term covers a wide range of ceramics including porcelain, earthenwares, creamwares, stonewares, celadons, majolica and delftwares. The term “china” has been used freely to describe ceramic articles of a range of compositions since the importation of porcelain from China in the 16th Century and some of the artefacts do not actually contain china clay (kaolinite).

Chinese porcelain: A generic term which historically describes the characteristic hard paste porcelain comprising *kaolinite* and *petuntse* as major raw materials fired at a high kiln temperature; termed “true porcelain” by some historians to differentiate it from “artificial porcelain” which is retained for the soft paste body variants which included synthetic components as raw materials, for example calcined bone ash, potash, lime and glass frit. During the early 18th Century, the Chinese perfected a soft paste body variant which has been recognised as a comparative rarity. Chinese porcelains have a range of compositions which varied both with their production sites geographically and also with the specified quality: the highest quality porcelains from the Imperial kilns at Jingdezhen were composed of an equal ratio of china clay (kaolin) and petuntse.

Cipher: See Monogram

Coat-of-arms: An officially registered heraldic device granted in Great Britain by Royal authority via the College of Arms in London, the Ulster King of Arms in Ireland and the Lord Lyon King of Arms in Scotland to members of a noble family or an arms-bearing organisation. Other European countries have their equivalent officers for the administering and granting of arms. This includes a shield (escutcheon) with dexter (right) and sinister (left) supporters, as viewed from the bearer of the shield—or by an observer from the front of the shield in a mirror image. The supporters are usually real or mythical animals or birds such as the lion, leopard, eagle, unicorn, dragon, wyvern and gryphon. Under the shield (base) a motto may be carried and also a badge of honour. On top of the shield (chief) can sit a coronet signifying the rank in the peerage of the bearer of the arms, a helm and visor accompanied by mantlings(lambrequins) and supporting a twisted silk torse upon which the crest sits. Atop the crest may appear a battle cry legend.

Crested china: A term reserved for a partial coat-of-arms depicted upon porcelain or earthenwares which may exhibit the crest solely, with or without its torse, and perhaps accompanied by a motto, usually placed in either the verge or reserve or centrally located on the artefact. It may be enamelled in polychrome pigments, or it is more usually found in gilt or occasionally and more rarely in a monochrome pigment.

Diplomatic gifts: The use of porcelain as a diplomatic gift in Europe can be traced back to the early 14th century with Nestorian emissaries from the Chinese Emperors: the first manufacture of armorial porcelain for this purpose can be credited to Francisco Medici, Duke of Tuscany, who in the mid to late 16th century used his own porcelain as diplomatic gifts to other European rulers. Porcelain was very highly regarded and was ideally suited for this purpose, giving it a status similar to that of religious relics, jewels and bullion. The rendering of armorial bearings on such porcelain gifts afforded them a personality that was specific to the receiver.

Electromagnetic spectrum: This describes the range of wavelengths which define the characteristics of electromagnetic radiation from the high-energy, low wavelength, high frequency cosmic and X rays, through the ultraviolet (200–400 nm) and then the visible spectrum (400–700 nm) from the violet (through indigo, blue, green, yellow and orange) to the red into the infrared, and thence through to the low-energy, high wavelength, low frequency microwave radiation. Analytical techniques probe different ranges of the electromagnetic spectrum to interrogate the molecular and elemental compositions of ceramic specimens and derive quantitative and qualitative data which can be correlated with the materials used in the ceramic bodies, glazes and pigments. The unit of the wavelength used analytically to define the electromagnetic spectrum is the nanometre, 10^{-9} m or nm. As the wavelength of the radiation increases, the associated energy becomes smaller, through from the Xray to the microwave region, according to the Planck-Einstein radiation law, $E = h\nu$, where E is the energy of the radiation, h is the Planck constant and ν is the frequency of the electromagnetic radiation, which is inversely proportional to the wavelength of the radiation, λ. Generally, the elemental composition is interrogated in the Xray region and the molecular composition is found from molecular energy transitions occurring in the visible and near infrared regions of the spectrum. Analysis of the elemental and molecular spectral data gives a complementary picture of the porcelain body, glaze and pigments used in the artefact.

Gilding: The application of a gilt decoration to fired porcelain involving 24-carat gold leaf in a carrier such as mercury, honey or an organic resin followed by burnishing, see under ***burnishing.*** Gilding was a highly specialist task in decorator's workshops, the best gilders being permitted to sign their work with a numeral on the underside of the piece or being mentioned specifically in the service commission documentation and gilding was an expensive component of the decorative process.

Glaze: The application of a coating to a fired porcelain biscuit body to seal the ceramic pores and to enhance the appearance with a glossy, smooth texture. Usually applied over the decorative pigments, it acted as a protectant for the decoration. Overglaze decoration, known technically as painting rather than enamelling in the ceramic workshops, was subject to paint loss and wear during usage. The glaze was usually applied as a slip, a thick suspension of clays, alkaline earths and other additives with lead oxide component (or a tin oxide additive, such as cassiterite, for a lead -free glaze). Fired at a lower temperature than the biscuit body, this final stage of preparation enhanced the beauty of the porcelain: it was essential that the glaze composition mechanically and chemically was compatible with the fired porcelain otherwise firing faults were created and an unsightly glaze "craquelure" could form, whereby the glaze physically separates from the porcelain substrate in contraction from the kiln upon cooling. Certain decorative pigments were favoured for their enhancement properties upon interaction with the glaze application, cobalt blue being one of these, and it was equally important that the pigment maintained its integrity and was able to resist "running" on application of the glaze.

Glost kiln: Used in the final stage of porcelain preparation before and/or after applying the enamelled decoration and involving the application of an alkaline "slip" containing china clay, soda, potash and a lead oxide component usually in the form of a powdered flint glass, which at lower temperatures will form a hard, transparent glaze coating. Occasionally, a glost kiln was used for the calcination of raw material components at lower temperatures and for the drying of raw materials in which variable amounts of water were found to occur. Initially, glazes were lead-based but in the first decade of the 19th Century, when the toxicity of lead compounds was fully appreciated, reversion to a less toxic, tin-based glaze was effected (as patented by John Rose at the Coalport China Works in 1820, using cassiterite, SnO_2, instead of lead oxide, litharge or massicot, PbO). China sold from manufactories "in the white" could be unglazed or glazed for decoration in ateliers, with subsequent glazing being undertaken by the enamellers.

Goss china: a type of armorial porcelain made *en masse* as souvenirs and mementoes for holidaymakers, tourists and visitors from the mid-19th century. The coats-of-arms of towns and places of historical interest featured on a variety of miniature bone china artefacts and numbering up to ten thousand different armorials on more than 2000 shapes.

Initials: see Monogram.

Instrumental Chemical Analysis: the use of instrumentation to detect qualitatively and to estimate quantitatively the presence of chemical elements and their compounds in ceramic bodies and glazes. The basis of instrumental analysis is the use of special signatures or properties of chemical species or chemical bonds such as silica, SiO_2, or the $Si = O$ bonds in complex macromolecular structures, such as silicaceous ceramic bodies, from their behaviour towards the scattering or absorption of electromagnetic radiation in the electromagnetic spectrum from the low wavelength (high wavenumber) Xray region to the high wavelength (low wavenumber) microwave region. When coupled with a microscope, an analytical instrument is capable of interrogating very small specimens or domain regions in a larger specimen of the order of cubic microns or cubic nanometres. Instrumentation that has been commonly used in the assessment of the constitution of porcelain and ceramic bodies, glazes and pigments includes scanning electron microscopy/energy dispersive and X Ray spectrometry (SEM/EDAXS), X Ray fluorescence spectroscopy (XRF) for the elemental oxide composition and X Ray diffraction spectrometry (XRD), laser Raman Spectroscopy (RS) and Fourier-transform infrared spectroscopy (FTIR) for mineralogical and pigment identification.

Monogram: Initials which are arranged in a design so that the removal of one initial compromises the design and renders one or more of the other initials indecipherable. A cipher is a design in which initials which are merely arranged together whereby the removal of one does not affect the others. Monograms, ciphers and initials all appear on applied ceramic decoration and when accompanied by an heraldic device such as a coronet or a crest then constitute an armorial representation. Otherwise, they are not strictly considered as armorial although historically they have often been included as such by many ceramics historians.

Moulding: The shape and size of porcelain items can be a reliable stylistic indicator of a particular factory's output and a clue to the attribution of unmarked pieces. Mouldings were generally unique to each manufactory and comprised individual and idiosyncratic cross-sections of flatwares such as saucers and plates, the shapes of cups, the presence and pattern of footrims, the shape of spill vases (e.g. cylindrical, fluted, tulip-shape, trumpet-shaped, flared), the impressed moulded and embossed verges of plates and the number and type of indentations at the rim. Many factories characteristically used the application of moulded flowers which were integral to the item and also applied reliefs such as feet, the heads of animals, floral decoration and beads, as well as knops or handles on tureens and comports.

Named service: A porcelain service which has been commissioned by a particular client or individual from the china factory or one of its agents, especially where factory records support the attribution and where occasionally a specific reason for its purchase may be stated, such as the *Duke of Cambridge* service in Nantgarw porcelain which was ordered from the Nantgarw agents, John Mortlock, Oxford Street, London, as a wedding present from George, the Prince Regent, for the marriage of his younger brother, Adolphus, Duke of Cambridge, to Princess Augusta, Landgravine of Hesse-Cassel, at Buckingham Palace in June, 1818. The term also includes services which have been purchased from the later sale of effects or estates and have thereby acquired the name of their new owners by association. Occasionally, the named services do not bear the name of the person who commissioned them but rather the place where they were to be used or perhaps referred to even the decoration adopted: an example of the former is the *Marino Ballroom* service in Swansea porcelain commissioned by John Vivian for his residence in Sketty Hall, Swansea, and of the latter, the *Garden Scenery* service of Swansea porcelain originally commissioned by Lewis Weston Dillwyn, proprietor of the Swansea China Works.

Pigments: In ceramics decoration, these are coloured minerals which are thermally stable at the operating temperatures adopted in the glost kiln after application of the glaze slip and firing up to 600°C or slightly greater. Mineral pigments are usually metal oxides, sulfates and sulfides such as haematite, gypsum and orpiment. Care needs to be taken in the interpretation of old recipes for the decoration of ceramics as minerals were often confusingly assigned an incorrect nomenclature, such as *minium*, which has been used historically to describe both red lead, trilead tetroxide, Pb_3O_4, and cinnabar, mercury sulfide, HgS, and red oxide of lead can mean both red lead ($2PbO.PbO_2$) and litharge (PbO). Analytically, the interrogation of pigments on enamelled porcelains is a fertile area of study as manufacturers often changed their pigment sources and the composition of their pigments with time and the presence of trace quantities of unusual elements such as zinc, niobium, rubidium and copper can indicate valuable information about the timelines of production and sources of the pigment by the manufactory concerned. For example, the presence of a minor elemental component of bismuth in a cobalt blue pigment on an Iznik glazed tile from the 16th /17th century indicated that the source of the cobalt ore was European (from the Erzengebirge mine in Saxony) rather than the more usually encountered Asian mines. Arsenic traces in the cobalt blue pigment on Chinese 18th century porcelains are indicative of the use of European cobalt ores in the preparation of the pigment rather than the Asian sources which are arsenic—free.

Porcelain types: Commencing with hard paste and soft paste porcelains historically (also termed rather confusingly as "true" and "artificial" porcelains), this inadequate description generated a further differentiation into the porcelain as hard paste (highly siliceous), phosphatic, bone china, magnesian, glassy and hybrid types. The categorisation of a porcelain type is dependent not only upon the body composition but also upon the process methodology and firing sequences adopted. However made, all types of porcelain must possess a degree of translucency to qualify for its appellation and inclusion in the category.

Raw Materials: The raw materials are components comprising the body and glaze recipes used in porcelain syntheses. Mostly these were natural minerals and rocks which were sourced from precise mines and locations which gave rise to several alternative names. Natural raw materials include feldspars, china clays, ball clays and soapstone. Synthetic raw materials include bone ash, pearl ash, soda ash, smalt, lime and magnesia.

Service: A porcelain service could be commissioned directly from the factory, from an appointed agent, bought at auction from "chinamen" or acquired through an import agency such as the Honourable East India Company who traded with Chinese porcelain agencies and their brokers or factors (Hongs) through the port of Canton (Guangzhou). A porcelain service was normally defined as a dinner, dessert or tea service, but combinations of dinner-dessert, tea-coffee and special breakfast services were quite commonly ordered. The usual dessert or dinner service comprised 42 pieces approximately, but special orders expanded this to over 100 pieces, with the inclusion of soup dishes, tureens, centre fruit comports and stands, ice pails, spill vases, muffin dishes, guglets (water bottles), vegetable dishes, egg-cups and carrier, sauce boats and trios (coffee can/cup, tea cup and saucer). There are instances of very large services being commissioned, such as those from Meissen in the 18th Century of over 1000 pieces with some 50 or 60 place settings. The service was usually decorated according to the specific requirements of the commission and the inclusion of a coat of arms or heraldic crest centrally or at the verge would confer upon it the description of an "armorial" service. An armorial service is especially valuable for provenancing and historical purposes as it facilitates the assignment of the commission to a particular person in an arms-bearing family and often enables one to attribute a definitive date for a commission, such as the granting of arms or perhaps a family succession to a title or elevation in the peerage which may have been celebrated accordingly with the ordering of a porcelain service.

Translucency: This is perhaps the greatest achievable asset of porcelain manufacture to which every manufacturer subscribed and hoped to attain in emulation of the Chinese "eggshell" wares which were imported to Europe in the mid-18th century. It describes the transmission of visible radiation, or light, through a solid object: at one end of the spectral scale is glass, which is usually *transparent* (100% transmission) and at the other is earthenware which is *opaque* (0% transmission).

However, see entry above for “Glass”, where the occurrence of a milk glass is noted in which the transparency is less than 100% and the glass is then termed translucent. Porcelain is measured by its translucency, which is the clarity for transmission when viewed with background lighting. An intermediate descriptor is “semi-opaque” or even “semi-porcelain”, which was applied to some china during the mid-19th century, which is rather indefinite and conveys little information about their true category. Other descriptors such as “vitreous china” and “ironstone china” likewise do little to assist in their classification. By definition, porcelain must be able to transmit light to some extent unlike stonewares, faience, terracotta and earthenwares.

Bibliography

Anderson, J. A. (2000). *Derby Porcelain and the Early English Fine Ceramics Industry*, Ph.D. Thesis, University of Leicester.
Bell, H. N. (1820). *The Huntingdon Peerage*. Baldwin, Cradock, & Joy.
Bemrose, W. (1898). *Bow, Chelsea and Derby Porcelain: Being Further Information Relating to These Factories, Obtained from Original Documents, Not Hitherto Published*. Bemrose & Sons Publishers.
Bentley, T. H. (Ed.). (2019). *'Chinese Porcelain, the East India Company and British Cultural Identity 1600–1800' by Pierson, S in Picturing Commerce In and From the East Asian Maritime Circuits, 1550–1800*. Amsterdam University Press.
Beurdeley, M. (1962). *Porcelain of the East India Company*. Barrie and Ratcliffe.
Bischoff, C. (2014). Women Collectors and the Rise of the Porcelain Cabinet. In J. van Campen & T. Eliëens. Waanders, (Eds.), *Chinese and Japanese Porcelain for the Dutch Golden Age*.
Brewer, J. & Porter R. (Ed.). (1993). *Consumption and the World of Goods* Routledge.
Brewer, J. N. (2005). *Delineations of Gloucestershire: Views of the Principal Seats of Nobility & Gentry* Stroud.
Cannon, J. (1984). *Aristocratic Century: The Peerage of Eighteenth-Century England* Cambridge University Press.
Carswell, J. (2000). *Blue and White Chinese Porcelain Around the World*. British Museum Press.
Cassidy-Geiger, M. (2007). Porcelain and Prestige: Princely Gifts and *White Gold* from Meissen, Chapter 1.In M. Cassidy-Geiger (Ed.), *Fragile Diplomacy: Meissen Production for Foreign Courts, ca. 1710–1763*, Yale University Press.
Chatterton, E. K. (2008). *A World for the Taking*. Fireship Press.
Church, Sir A. (1886). *English Porcelain: A Handbook to the China Made in England*. Published for the Committee of Council on Education by Chapman & Hall, Ltd.
Cocks, A. S. (1989). The Nonfunctional use of Ceramics in the English country House During the Eighteenth Century. In G. Jackson-Stops, et al. (Eds.), *The Fashioning and Functioning of the British country House, Studies in the History of Art 25*.
Colomban, P., Milande, V., & Lucas, H. (2004). On-site Raman analysis of Medici porcelain. *Journal of Raman Spectroscopy, 35*(1), 68–72.
Dawson Dobson, S. (1795). *Historical Anecdotes of Heraldry and Chivalry: Tending to Shew the Origin of Many English and Foreign Coats of Arms, Circumstances and Customs,* Holl and Brandish, Worcester.
Dawson, A. (2010). *English and Irish Delftware: 1570–1840*. British Museum Press.
Day, I. (2000). *Eat, Drink and Be Merry: The British at Table, 1600–2000*. Philip Wilson Publishers, Ltd.
Earle, P. (1989). *The Making of the English Middle Class: Business, Society and Family Life in London 1660–1730*. University of California Press.

R. L. Denyer et al., *Armorial Porcelain*, https://doi.org/10.1007/978-3-031-63745-2

Easthaugh, N., Walsh, V., Chaplin, T., & Siddall, R. (2004). *Pigment Compendium: A Dictionary of Historical Pigments*. Elsevier Butterworth-Heinemann.

Estabrook, C. (1998). *Urbane and Rustic England: Cultural Ties and Social Spheres in the Provinces* 1660–1780.

Ferguson, J. (2021). *Visualising Protestant Monarchy: Ceremony, Art and Politics after the Glorious Revolution (1689–1714*). Boydell.

Ferguson, P. (2016). *Setting the Table at Gibside: The Bowes Family of County Durham and their Ceramic Acquisitions in the Eighteenth Century*. Transactions of the English Ceramic Circle.

Galinou, M. (Ed.). (2004). *City Merchants and the Arts 1670–1720*. Oblong for the Corporation of London.

Gettens, R. L., & Stout, G. L. (1942). *Painting Materials: A Short Encyclopaedia, D*. Van Nostrand.

Glanville, P., & Young, H. (Eds.). (2002). *Elegant Eating: Four Hundred Years of Dining in Style*. V&A Museum Publications.

Graca, J. (1977). The Portuguese Porcelain Trade with China. *Arts in Asia, 7*, 45–47.

Grandjean, G. (1999). The Porcelain of Rouen. In B. Rondot (Ed.), *Discovering the Secrets of Soft Paste Porcelain at the St Cloud Manufactory, ca* (pp. 1690–1696). The Bard Graduate Center for the Studies in the Decorative Arts.

Hanscombe, S. (2008). Armorial Porcelain Decorated in the London Workshop of James Giles. *Transactions of the English Ceramic Circle, 20*. Part, 1, 124.

Harley, R. (1982). *Artists' Pigments ca. 1600–1835 : A Study in English Documentary Sources,* (2nd ed.). Archetype Publications.

Holdridge, D. A. (1956). *Pottery Science – Materials, Processes and Products,* Ellis Horwood, Chichester.

Habakkuk, H. J. (1994). *Marriage, Debt, and the Estates System: English Landownership 1650–1950* Clarendon Press.

Hall, L. J. (1983). *The Rural Houses of North Avon and South Gloucestershire, 1400–1720*. City of Bristol Museum & Art Gallery.

Hancock, D. (1995). *Citizens of the world: London Merchants and the Integration of the British Atlantic Community, 1735–1785*. Cambridge University Press.

Hubback, J. (1845). *A Treatise on the Evidence of Succession to Real and Personal Property and Peerages*. T. & J.W. Johnson.

Impey, O. (1977). *Chinoiserie: the Impact of Oriental Styles on Western Art and Decoration* Scribner's.

Jewitt, L. (1883). *The Ceramic art of Great Britain* (2nd ed.). J. S. Virtue and C. Ltd.

Kennedy, R. (1999). *Between Bath and China: Trade and Culture in the West Country 1680–1840*. Museum of East Asian Art.

Kerr, R. (2011). Mengoni, L, & Wilson, M., *Chinese Export Ceramics Harry N Abrams*. V & A Publishing.

Kerr, R., & Wood, N. (2004). *Science and Civilisation in China*, Vol. 5, Part 12: Ceramic Technology, Series editor J. Needham, Cambridge University Press.

Kroes, J. (2007). *Chinese Armorial Porcelain for the Dutch Market Waanders BV*, Uitgeverij, Zwolle.

Lane, A. (1954). *Italian Porcelain*. Faber & Faber.

MacGuire, B. (2023). *Four centuries of blue and white: The Frelinghuysen Collection of Chinese & Japanese Export Porcelain*. Paul Holberton Publishing Ltd.

McKendrick, N., Brewer, J., & Plumb, H. J. (1982). *The Birth of a Consumer Society: The Commercialization of Eighteenth Century England*. Indiana University Press.

Morse, H. B. (1926). *The Chronicles of the East India Company, trading to China 1635–1834* Clarendon Press.

Pearce, J. (2001). Ceramics for Refined Taste: Chinese Porcelain from London Excavations. *Oriental Art, XLVII,*(2): 56–58.

Pierson, S. (2007). *Collectors, Collections and Museums : The Field of Chinese Ceramics in Britain 1560–1960*. Peter Lang.

Pierson, S. (2013). *From Object to Concept: Global Consumption and the Transformation of Ming Porcelain* Hong Kong University Press.

Ramsay, W. R. H., & Ramsay, E. G. (2008). A Case for the Production of the Earliest Commercial Hard Paste Porcelain in the English-Speaking World. *Proceedings of the Royal Society Victoria, 120*, 236–256.

Ramsay, W. R. H., & Ramsay, E. G. (2017). *The Evolution and Compositional Development of English Porcelains from the 16th Century to Lund's Bristol c.1750 and Worcester c/1752—The Golden Chain*, Invercargill Press.

Redford, B. (Ed.). (1992). *The Letters of Samuel Johnson, 5 Volumes*. Princeton University Press.

Rissik Marshall, H. (1951). *Coloured Worcester Porcelain of the First Period, 1751–1783*. Ceramic Book Co.

Robson, T. (1830). *The British Herald, or Cabinet of Armorial Bearings of the Nobility and Gentry of Great Britain and Ireland* (Vol. 2). Turner & Marwood.

Roy, A. (2007). Cobalt Blue. In B. H. Berrie (Ed.), *Artists' Pigments: A Handbook of their History and Characteristics* (Vol. 4, pp. 151–178). Washington & London.

Savage, G. (1952). *Eighteenth Century English Porcelain*. Spring Books.

Scott-Giles, C. (1954). *Boutell's Heraldry*. Frederick Warne & Co.

Solon, M. L. (1905). The Rouen Porcelain. *The Burlington Magazine for Connoisseurs*, 7(26), May: 116–124.

Stobart, J., Hann, A., & Morgan, V. (2007). *Spaces of Consumption: Leisure and Shopping in the English town, c. 1680–1830*. Routledge.

Stone, J. F. (1995). *An Open Elite? England 1540–1880* Clarendon.

Strong, R. (2003). *Feast: A History of Grand Eating*. Houghton Mifflin Harcourt.

Styles, J. (2001). Georgian Britain 1714–1837: What Was New? In M. Snodin & J. Styles (Eds.), *Design and the Decorative Arts: Britain 1500–1900* V&A Publications.

Thurley, S. (2003). *Hampton Court: A Social and Architectural History*. Yale University Press.

Verey, D., & Brook, A. (1999). *The Buildings of England: Gloucestershire I: The Cotswolds*. Penguin.

Walton, G. (1992). The Era of Louis XIV: A Turning-Point in the History of Diplomatic Gifts. In H. C. Collison (Ed.), *Versailles: French Court Style and Its Influence*. University of Toronto Press.

Woodcock, T. (1988). *The Oxford Guide to Heraldry*. Oxford University Press.

Woolf, D. (2003). *The Social Circulation of the Past: English Historical Culture, 1500–1730* Oxford University Press.

Wyrley, W. (1853). *The True Use of Arms*, 1592, reprinted by J.G Bell.

Watanabe-O'Kelly, H., & Morton, A., (Ed.) (2017). *Queens Consort, Cultural Transfer and European Politics c. 1500–1800*. Routledge.

Weatherhill, L. (1986). A Possession of One's Own: Women and Consumer Behaviour in England 1660–1740' *Journal of British Studies, 25*(2) April: 131–156.

Wilson, C. A. (1991). *Banquetting Stuffe' The Fare and Social Background of the Tudor and Stuart Banquet* Edinburgh University Press.

Index

R. L. Denyer et al., *Armorial Porcelain*, https://doi.org/10.1007/978-3-031-63745-2

GPSR Compliance

The European Union's (EU) General Product Safety Regulation (GPSR) is a set of rules that requires consumer products to be safe and our obligations to ensure this.

If you have any concerns about our products, you can contact us on ProductSafety@springernature.com

In case Publisher is established outside the EU, the EU authorized representative is:

Springer Nature Customer Service Center GmbH
Europaplatz 3
69115 Heidelberg, Germany

Batch number: 10726135

Printed by Printforce, the Netherlands